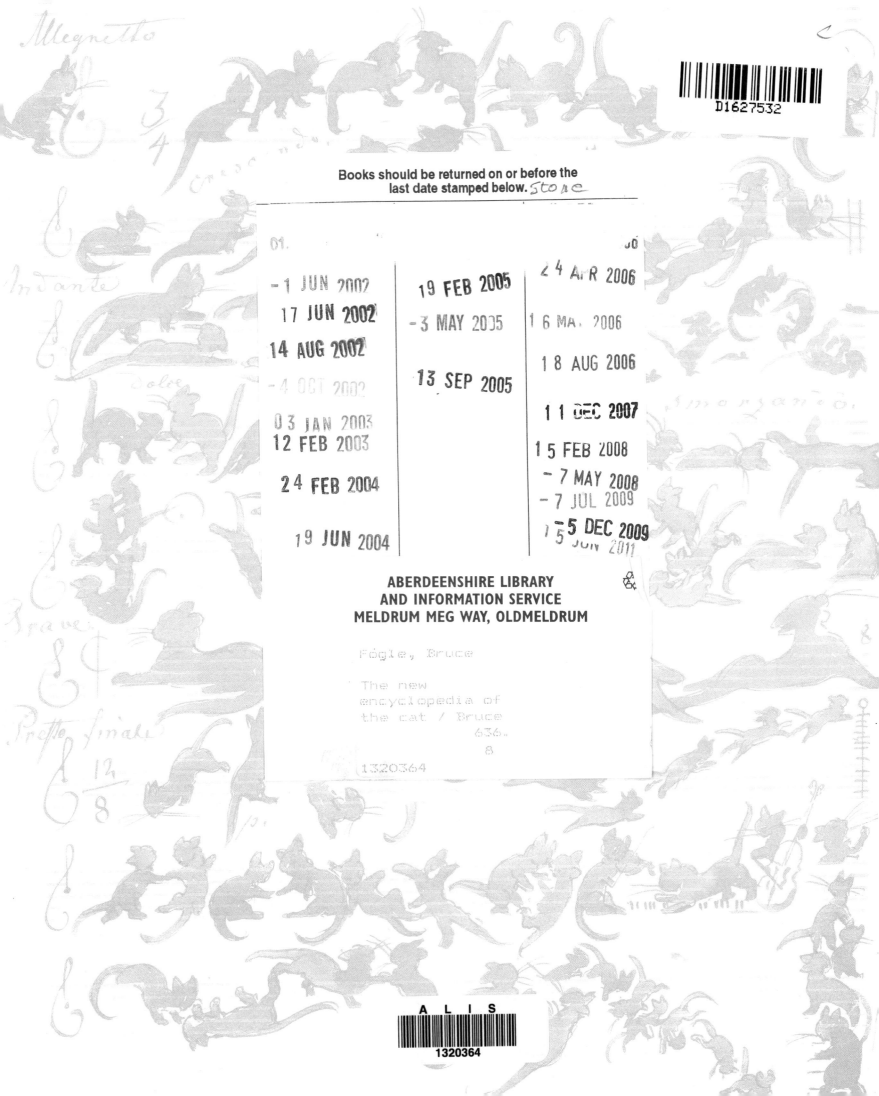

The NEW ENCYCLOPEDIA *of the*
CAT

The NEW ENCYCLOPEDIA *of the*

CAT

DR. BRUCE FOGLE

DORLING KINDERSLEY

LONDON ■ NEW YORK ■ SYDNEY ■ MOSCOW

DORLING KINDERSLEY

LONDON, NEW YORK, SYDNEY, DELHI,
PARIS, MUNICH AND JOHANNESBURG

Original edition (1997)

Project Editor CANDIDA FRITH-MACDONALD

Art Editor URSULA DAWSON

Editors TRACIE LEE DAVIS, SHARON LUCAS,
SARAH WILDE

Designer HEATHER DUNLEAVY

Managing Editor FRANCIS RITTER

Managing Art Editor DEREK COOMBES

DTP Designer SONIA CHARBONNIER

Production Controllers RUTH CHARLTON,
ROSALIND PRIESTLEY

Picture Researcher SAM RUSTON

Breed photography TRACY MORGAN, CHANAN,
MARC HENRIE, TETSU YAMAZAKI

This edition

Project Editors CANDIDA FRITH-MACDONALD,
SANDRA JENKINS

Art Editor JANICE ENGLISH

Managing Editor DEIRDRE HEADON

Art Director CAROLE ASH

Production Controller MANDY INNESS

Picture Researcher ANNA GRAPES

Picture Librarian HAYLEY SMITH

Consultant JENNIFER ANNE HELGREN

Jacket Design NATHALIE GODWIN

This edition first published in Great Britain in 2001
by Dorling Kindersley Limited
80 Strand, London WC2R 0RL

CONTENTS

INTRODUCTION

Early evidence
The cat probably reached Britain with the Roman Empire: this terracotta tile from Reading dates from the first century AD.

As we begin the 21st century, the domestic cat has become the most popular household companion in North America and many parts of Europe. Yet in these regions there are still as many feral cats, surviving on their own as there are individuals living as family pets. Of all the domestic species, the cat is the most independent and least understood. Our lack of understanding has simple origins. We are a sociable species and so too are other domesticated species – dogs, goats, sheep, cattle, horses, pigs, even chickens. Not so the cat, which marches to its own drumbeat.

Descended from the North African wildcat (*see page 18*), the domestic cat evolved to be a solitary hunter. It does not come from a co-operative culture, where individuals live and hunt together, as do some larger cat species such as the lion. With small rodents, amphibians, and birds as its food sources, the wildcat's meal is only

Religious acceptance
The role of the cat in Christian religion has varied. It appeared favourably in the Lindisfarne Gospels (see left), but was later associated with heresy.

large enough to fill its own stomach, so it developed natural independence, being content to live and hunt alone, and to meet other cats only during territorial disputes or mating. When cats moved into new human communities, those that were cared for discovered a source of ample food. Natural selection favoured more sociable individuals, slowly transforming the wildcat into the domestic cat that we know today.

Witch's cat
By the 14th century, Europe was gripped by religious fervour. The cat's nocturnal habits condemned it as evil in the eyes of the Church.

FORM AND FUNCTION

Since that domestication, over 4,000 years ago, the cat has been both deified and persecuted in turns. Some cultures have been fascinated by its natural fertility, while others saw in its behaviour the signature of the devil. A misunderstanding of the cat's ability to survive on its own as a nocturnal hunter is at the root of many of the misconceptions that developed, and some survive even today. In fact, the cat is one of nature's most successful products. All aspects of its anatomy result from evolutionary logic. Its eyes dilate fully for night vision but shut down to slits in bright sunshine. Touch-sensitive whiskers help the cat to "feel" its way

Cartoon favourite
Cartoon cats have shown enduring popularity. Felix, one of the first, became a U.S. Air Force emblem and is still well-loved worldwide.

Artists' models
Théophile Alexandre Steinlen (1859–1923) became well known for his brilliantly accurate portrayals of cats. Many other artists have found a highly successful niche by specializing in felines.

when it hunts in low light, while its delicate hearing lets it hear the tiny, high-pitched sounds of a mouse squeaking. An accurate sense of balance permits it to climb away from danger, while its righting reflex allows it to regain balance if it accidentally falls. All of these characteristics create an animal that is appealing to our eyes and minds.

THE SOCIABLE CAT

Although biologically predisposed towards a solitary life, in certain circumstances, when food is plentiful, the cat can alter its solitary behaviour. Female groups share maternal responsibilities such as feeding and protecting the young. As the male kittens mature, fights with female relations intensify until they leave the colony, forming an informal brotherhood with other male cats. These patterns are seen clearly in colonies of feral cats, and are also found in household cats. During the 20th century, the cat moved from a minor role in human society to a more commanding position as one of the most popular of all animal companions. By the 1980s, it was the most popular animal companion in North America, a position it also reached in many European countries

Naive art
Folk art has incorporated many motifs from everyday life. This feline figurine is an example from the rich tradition of American folk art.

in the mid-1990s. The popularity of purebreds and the creation of new breeds is increasing everywhere. This is not necessarily good news for cats. Selective breeding within breeds already has created previously unseen genetic conditions. Equally worrying, traits such as hairlessness, dwarfism and miniturization are being selectively bred in order to create new breeds.

The cat is a perfect animal companion for us. Quiet, self-cleaning, content to snooze whenever the opportunity arises but capable of astounding episodes of activity when its brain is engaged, it is also aesthetically pleasing. Perhaps the cat's most important characteristic is its independence. A cat may live in your home as a result of your choice or its own, but if left to its own devices, its chances of surviving remain the same. The greatest value of the cat is that by living with us it gives us an opportunity to view the natural world in our own homes. In the future development of our relationship with the cat, that integrity is what we should strive to maintain.

One of the family
As it spread eastwards, the cat basked in the reflected glory of the tiger, which held symbolic importance in the Orient.

CHAPTER ONE

THE CAT FAMILY

MILLIONS OF YEARS AGO, MEAT-EATING mammals evolved: species with unique shearing and cutting carnassial teeth. The cats' dependence upon eating the flesh of other animals has been at the core of their evolution. They developed matchless hunting capabilities, making them the most efficient land-based carnivores. Some of the large cats worked co-operatively to capture food. Most of the smaller cats evolved as lone hunters, preying on animals large enough to fill only a single stomach. Three million years ago, a species not unlike the modern domestic cat first emerged, but only within the last few thousand years that members of this species chose to live in close proximity to man.

ABOVE MALE RED TABBY AMERICAN SHORTHAIR

LEFT FEMALE INDIAN TIGER

ANCESTORS OF THE CAT

TODAY'S CAT FAMILY COMES FROM THE SAME ROOTS AS ALL OTHER predator mammals. The world's first mammals were small, with limited brain capacity, but in the evolutionary process these were replaced by newer species with more refined abilities. Over tens of millions of years, primitive cats evolved, adapting to climate, to available prey, and to competition from other evolving carnivores. Successful adaptations have occurred, been lost, and then recurred as the environment altered. Educated guesswork is used to map evolution, but recently genetic fingerprinting has been employed to calibrate a "molecular clock" for the evolution of today's species.

The history of modern mammals, including the cat, is traced from fragments of ancient bone and teeth. These shards of evidence do not, however, give us any clues about the two most fundamental traits of mammals, their fur and their skin glands. Fur offered insulation while skin glands allowed mammals both to sweat and to manufacture nourishment for their young in the form of milk. These are simple adaptations, but are at the root of the mammal's success in a variety of environments.

THE EARLIEST ANCESTORS OF THE CAT

After the dinosaurs died out 65 to 70 million years ago, they were survived by, among others, the mammals. The earliest mammals were the fish-eating creodonts, and one of their forms evolved into smaller, forest-dwelling miacids. For unknown reasons, the creodonts became extinct and the miacids thrived. It has been assumed that this evolutionary change occurred because the miacids had larger brains, although there is no fossil evidence that this is so.

The early miacids were long-bodied, short-legged animals that evolved into the variety of modern families of carnivores we know today. Each family fits into its own ecological niche, from the bears at the largest end of the scale to the civets and mongooses at the smallest, and the cat family spans the range from tigers and lions to tiny sand cats. Among the carnivore families, the cats are most closely related to the civets. Palaeontologists look at structures at the base of the skull, in particular the structure which acts as a bony case surrounding the

The spread of the cat
Australia and South America separated from the ancient land mass called Gondwanaland before feline evolution began. The cat family evolved in Eurasia and North America, and when South America linked with its northern neighbour, cats entered that continent. With no land bridge available, the cat did not enter Australia.

middle ear, to help differentiate one species from another. Unfortunately the fossil record for small cats is sparse, in part because many ancestral species inhabited forested regions where conditions for preserving fossils were poor. Fossil records for larger species are more abundant and show that cats lived worldwide.

THE FIRST CAT-LIKE CARNIVORES

The early "cats" are classed as *Proailurus*. They shared many characteristics of the *Viverridae* family, including walking plantigrade, or flat-footed, and can be considered half-cat, half-civet. About 20 million years ago, *Proailurus* gave way to *Pseudaelurus*, walking digitigrade, on the tips of its toes, and possessing stabbing canine teeth. Experts consider this group of animals as the first members of the modern cat family. Within this group there were both larger and smaller species. They were light, agile, and varied in sizes between the modern puma at 100 kg (220 lb) and the lynx at 29 kg (64 lb).

Out of the *Pseudaelurus* group emerged the direct ancestors of the modern cat family, *Felidae*, as well as other "dead-end" branches, such as *Smilodon*, the sabre-toothed cats that survived until less than a million years ago in Europe, Africa, Asia, and the Americas. Sabre-toothed species evolved independently at least four times, as creodonts, as marsupials, and twice as ancient species of cats. Their jaws opened 50 per cent wider than other cats', and the long teeth were flattened and serrated for more efficient cutting. The species evolved to hunt large grazing mammals and probably died out when their food supply diminished, having become too selectively adapted for other prey.

EARLY CAT DISTRIBUTION

The early big cats inhabited most parts of the world. Lions lived not just in Africa, but in Europe, throughout Southeast Asia, across Siberia, and throughout North and Central America down to northwest South America.

Cat evolution
The evolution of modern cats does not follow a simple plan. During 60 million years of carnivore development, evolution created and rejected numerous varieties of "proto-cats". Fossil evidence is used to track evolutionary developments, but it can be misleading. Fossils from arid regions tend to remain intact, while those from damper regions disintegrate. One consequence is that experts disagree on the details of cat evolution, although they do agree on the broad picture. Feline evolution has taken many dead ends, one of the most spectacular being the sabre-toothed cats (Smilodon), but it has lead inexorably to the variety of cat species that survive today.

Creodonts
The predominant carnivores 53 to 60 million years ago, creodonts were rather small, no more than 30 cm (12 in) high at the shoulder. They had thick necks and long bodies in relation to their legs. All of today's carnivores evolved from creodonts.

Miacids
Miacids evolved around 60 million years ago, and survived the creodonts. They were up to twice the size of the creodonts, with longer legs and slender heads.

PALAEOCENE   EOCENE

Leopards and jaguars lived throughout western, central, and eastern Europe, Africa, Asia, on the island of Java, and throughout the Americas. The lion and the lynx are thought to originate from Africa while the leopard may have evolved in India. Jaguars evolved in Eurasia and spread into the Americas. Tiger fossils two million years old have been found in China; the oldest fossils of cheetahs were found in France, and of pumas in North America.

Small cats never populated North America. The small predator's niche was already filled by carnivores from the Mustelid family, including the mink, marten, wolverine, and skunk. The domestic cat's extinct ancestor, Martelli's wild cat (*Felis lunensis*), evolved in Europe or the Middle East.

NEW EVOLUTIONARY EVIDENCE

Traditional classifications of cats have been based on the geographical distribution, size and shape, and behaviour of species. Cats are divided, on the basis of where they are found today, into Old World and New World species. Small cats that did not roar were further classified as *Felis*, while big cats that did roar were classified as *Panthera*: an arbitrary but useful division, with only a couple of misfits – the cheetah (*see page 12*) and the clouded leopard (*see page 13*). However, recent studies of DNA and the ancient viruses found in body cells have cast this system into confusion.

The old geographical divisions have been undermined as cats from different continents are found to be closely related. Even more surprisingly, cats that appear very different physically have been shown to share almost identical DNA and virus-resistance patterns. Several cats currently classified as *Felis*, including the serval and golden cats (*see page 16*) are now thought to belong in the *Panthera* genus, despite their inability to roar.

In the future, the traditional classifications will be replaced with new, accurate groupings that will tell us more about the history of the cat family. For now, however, the old names remain in use, in this book as elsewhere, to avoid a confusing variety of classifications.

The sabre-toothed cats
Skeletons recovered from the Rancho La Brea tar pits in California show that the Smilodon had canine teeth over 15 cm (6 in) long. While deadly in appearance, the teeth were too fragile for use in attack and probably were used to subdue and kill prey after it was brought down.

AUSTRALIAN "CATS"

Australia, New Guinea, and New Zealand separated from the major land mass called Gondwanaland about 85 million years ago, long before cats evolved. In these regions, animal evolution was isolated from the rest of the world. While no indigenous mammals had arrived in New Zealand, marsupials – mammals that give birth to young which develop fully in a pouch in the mother's body – existed in Australia. In the absence of true cats, a marsupial evolved to fill their predatory niche. The tiger cat, *Dasyurus maculatus*, became one of Tasmania's and eastern Australia's small native predators.

Carnivorous marsupial
In Australia, the absence of true cats allowed this marsupial carnivore to fill the cat's environmental or ecological niche.

HOPLOPHONEUS

NIMRAVINUS

DINICTIS

SMILODON

MACHAIRODUS

ACINONYX

Pseudaelurus
By 20 million years ago, carnivores that looked like modern cats had evolved. The Pseudaelurids were larger than miacids, typically feline, with flexible shoulder blades, spines, and tails. These species walked almost on their toes and inhabited open plains in North America and Europe.

FELIS AND PANTHERA

Felis lunensis
Felis lunensis, a species that lived in Europe about 12 million years ago, is thought to be the direct ancestor of today's wildcat family, *Felis silvestris*. They were smaller than their predecessors, and considering their European habitats, probably had thick coats with spotted tabby markings.

Modern cats
Cats radiated into two groups, Old World and New World, between 8 and 12 million years ago. Scientific developments have lead to genetic fingerprinting, which reveals the close relationship between the domestic cat and other species like the Pallas' cat.

MIOCENE

PLIOCENE

PLEISTOCENE

BIG CATS WORLDWIDE

THE BIG CATS SHARE THEIR ORIGINS AS WELL AS MANY ASPECTS OF THEIR behaviour with their smaller relations, although they are mostly members of the *Panthera* genus, while the small cats are members of the *Felis* genus. There is, however, a fundamental distinction between the two groups: the big cat has the ability to roar, but cannot continuously purr, while small cats can purr continuously, but do not roar. Members of the big cat group naturally inhabit Africa, Asia, the Indian subcontinent, and South America. With the exception of the Masai subspecies of lion, these relations of the domestic cat are classified as vulnerable or endangered species.

The classification of the big cats still varies according to different authorities, but with the development of genetic fingerprinting the genetic identity of each species is now being investigated. Without doubt, the results will lead to some reclassification of the species.

Like the domestic cat, all big cats are born blind and helpless, covered with fur, in litters ranging from one to six individuals. Unlike other carnivores such as bears or wolves, which supplement their meat diets with vegetation, cats feed exclusively on vertebrate prey. They cannot survive without eating animal protein and fat, and each species of big cat has evolved its own technique for capturing its meal. The lion uses co-operative social behaviour and brute force to catch and kill mammals that are often considerably larger than itself, the cheetah employs speed and an under-the-throat death bite to kill, while the tiger, leopard, and jaguar all stalk and pounce on their prey.

LIVING WITH HUMANS

To humans, the big cats have often held social and religious significance. Tigers and jaguars have been worshipped in their native lands, while lions and leopards have been used by many nationalities as symbols of royalty. Other than the domestic cat, the cheetah has the longest history of living alongside people. From Ethiopia through the Arabian peninsula to Mogul India, they were used for centuries as hunting companions, raised from kittenhood to hunt antelope and other herbivores.

In India, the cheetah's role was similar to that of the falcon in Europe: it was taken, blindfolded, to the hunting area and freed when the quarry was sighted. India's other big cat, the tiger, has never been tamed. Although tigers rarely make contact with humans, in the absence of other prey some do learn to kill people, and the man-eating tiger has become embedded in Indian folklore. Labelled a "man-eater", it was hunted extensively in the 19th century and is now an endangered species.

Leopard *Panthera pardus*
Most leopards, found from Africa through to the Far East, have black spots on a fawn coat. The black panther, previously thought to be a distinct species, is now known to be a black-coated variety of leopard. This natural variation in coat colour has been deliberately increased in domestic cats.

Cheetah *Acinonyx jubatus*
Perhaps the most distant of all the domestic cat's living relations, the cheetah is the fastest animal on land. It shares the hunting territory of lions and leopards, but prefers to hunt for its meals in the morning and early afternoon, when other predators rest. To assist with rapid acceleration, its claws, unlike those of other cats, do not retract into sheaths.

Jaguar *Panthera onca*
The only big cat to be found in the Americas is a New World equivalent of the leopard. Although classified as a big cat, it grunts and growls but seldom roars. Jaguars are lone hunters, preferring dense forests and swamplands with easy access to water. Now exterminated in the United States, subspecies exist from Mexico to northern Argentina.

LION AND TIGER HYBRID

Lions and tigers have been cross-bred in captivity, producing new hybrids referred to as "tigons" or "ligers". The viability of matings shows how closely, in evolutionary terms, these two distinct species are related, but the fact that the offspring are usually sterile indicates the degree of divergence that has occurred within the cat family.

Infertile offspring
Like matings between the horse and the donkey that produce the sterile mule, the lion and tiger hybrid cannot breed.

Tiger *Panthera tigris*
Found in isolated pockets from India to Siberia, the tiger's habitat ranges from tropical forests to rocky mountainsides. A male's territory covers that of four or five females but never another male's. When prey is plentiful, tolerance of other tigers increases, but when it is scarce, tigers live alone and meet only to mate.

Clouded leopard *Neofelis nebulosa*
Found from Nepal to Taiwan, the clouded leopard hunts in trees, preying on monkeys, squirrels, and birds. Its rigid hyoid bone is like that of the small cats and prevents it from roaring, but its resting posture is that of a big cat.

Lion *Panthera leo*
The most sociable members of the cat family, lions live in "prides" of related females, a characteristic also found in the domestic cat. The lion is restricted to Africa, although a small population of Asiatic lions exists in the Gir Forest of northwest India. Expansion of agriculture has forced several subspecies into extinction, although lions recently found in the Addis Ababa zoo in Ethiopia may be survivors of the extinct Barbary lion of North Africa.

Diminishing habitat
The broad geographical distribution of big cats, living in so many varied habitats, gives an impression of stability and success, but big cat numbers are declining almost everywhere. This is partly due to hunting, for pelts and the Far East animal-products trade, but also through loss of natural habitat. A hundred years ago, tigers ranged from Turkey to Korea and Bali. Without active human intervention, the habitat of the big cats will shrink even more.

SMALL CATS IN THE AMERICAS

ANIMALS ADAPT AND EVOLVE SO THAT THEY MAY SURVIVE IN THEIR environment. Due to the rich diversity of topography and climate in the Americas, a wide variety of species in the genus *Felis* has developed there. The largest of the small cats, the puma, has been successful throughout all the Americas, while the medium-sized lynx and bobcat found a niche in North America, although their future is not certain. In South America, a number of smaller cats evolved. Many of these are similar in size and appearance to the domestic cat, but they are intrinsically wild. None has the potential tameability that the ancestors of the domestic cat possessed.

Evolving cat populations became separated when continental drift created land masses that were surrounded by ocean. Because this effectively halted the spread of animals to some areas of the world, regions such as Madagascar and Australia have no indigenous cats (*see page 11*). The Americas have only one big cat, the jaguar (*see page 12*), but a diverse range of small cats in the genus *Felis* evolved. Scientists think that many of these cats developed in North America and then migrated south when continental movement brought South America back into contact with North America. In South America, cats successfully diversified into the variety of species that exist today.

The domestic cat's American relations are of similar size but of shy temperament. Few have been studied in any detail, and very little is known of their biology or social behaviour. Many of these small cats prefer to live in dense underbrush or in trees, where they hunt at night, and these habits make them far more difficult to observe than cats that live on open plains and hunt by day. Genetic tests of these American small cats have not been performed, but it is very probable that many are closely related to each other. The ocelot, margay, tiger cat, kodkod, and Geoffroy's cat are now all thought to be strains of one species. The bobcat and lynx are often classified as another single species.

HUMAN CONTACT

Due to the secretive lifestyles of the smaller cats, there are few native stories about them. The one exception is the jaguarundi, which, according to some authorities, was tamed by local Indians before the Spanish arrived in South America. Both the jaguar and the puma are thought to have played significant roles in the mythology of pre-Columbian civilizations. Artefacts left by Aztec and other societies suggest that these cats were respected, even venerated. In North America, the bobcat was characterized as a trickster by the Native Americans, making appearances as such in their folklore, but it was never tamed.

Lynx *Felis lynx canadensis*
The only member of the cat family that lives on both sides of the Atlantic, the lynx is widespread throughout its native habitat. While European lynxes are large (*see page 17*), some adult American lynxes are no bigger than domestic cats. A taste for snowshoe hare is detrimental to the species: fluctuations in the hare population directly affect lynx numbers.

Bobcat *Felis lynx rufus*
A subspecies of the lynx family, the bobcat or red lynx inhabits vast areas of North America from Nova Scotia extending down into Mexico. Its black-spotted, brown coat is camouflage in the rocky, bushy terrain where it hunts. Slightly larger than the lynx, with a longer tail, the bobcat has large ears, but less dramatic ear tufts than its relation.

Ocelot *Felis pardalis*
Once widely distributed, the ocelot has suffered the misfortune of having a handsome orange-yellow coat, black-striped and spotted, for which it has been hunted to extinction in parts of North America. A superb climber, it hunts in trees and sleeps during the day.

Jaguarundi *Felis yagouarundi*
Sometimes called the ottercat or weaselcat, this feline inhabits dense undergrowth from Arizona to Argentina. Because its fur has never been popular, it has escaped the hunters, but little is known about its behaviour. Most active in the morning, jaguarundis eat rodents and rabbits but also frogs and birds. South American Indian stories tell of taming jaguarundis for use in rodent control.

Puma *Felis concolor*
Ranging from southern Canada to Patagonia in South America, the puma is also known as cougar, panther, and mountain lion. The largest member of the genus *Felis*, its prey ranges from rodents to deer. Because it takes domestic livestock, it has been hunted to the point where both the Florida panther and the Eastern puma are now endangered species.

Threat from humans
The most serious threats to small cats are the fur trade, the exotic-pet trade, and loss of habitat. Public awareness of how wild animals are killed has led to a dramatic reduction in the trapping of wild cats, and legislation now protects endangered species, but both came too late for some species. For fashion items, skins needed to match: once there were not enough skins from one species, hunters simply moved on to the next species. The result is that some populations of spotted cats may never recover. In the face of habitat loss due to agriculture, the remaining populations must adapt or decline further.

Tiger cat *Felis tigrinus*
Also known as the little spotted cat, ocelot cat, and oncilla, the tiger cat is often mistaken for its close relative the margay. Its habitat extends from Costa Rica to northern Argentina, but it is rare because it is hunted for its coat, which varies in colour from cream to rich ochre. Although a forest dweller, it does not live in trees as much as the margay. It preys on birds, mammals, lizards, and insects, but little is known of its social behaviour.

Margay *Felis wiedi*
The margay, sometimes called the long-tailed spotted cat, looks much like a small ocelot. Although a close relative, it is considerably smaller – about the size of a large domestic cat. Its habitat extends from Mexico to Argentina, although it has become extinct in vast areas of the Americas.

Mountain cat *Felis jacobita*
This rare cat inhabits the high areas of the Andes. It is the same size as the average domestic cat, although males can be slightly larger. Its silver-grey coat, spotted or striped with brown or orange, is dense, and makes it well adapted for the bitter weather of its high-altitude habitat. It has been seen at altitudes of up to 5,000 m (16,000 ft) above sea level.

Pampas cat *Felis colocolo*
Also called the "gato pajero" or grass cat, because of its favoured habitat, the pampas cat ranges from Ecuador to northern Chile and Argentina. It is a nocturnal predator, and preys on ground-nesting birds and small mammals. The size of a domestic cat, the pampas cat arches its back dramatically and its fur stands on end when it is excited or aggressive.

Geoffroy's cat *Felis geoffroyi*
Sometimes called Geoffroy's ocelot, this tiny cat is closely related to the kodkod and the much larger margay. An excellent climber and swimmer, it preys on small mammals and birds. Its habitat extends from Brazil and Bolivia south into Argentina. The colour of the coat varies from ochre to silver, but spotted grey is the most common.

Kodkod *Felis guigna*
The smallest of the American cats, the male kodkod, or guiña, is never more than 3 kg (7 lb) in weight and 52 cm (21 in) long. This spotted, grey-brown tree-dweller is found mostly in central and southern Chile, but there are also populations in western Argentina. Very little is known about its social behaviour or hunting habits, although farmers claim that it attacks poultry.

SMALL CATS IN AFRICA AND EURASIA

MOST AUTHORITIES NOW CLASSIFY THE SMALL AND MEDIUM-SIZED cats of Africa and Eurasia as members of the genus *Felis*, the domestic cat's close family. There are many similarities between the sizes, looks, activities, and even, when they are known, the genetics of these cats. Like the domestic cat, the lifestyle of its close relatives is one of prolonged inactivity punctuated by bursts of activity when in search of food. Little is known about the social behaviour of these species, but those that have been observed have displayed habits similar to those of the domestic cat. Few, however, carry the potential for domestication in their natural form.

The individual anatomical and behavioural characteristics of each species of wild cat are the result of evolutionary adaptations to take advantage of available sources of food. Some species, such as the fishing cat, have adapted to life on the edge of water, where they feed on shellfish, snakes, and rodents. Others, like the Chinese desert cat, have had to adapt to life in an arid climate, and more typically prey upon small mammals.

All cats have developed coat colours and markings that camouflage them from their prey in their particular environments. Generally speaking, spotted cats live in forests, while less vividly patterned cats live and hunt in tall grass or open territory. Regardless of their habitat, however, all of the domestic cat's close relations share common characteristics. In the wild, most cats are solitary and keep to their own territories, which they mark with scent. A male's territory may overlap the territories of several females but they rarely meet, except to mate after engaging in short periods of courtship.

SMALL CAT HYBRIDS

Small cats can successfully interbreed, although cross-species matings are unlikely to take place in the wild. Just as a lion mated to a tiger produces a "liger" or "tigon" (*see page 13*), hybrid small cats are born. The difference between these matings and those of the big cats is that, with the correct cross-species alliance, the resulting kittens are fertile, capable of perpetuating the new hybrid. Some species have been bred with the domestic cat, creating handsome hybrids. However, the resulting hybrids are often intractable, especially in early generations. Tameness, developed through generations of breeding, is quickly diminished by breeding with species that have not undergone the biofeedback or hormonal changes associated with tameness (*see page 58*).

Jungle cat *Felis chaus*
This cat is misnamed, for it prefers open country, reedy marshes, and woodland, and is also known as the reed cat, swamp cat, and marsh cat. Mummified jungle cats have been found in Egypt.

Sand cat *Felis margarita*
This cat's territory stretches from Saharan Africa to the deserts of Baluchistan in Asia. It has thick hair between the pads of its feet to prevent it sinking into the sand, and it is able to survive without water, obtaining moisture from its prey of desert rodents.

Caracal *Felis caracal*
A member of the lynx family that has adapted to the heat of the tropics, the caracal has distinctive ear tufts. Its habitat ranges from Africa to the Arabian Peninsula and India.

African golden cat
Felis aurata
This rare cat inhabits the high forests of Africa, although it is not a tree-dweller. Its prey includes rodents and birds, which it hunts by night and day. Its coat varies from brown to grey.

Serval *Felis serval*
A native of the African savannahs, the serval's coat is tawny with black spots. It can leap twice its own length to bring down birds.

Black-footed cat
Felis nigripes
The smallest of the African cats, this spotted cat is found in the savannahs of southern Africa, where it preys on lizards and small rodents.

Rusty spotted cat
Felis prionailurus rubiginosus
The smallest of the Asian cats, this tiny feline, which is smaller than a domestic cat, inhabits the scrub and grassland of southern India and the humid mountain forest of Sri Lanka. This cat is nocturnal and hunts small mammals, birds, and insects.

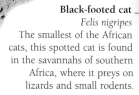

Lynx *Felis lynx pardina*
The largest of the European cats, the lynx can be up to five times heavier than the domestic cat. Ranging from Scandinavia to Siberia, lynx populations vary. Extinct in France and Italy, the smallest population is found in the desert area of central Spain, where around 1,000 survive.

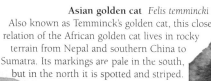

Pallas' cat *Felis manul*
From Iran to western China, the Pallas' cat inhabits mountainous and rocky terrain, where it preys upon rodents. It has evolved with small ears and a very dense coat in order to survive subfreezing temperatures.

Asian golden cat *Felis temmincki*
Also known as Temminck's golden cat, this close relation of the African golden cat lives in rocky terrain from Nepal and southern China to Sumatra. Its markings are pale in the south, but in the north it is spotted and striped.

Asian leopard cat *Felis prionailurus bengalensis*
Similar in size to the domestic cat, the leopard cat inhabits an area from northern China to the islands of Indonesia – where its success is aided by its ability to swim and catch fish. It has been bred with the domestic cat to produce the Bengal (*see page 208*).

Marbled cat
Felis pardofelis marmorata
The marbled cat is found from the eastern Himalayas to Borneo. Very rare, it is now protected in India and Thailand. Its coat resembles that of the clouded leopard, but it is considerably smaller in size.

Fishing cat
Felis prionailurus viverrinus
With slightly webbed toes and less retractile claws than other cats, the grey-brown, spotted fishing cat lives in the mangrove swamps of India, Sri Lanka, China, and Myanmar. It preys on fish, shellfish, snakes, and small mammals.

Flat-headed cat
Felis prionailurus planiceps
Also known as the marten cat, this red-brown cat's most distinctive feature is its long, flat head. Throughout its habitat in India, Sri Lanka, Indonesia, and southern China it prefers proximity to water, where it hunts small mammals, birds, and fish.

Iriomote cat
Felis prionailurus iriomotensis
Discovered in 1967 on the Japanese island of Iriomote, this small nocturnal cat lives by the water's edge, eating birds, crabs, and small mammals. Probably fewer than 100 survive, each hunting in its own territory.

Chinese desert cat *Felis bieti*
Inhabiting the steppes and mountains of western China and southern Mongolia, this domestic cat-sized species has a dense coat, and thick hair on the soles of its feet. Its yellow-brown coat offers camouflage in its dry habitat.

Bay cat *Felis pardofelis badia*
Extremely rare, this close relative of the marbled cat is also known as the Bornean red cat. It has a chestnut-brown coat with spotting on the underbelly. Most rare are individuals with bluish fur. It inhabits rocky, forested regions of Borneo, where it preys on small mammals and birds.

Vanishing species

As do many of their relatives, a number of the small cats of Africa and Eurasia face an uncertain future. Fortunately, species such as the caracal, serval, African golden cat, and Pallas' cat exist in viable numbers in secure wildlife refuges. Others, like the Iriomote cat, the Asiatic golden cat, bay cat, and flat-headed cat, are vulnerable and in danger of extinction. Governments realize the value of "ecotourism" and now invest in protecting the habitat of many of these species, but unfortunately for both the ecotourist and the dedicated biologist, many of these cats are too wary of human presence to be seen. Wildlife films continue to be the only way of seeing many of these relations of the domestic cat.

CLOSEST RELATIONS

THERE IS A GROUP OF SMALL, INDEPENDENT WILDCATS SCATTERED ACROSS Africa, Europe, and Asia that are members of the genus *Felis*. They are so similar in appearance that experts are unable to agree whether each is a distinct species, or simply a regional race or subspecies adapted for its environment. These wildcats have proved to be the domestic cat's closest relations: recent genetic studies suggest that the domestic cat is virtually identical to the African wildcat, and that these two varieties are appreciably different from both European and Asian wildcats. All of these cats can be interbred, producing fertile offspring.

The wildcat, *Felis silvestris*, is the domestic cat's closest relative. Until recently, taxonomists have failed to agree on the significance of differences within this species of small, lone hunters. There are visual differences between varieties, with coat colour and texture varying according to habitat, and there are also differences in temperament between the wildcats of Africa and those of Europe. In Somalia, for example, African wildcats live close to human settlements, even scavenging within the boundaries of villages. In Britain, Scottish wildcats are virtually never seen, although some have adapted to hunting the plentiful supply of rabbits that are found on agricultural land.

Recent studies carried out by Professor Eric Hurley at Cape Town University in South Africa have indicated that there are distinct genetic differences between the European wildcat and the African wildcat. The study suggests that although physically similar, they diverged into two different species hundreds of thousands of years ago. Genetic studies are likely to reveal that all the regional varieties of European wildcat are isolated populations of a single species.

DISTINCT DESCENDANT

The studies of the African wildcat have revealed that it has almost identical "genetic fingerprints" to the domestic cat, giving sound evidence that the domestic cat is an emigrant from Africa.

Sufficient subtle differences remain between the genes of African wildcats and domestic cats to identify an individual as one or the other, even when populations are mixed. Experts have disagreed whether the scavenging African wildcats and adapting Scottish wildcats are full-blooded wildcats or wildcat-domestic hybrids: genetic testing may answer this question.

The wildcat family

Wildcats have thrived in climates ranging from the harsh winters of Scotland to the burning scrub of Africa. They have evolved into a range of types, with differences subtle enough for some to argue that they are closely related, diverse enough for others to see them as separate species.

Scottish wildcat *Felis silvestris grampia*
Once prevalent throughout Britain, wildcats had disappeared from England by the 1800s, retreating to Scotland, where they survive north of a line drawn from Glasgow to Edinburgh. With protection, their numbers have begun to recover. The Scottish wildcat has a broader head and shorter tail than the domestic cat.

Spanish wildcat *Felis silvestris iberia*
Looking like heavily built domestic tabbies, the wildcats of Iberia are a remnant of the wildcats that once lived throughout Europe. Hunted for their pelts and persecuted because they are believed to prey on livestock, they survive today only in isolated regions.

European wildcat *Felis silvestris europeus*
After declining almost to extinction, the populations of European wildcats are now stable in many countries. Intractable and virtually untamable, this indigenous wildcat is now a protected species in many regions, including Germany, the Czech Republic, and Slovakia.

Indian desert cat *Felis silvestris ornata*
Also called the Indian desert wildcat, the habitat of this cat ranges south from Russia down to central India. It is smaller than its African and European relations, and its coat is spotted rather than striped.

African wildcat *Felis silvestris lybica*
The African wildcat, like its European relatives, has a larger brain than the domestic cat, which is probably due to the demands of survival. Unlike European wildcats, it can be relatively tame if raised in captivity. Genetic investigations indicate that this is the true ancestor of the domestic cat.

HYBRIDIZATION OF THE WILDCAT

Wildcats in Europe have been threatened by human persecution and the spread of agriculture. Less obviously, they may also have been endangered by hybridization with the domestic cat. It is possible, even likely, that male wildcats have fathered kittens with female domestic cats. At adolescence, these hybrid kittens exhibit the father's wild ways and move away from human settlements. As a result, wildcats are unlikely to play a role in the development of the domestic cat, but domestic cats may play a role in the evolution of the wildcat. Genetic testing suggests that hybridization between the African wildcat and the closely related domestic cat is uncommon. Although tests have not yet been carried out on the European wildcat, its natural reclusiveness suggests that hybridization is not a serious threat to its survival.

Truly wild
Unlike the African wildcat, which can be domesticated if raised from kittenhood, the wildcats of Europe appear to be innately fearful of humans. This trait is so strong that even a hybrid of wildcat and domestic cat, such as this one, will be a lifelong "spitfire", unable to adapt to life with people.

FROM WILDCAT TO DOMESTIC CAT

UNLIKE OTHER DOMESTIC ANIMALS, THE CAT IS SELF-DOMESTICATED: it chose to live in close proximity to people because it was in its interests to do so, a natural selfishness that remains at the core of domestic cat behaviour to this day. This self-domestication was relatively recent, occurring long after we had domesticated the dog.

The cat's ancestors needed specific psychological and physiological traits to help them to infiltrate the new and unique ecological niche of the human settlement. Under the altered environmental pressures, both physical and behavioural changes occurred, creating the most successful ever of all the species of cat: the domestic cat.

The history of the cat is best documented in ancient Egypt, but Egypt was not necessarily its first domestic home. A cat's tooth from 9000 BC has been found in a settlement site in Jericho, Israel. Wildcats did not inhabit Cyprus, so feline remains from before 5000 BC found on the island (see page 24) suggest that cats were taken there, perhaps even as pets. Remains from before 4000 BC have also been found near Harappa, in the Indus valley.

The earliest remains from Egypt that show a possible affectionate relationship between cats and people are as old as those from Harappa. At a burial ground in Mostagedda, Middle Egypt, a man was buried with a gazelle near his head and a small cat by his feet; Egyptologists believe that the gazelle was intended to serve as his funerary meal, and the cat was his pet.

Egypt's first permanent human settlements, with their granaries and silos, appeared around 4000 BC. This unique development offered concentrations of rodents and edible waste, together with abundant shelters and safety from larger predators, to any cat able and willing to take advantage of them. Two small cats, the jungle cat *Felis chaus* (see page 16) and the African wildcat *Felis silvestris lybica* (see page 19), lived in this region. With a less fearful nature, the African wildcat was prepared to enter this new environment in search of its favoured rodent prey. A further minor genetic mutation, perhaps involving the hormonal control of emotions (see page 58), created a population with a unique survival advantage: an ability to live and breed in human settlements.

Birth place of the domestic cat
Although there is evidence that cats lived alongside humans in other places, the floodplains and delta of the Nile have yielded the most abundant records of feline domestication.

Records of domestication
The development of settlements and agriculture along the fertile floodplains of the Nile provided the indigenous wildcat with the potential of a new environmental niche. From this time onwards, burials and paintings provide tantalizing scraps of evidence for a growing relationship between felines and humans. The absorption of the domesticated cat into Egyptian society seems to have been complete by about 2000 BC, and was followed by its recognition as the symbolic embodiment of a deity.

The African wildcat's natural lifestyle
In its natural environment, for which it has evolved over millennia, the African wildcat has very little stress from overcrowding, but it faces fierce competition for mates, prey, and territory: few kittens live to adulthood. Domestication suits stress-tolerant animals that can accept more crowding.

Living in two worlds
The development of agriculture around 4000 BC (see right) opened up a new world of possibilities to the wildcat; here was an unprecedented abundance of ready prey gathering around crop fields and granaries. To take advantage of this bounty, however, the cat had to be able to tolerate living not only at higher densities of population, but also in close proximity to other species. The first feline remains in a human tomb, found at Mostagedda, date from around this significant time. Others exist from the following millennia, although it is hard to say exactly how domesticated these cats were.

ADAPT AND SURVIVE

Under these new environmental pressures, the wildcat evolved into the domestic cat. There were enduring changes in its character. Placidity became an increasing asset, as only the most placid and least fearful could survive in close proximity to so many people and animals. Sympathetic people became substitute parents for motherless kittens. Uniquely, compared with other wildcat subspecies, kittens raised by people during the critical first seven weeks of life did not develop a fear of them on reaching adulthood; they remained willing companions.

There were also physical changes. Camouflage was no longer needed, so coat pattern and colour mutations that would not have survived in the wild were perpetuated in the settlement cats. The gastrointestinal system changed to cope with a more varied diet; the domestic cat's intestines are longer than the wildcat's. The brain became almost 30 per cent smaller because the cat was no longer so dependent upon its senses for survival.

THE ROAD TO SUCCESS

People recognized the value of the newly domesticated cats as vermin-killers, but they probably also served a less well-known purpose. Rats and mice were devastating pests, but poisonous snakes, both indoors and outdoors, were lethal. There was little that people could do to safeguard themselves from cobras and vipers, so supreme, fearless hunters such as wildcats were tolerated or even openly welcomed in settlements for their willingness to hunt snakes as well as rodents. The cat became overwhelmingly popular because of its humble origins and its doubly protective role.

The ancient Egyptians developed a deep respect and understanding of cats and their behaviour, in time even turning them into symbols of deities. Some researchers claim that the cat itself eventually became a deity of sexuality or fertility in Egypt, but such direct worship of felines is unlikely. What is certain is that the cat's natural behaviour ensured that it became a prominent symbol in religion and superstitions, and its image became incorporated into a huge variety of objects. The domestic cat was now exportable: at first, only stories and images of the domestic cat were carried around the Mediterranean, but these reports were soon to be followed by the cats themselves.

Direct descendant
Today's pets remain close to the ancient stock of self-domesticated wildcats – at heart, virtually nothing has been either added or taken away. Without socialization in kittenhood, even the most selectively bred pedigree cat will turn its back on countless generations of domestication; given the chance and a little practice, it will hunt and defend its territory just like a wildcat.

In the heart of the home
Those cats that could overcome natural timidity and solitary habits found the human home to be a safe and comfortable niche. By 2000 BC, the cat was well established, appearing in hieroglyphics from El Lisht. The first artwork to show a cat in a domestic setting dates from c. 1950 BC and was found at Beni Hasan. In keeping with the elaborate coding of later Egyptian art, the mistress of the house is indicated by the cat crouching beneath her chair (see left). It was in the Ptolemaic period, from 333–30 BC, that cat worship reached its peak, centred on Bubastis in the Nile delta.

HYBRID THEORIES

Among the many mummified feline remains found at Beni Hasan, thought to have been offerings to Bastet (see page 26), are several large skeletons of the marsh-dwelling jungle cat *Felis chaus*. This has led some to speculate that the domestic cat is in fact an evolutionary hybrid of the African wildcat and the jungle cat. However, Professor Eric Hurley's genetic studies of the cat, which use minute examinations of gene sequences, suggest that the domestic cat is genetically far too similar to the African wildcat to be a hybrid with any other species, including the jungle cat.

Discredited ancestor
The role of Felis chaus *in the development of the domestic cat has long been disputed: the first scientific evidence appears to rule it out.*

CATS
AND MAN

HISTORICALLY, CATS HAVE PLAYED A remarkably ambivalent role in human society. Worshipped as beneficent gods in some cultures, they were castigated as the agents of evil in others, including Western Christianity. In the last 150 years our relationship with the cat has changed dramatically. Although in literature and art it is sometimes still equated with witchcraft and immorality, in the modern culture of advertising and entertainment the cat now represents sensuousness and refinement. The cat's remarkable success in surpassing the dog as man's most popular animal companion in North America, Britain, and elsewhere is ample evidence of the universal appeal it has to modern sensibilities.

ABOVE FEMALE USUAL SOMALI

LEFT MEDIEVAL MANUSCRIPT PORTRAYING CATS AS HUNTERS

THE SPREAD OF THE CAT

THE DOMESTIC CAT HAS BEEN A REMARKABLY SUCCESSFUL MIGRANT; in a relatively short space of time, it has spread from the Near East and Africa to all the continents except Antarctica, and on to many islands. Early in their dispersal, cats may have been transported simply to satisfy people's curiosity, but the cause of their continued success was the spread of the mouse and rat. Cats became a valuable commodity in the war against rodents, and it was in this "terminating" role that they travelled across the world's trade routes, eventually reaching the New World. In total, the domestic cat population currently numbers hundreds of millions worldwide.

Language studies offer interesting clues about where cats originated and when they travelled. In his classic work, *A History of Domesticated Animals* (1963), naturalist Frederick Zeuner claimed that the cat came from the Near East and Africa, and discussed the origins of its various names. He theorized that the word "cat" was of North African descent and came from the Berber word *kadiska*, which corresponds with biological evidence that points to this region as the origin of the domestic cat. In other parts of North Africa and the Middle East, the domestic cat was known as *kadis* in Nubian, *qato* in Syrian, and *quttah* in Arabic.

FELINE NOMENCLATURE

The ancient Egyptians used the onomatopoeic word *miou* for the domestic cat, now reinstated in the modern breed name Egyptian Mau. When cats first arrived in Greece they were called *gale* (the Greek name for polecats), while the Latin word *felis* was used to denote any yellowish carnivore, which included the weasel, the polecat, and the cat. By AD 100, the word *katoikidios* was attributed to the cat in Greek texts, and 200 years later *cattus* had supplanted *felis* in Latin. Variations of these words entered many other languages including French with *chat*, Spanish and Portuguese with *gato*, Italian with *gatto*, Swedish and Norwegian with *katt*, Danish and Dutch with *kat*, German with *katz*, Czech with *kocka*, and Russian with *koshka*.

CONQUERING THE WORLD

In mapping the spread of the cat, one of the most significant finds comes from a tomb in Cyprus, thought to be the remains of an African wildcat, *Felis libyca*, dating from about 5000 BC. This cat, which must have been imported, was a domestic animal but still as large as a wildcat.

It seems that the cat reached India, probably with Phoenician traders, by 500 BC, and in all likelihood even earlier: the two Indian epics, the *Mahabharata* and *Ramayana* both mention the cat, and are dated *c.* 300 BC. It is, however, much harder to map the arrival of the domestic cat in countries of the Orient. Dates vary from 2000 BC to AD 400 for the cat's arrival in China and Southeast Asia, while the traditional date of AD 999 for Japan (*see page 30*) does seem rather late considering the proximity of these lands. The comparatively late arrival of the cat in Europe was undoubtedly due to the embargo that the Egyptians placed on this sacred animal. But with the advent of Christianity, the cat travelled north with the expanding Roman Empire, eventually reaching southern Russia and northern Europe around AD 100. A cat skeleton excavated from a Roman villa at Lullingstone in southeast England supports this theory. Around the same time, cats arrived in Norway under a different escort, possibly accompanying returning mercenaries from Byzantium (now Istanbul). Surprisingly, it took a further 400 years for the cat to reach Latvia, by which time cats were being traded freely throughout most of Europe.

There was then a very long period before the cat embarked on its voyage to the New World. Although French Jesuits took cats to Quebec in the 1500s, and at least one cat accompanied the Pilgrims to America on the *Mayflower* (1620), it was not until the 1700s that cats appeared in the New World in appreciable numbers, when settlers in Pennsylvania imported them to control a plague of rodents.

Domestic adaptation

The majority of skulls from Egyptian cat cemeteries resemble Felis libyca *(see page 20). Since the first cats were domesticated, subtle, gradual changes in coat colour have arisen through climate, terrain, and artificial selection. American and Australian coat colours show that cats travelled here from British ports in the 1700s and 1800s respectively. Cats with an almost white "Van" bi-colour originate from Byzantium (later known as Constantinople, now as Istanbul), but became isolated in Scandinavia after being taken there by returning traders and mercenaries.*

Spiritual effigy

This wooden feline effigy, dating from around 1450, was found at Key Marco in southwestern Florida. It is believed to be a ritual item symbolizing a spiritual connection between the human and animal kingdoms. The Key Marco feline could be mistaken for a domestic cat, except that it predates the cat's arrival in the Americas and is more likely to be a Florida panther.

NORTH AMERICA
18TH CENTURY

French funerary stele

Similar to modern-day tombstones, funerary steles were once erected to commemorate the dead. One theory suggests that it was customary for the Romans and Gauls, who buried their dead children with their toys, also to show them with their pets – such as in this stele of a boy called Laetus (*c.* AD 100). The pet, in this case a cat, was thought to assist in a safe journey to the afterlife.

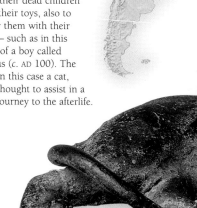

Scandinavian figurine

This rudimentary cat, carved in amber, was found in an archaeologial excavation of a site in northern Sweden. It is thought to date from the 9th century. Such artefacts indicate that the domestic cat was established, if not commonplace, across Scandinavia by this time.

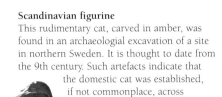

Canine confrontation

When cats first settled in mainland Greece they were probably viewed with curiosity. This Greek marble relief (c. 500 BC), shows a cat and a dog being introduced. The interest shown by the observing men indicates that the intention was to see the outcome of this controlled confrontation, while the use of leads indicates that the animals were regarded as personal property.

Middle-Eastern statuette

Found in Lachish, Israel, this ivory cat carving dates from around 1700 BC. Although the Egyptians forbade the exportation of their most revered animal, regular commercial traffic between Egypt and other Middle-Eastern ports would have assisted the cat's eventual migration.

RUSSIA
AD 700

BRITAIN
AD 100

JAPAN
C. AD 600

INDIA
500 BC

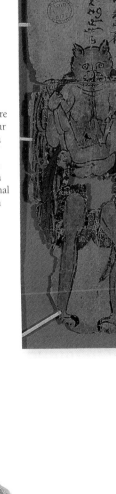

Protective feline spirit

The inscription on this Buddhist scripture warns, "If you dream of a kitten and your small child lolls out its tongue, then you may know that this spirit is the cause of the trouble. Sacrifice to it and all will be well". This script was originally part of a set depicting 16 female spirits with animal heads, all of which were concerned with protecting the young, from Dunhuang, Chinese Central Asia (9th century).

AUSTRALIA
19TH CENTURY

Turquoise pottery

Only this tiny fragment remains from a piece of glazed Persian pottery, dating from around the late 13th century. Although the nose is missing, the embossed head is recognizably that of a feline, and positioned slightly above it is the figure of a bird.

Fertility talisman

The increase in the popularity of the cat deity Bastet (see page 30) in late Egypt (c. 600–200 BC) was reflected in the large numbers of cat talismans which were produced at that time. This bronze female cat with kitten may have been presented to Bastet as an expression of gratitude or in expectation of future favours, especially in connection with matters such as fertility and child care.

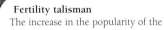

CAT-VENERATING RELIGIONS

ALTHOUGH IT IS NOT AS COMMON IN MODERN CULTURES, ANIMAL worship was once widespread. Deities who were not depicted in animal form usually had a sacred animal that typified their attributes, and cats of all sizes were among the animals that played such a role. The big cats, such as lions and jaguars, were often associated with protective forces, whereas the domestic cat symbolized more "feminine virtues", especially fertility. Because of its natural nocturnal inclinations, the cat also came to be seen as a guardian of the night and sometimes death's companion, which may have led to its later associations with magic and the devil.

Many religions have incorporated cat worship in one form or another. It is generally agreed that the domestic cat originated in North Africa, which is probably why the ancient Egyptians were the first to recognize its religious potential. In South America, where big cats were prevalent, early societies similarly incorporated the jaguar and cougar into their pagan rites.

Religious charm
Small cat amulets, such as this one (c. 305–30 BC), were probably worn as jewellery by Egyptian worshippers of Bastet.

EGYPTIAN CAT CULTS
Cult activities reflecting the ancient Egyptian fascination with felines focused chiefly on two anthropomorphic deities – the lion-headed goddess Sekhmet and her sister, the cat-headed Bastet. Religious or royal associations with the "King of Beasts" were common worldwide, but a cult based on the domestic cat was unique.

Bastet was a contradictory figure because she held the dual associations of life and death. This connection was taken to the extreme with the mummification of large numbers of domestic cats. During the late 19th century, excavators at the temple of Bast at Beni Hassan, Egypt, discovered over 300,000 mummified cats at a single site – unfortunately they were shipped to Britain, ground down, and sold as fertilizer. This extraordinary find shows the prevalence of cat worship.

After the mummies had been offered to the gods, priests gathered them up, and buried them in nearby cat cemeteries. When a cat died, its owners would try to give it a decent burial. They went into deep mourning and shaved off their eyebrows. Temple priests even killed and preserved specially bred cats, then sold them to pious pilgrims as cheap versions of the highly desirable cat statues cast in bronze.

In Egypt, to kill a cat, even unintentionally, was punishable by death. Historian Herodotus (484–424 BC) told the tale of an unfortunate Greek who did just that. Extreme veneration for the cat contributed to the eventual downfall of the Egyptian empire.

WILD FELINE DEITIES
Domestic cats did not reach the Americas until after the arrival of the European immigrants, but the jaguar and other cats of the New World had their place in the pantheons of its early cultures. Beliefs among South American peoples included the idea of medicine men turning into jaguars after death, and the concept of the cougar as *amigo del christiano*, or the Christian's friend, after a young Spanish girl's life was said to have been saved by a cougar.

Egyptian funeral stele
On this tomb stele, or stone, from about 1250 BC, two cats are stationed over a man and his wife. The inscription is a hymn to the "great cat" Atu, and the "perfect god", known as the sun-god, Ra.

Mediterranean painted vase
Domestic cats and their habits were undoubtedly well known in the classical world. On this vase (c. 600 BC), two figures – the seated one may be Aphrodite – sport with a ferocious-looking cat and a white pigeon.

CLASSICAL IMAGES
Greek mythology reveals a wealth of symbolism concerning the lion, but very little involving the domestic cat. But there is some ambiguity in the mythical tales concerning the multifaceted goddess, Artemis. She was originally regarded as a lioness, but was credited with creating the cat, and later fled to the moon in cat form.

The Roman equivalent of Artemis was Diana. Both were revered as great hunters as well as moon goddesses – the moon is said to rule fertility, which links these deities with Bastet (*see page 30*). The Romans imported many of their cultural inspirations from Greece: Venus, goddess of love and sexual pleasure, also equates with the Greek Aphrodite, Egyptian Bastet, and Norse Freya. Venus is at once an affectionate little kitten and a savage cat armed in tooth and claw, an embodiment of maternal emotions. Paintings of Venus as goddess of love frequently depict her with a cat.

Fertility instrument
Bastet, a goddess of pleasure, music, and dance, is linked with the ancient musical instrument called the sistrum, a form of rattle consisting of four beaded rods held in place by a loop, often used in fertility worship.

CATS IN PRE-CHRISTIAN EUROPE

Celtic and Nordic cultures integrated cats into their customs and folk tales, which in turn became part of the early Christian faiths. For example, the Welsh pagan goddess Cerridwen (c. 800 BC) has a link with the fertility cat cult through her son, Taliesin, who changed himself into a grain of wheat to hide from her wrath. Cerridwen swallowed the wheat grain and nine months later her son was reborn. During a description of his other fantastic incarnations, Taliesin said, "I have been a cat with spotted head upon a tripod".

In Nordic legend, the fertility goddess Freya arrived in a chariot drawn by a team of cats. These legendary felines symbolized the twin qualities of their Scandinavian mistress, namely fecundity and ferocity. Like real cats, they were normally affectionate and loving, but could be fierce when aroused. Far up in the north, Finnish people also believed in a cat-drawn sledge, but this time it was sent to fetch their souls. An account of this can be found in the epic Finnish poem *The Kalevala*.

Not all cat associations have been benign. After all, the giver of life was equally connected with the realms of the dead. In the Germanic myth of Thor and the Midgard Serpent, we get a glimpse of the cat's more sinister associations when the Midgard Serpent, who is bent on destroying the earth, is disguised as a cat. In Christianity, the serpent represents evil and betrayal. This story may have been a forerunner to the many tales of perfidy and evil concerning the cat in later religions (*see page 28*).

ISLAMIC BELIEFS

Neither the ancient Sumerian nor Babylonian religions had records of a cat cult as such, although an ancient Persian legend maintained that the cat was born from the lion's sneeze.

Widespread cat appreciation came later, due to a variety of tales associating cats with the Islamic prophet Mohammed. One legend tells of a cat that saved the Prophet from almost certain death from a snake bite. Another, tells of a cat, Muezza, that lived with Mohammed in Damascus. On completing his work one day, Mohammed saw that Muezza had fallen asleep on the flowing sleeve of his garment. Out of respect for Muezza, Mohammed cut away the

Babylonian lion relief

The lion is an ancient symbol of royalty. For example, this lion relief – dated between 604 and 561 BC – was excavated from the Babylonian city of Ishtar and is now on display in the Archaeological Museum in Instanbul.

Sacred golden puma

From about 600 BC onwards, Peruvian peoples have worshipped a puma god, which their early artists made in a variety of stylized forms.

sleeve, rather than disturb his sleeping companion. To this day, because of this story, cats are allowed to enter all mosques freely. Stories such as this were brought from the Holy Land to Europe by crusaders from France, Italy, Germany, Spain, and Britain. The cat became associated with Islam, and this is one reason why the Catholic Church was so successful in its efforts to equate cat ownership with heresy (*see page 28*), and later with evil.

SACRED CATS OF THE ORIENT

The Hindu and Parsee religions are known for their respect for all living creatures. In theory, at least, each orthodox Hindu household is expected to feed and house a cat. The Parsees – established from a group of Persians that settled in India during the 8th century – considered it a serious crime to kill a cat, although it did not attract the death penalty as it had in Egypt.

In the original canons of Buddhism, the cat is excluded from the list of protected animals, because of an incident that took place at the time of the Buddha's death. All the animals gathered, weeping, around the sacred remains; only the snake and the cat remained dry-eyed.

Venerable temple cats

In Buddhist temples across Asia, cats are kept as mousers. The Siamese breed is said to be descended from these temple cats, and many show the typical pointed pattern.

Then the cat pounced on the rat and ate it, proving its disregard for the solemn event. Today, cats can achieve nirvana (the attainment of ultimate serenity) like all other beings.

Early Chinese Buddhists revered the cat's self-containment. Its tendency to "meditate" gained it respect and protection, and its ability to see in the dark was thought to be of value in warding off evil spirits. Throughout China, cat figurines and amulets were used to keep both evil spirits and rats away from homes. In Japan, an analogous Buddhist tradition evolved. When a cat died, it was buried in, or near, the owner's temple. On the same day, a painted or sculpted likeness of the cat was offered at the temple altar. This action assured the cat's owner of a lifetime of tranquillity and good fortune.

In Thailand, cats were ritually entombed in burials of members of the royal family. As recently as the 1920s, a cat was included in the king's coronation procession to enable the late king to witness the installation of his successor.

CATS AND CHRISTIANITY

MANY EARLY CHURCHES OCCUPIED PAGAN SITES, AND RELIGIOUS FESTIVALS often incorporated pagan symbols, particularly at the time when Christianity first spread across to Rome and northern Europe. The cat had many roles in the old religions, and these were sometimes incorporated into the new faith. The Church's ambivalent attitude to cats has its roots in these beginnings. Some parts of the Church ignored their presence; others, such as the Russian Orthodox Church, valued and respected cats, while there are records of early Catholic Popes who treasured their cats. But in 1484 the Church turned upon cats, and for hundreds of years they were persecuted.

As independent and nocturnal creatures that lived alongside man, cats became the subject of myth and folklore in many cultures. Deified in Egypt, in other parts of the world they were associated variously with the supernatural, evil, and fertility (*see pages 30–32*). It is perhaps unsurprising that when looking for enemies within, the Christian Church turned on the cat.

EARLY ACCEPTANCE

Although African wildcats are known to have lived in Israel thousands of years ago, there is no mention of cats in either the Old or New Testament. Christianity is unlikely to have inherited a position on cats from Judaism, which made no reference to them until around AD 500, when the Talmud briefly praised feline cleanliness. It was as Christianity spread to the

Middle East, North Africa, and the European continent that cats and other animals became part of the religious symbolism. Some pagan symbols were adopted by the early Church, while others were demonized. Jesus became known as "The Lion of Judah" while the snake, which was previously venerated in some North African religions, became the image of evil. One of the first Christian sects to evolve were the Egyptian Copts, in the first century AD. Their gospel included cats that were judges after death. This may be linked to the fact that the Egyptian god Osiris, the Lord of the Dead, was often represented in feline form.

In northern Europe, the early Celtic Church encouraged conversion to the new faith by accepting and adapting pagan traditions that did not conflict with the new teachings. The pagan reverence for the natural world was often incorporated into early Church teachings, and several saints from this period are associated with cats. For example, St. Gertrude of Nivelles is often depicted with a cat and is the patron saint of cats, gardeners, widows, and travellers. In France, 5 February was St. Agatha's feast day, when women were supposed to refrain from working. Those who were caught doing so were said to be confronted by the saint in the form of an angry cat. Another story that originated in France has it that the cat was the only creature who was privileged with knowing the path back to the Garden of Eden after Adam and Eve's expulsion, and this cat later took Jacob to the gates of heaven.

MEDIEVAL PERSECUTION

The earliest Christian association of cats with evil began in the AD 600s, when the Gnostics preached dualism. Dualism acknowledged equal importance to the teachings of Jesus, Buddha, and Zoroaster. The Gnostics were considered heretics by the Church and were accused of placating the devil in the form of a black cat.

Feline martyr

Artists often used cats to represent evil. In this 1554 engraving, a cat dressed as a priest is hanged at an anti-Catholic demonstration in London.

Philosophical cat

St. Jerome is sometimes portrayed with a cat, rather than a lion, because cats were the traditional companions of philosophers. This 16th-century Flemish roundel is in Begbroke Church, near Oxford.

Five hundred years later, similar accusations of heresy were becoming common in Europe. In 1232, Pope Gregory IX established the Inquisition in order to detect heretics who, it was declared, worshipped the devil in the form of a black tom cat. In 1344, during an outbreak of St. Vitus' dance (a disease of the nervous system) in the French town of Metz, cats were blamed and publicly burned. In 1347–1348, the Black Death (bubonic plague) swept across Europe, reducing the population by a third. The disease was spread by rats, but cats were blamed. The Lord Mayor of London was one of many officials who ordered all cats to be destroyed, unwittingly removing the greatest barrier to the continuing spread of the disease.

Mother and child

Cats often feature in art and sculpture with the Virgin Mary. This may be due to the Italian legend about a cat whose kittens were born in the stable at the moment Jesus was born. Their descendants all have crosses on their backs.

Holy decoration
Heathen temples were often adorned with images of cats and birds. In the 20th century, these images feature in Christian places of worship. This carving appears in a new cathedral in Washington DC.

In the early 15th century, the Knights Templar, founded 250 years earlier to protect pilgrims on their way to the Holy Land, were accused of worshipping the devil, who appeared in the form of a black cat. The Church's persecution of people with principles that varied from orthodox Christian beliefs, and the association of these people with cats, led inevitably to the cat's association with paganism and witchcraft.

The shrieks of cats as they mated were interpreted as the cries of innocent people consumed by the "familiars". In Hungary, it was said that all cats automatically become witches between the ages of seven and twelve years unless a cruciform incision was made in their skin at birth. Old women were accused of turning themselves into black cats at night to slip into stables to harm cattle. The witch hunts spread through Europe and across the Atlantic to the American colonies, where they culminated, in 1692, in witch trials in Salem, Massachusetts, where 150 people were accused of witchcraft and 20 were executed. In Europe, the association of cats with heresy and witchcraft led to cat "festivals" in which cats were tortured and killed. The Ypres Cat's Parade in France, which originated around AD 962, is an example. Cats were thrown from the belfry for being associated with witches.

Today, the rehabilitation of the cat is so well established that the Ypres Cat Parade is a popular attraction for cat-lovers, who come to see models of cats paraded through the streets and cloth toy cats thrown from the belfry.

ACCEPTANCE OF CATS
During that period of intense persecution, cats were also loved companions. In Russia, cats were the only animals that were permitted to live in nunneries and monasteries. Throughout Europe, many individuals, including members of the Church, kept cats, although most had to do so secretly for fear of persecution. With the Act of 1736 abolishing all witchcraft laws, the Catholic Church's antipathy towards cats began to diminish. In the 19th century, Pope Leo XII's cat, Micetto, was born in the Vatican, and later Pope Pius IX's cat sat on his knee while he gave audiences.

Demon cat
Flauros, a feline-headed demon, was invoked by those who wished to wreak revenge on other demons. This illustration is from the 1864 edition of Dictionnaire Infernal by occultist Collin de Plancy.

CATS AND WITCHCRAFT
In the 15th century, in Germany and elsewhere in Europe, there was a revival in pagan rites. The cult of Freya (*see page 26*) honoured the fertility of cats. Cultists were first accused of heresy, but accusations of witchcraft soon followed. In 1484, Pope Innocent VIII decreed that witches worshipped Satan and that they took on the form of their animal helpers, called "familiars". The usual "familiar" was a cat.

Throughout Roman Catholic Europe, people were tortured and killed because they owned or cared for cats. A lack of understanding of cat behaviour perpetuated the myths. Cats slept, or pretended to sleep, all day so that they could wake at night to guide evil spirits.

Peruvian fiesta (left)
This 18th-century drawing depicts a saint's day celebration in South America, in which the local inhabitants adorn themselves in various forms of feline-based costume and parade through their village.

St. Cadoc's cat (right)
A common European myth relates how the devil constructed a bridge and claimed for himself the first living being to cross over it. Here St. Cadoc is seen duping the devil, presenting him with a tiny cat instead of the human being he had hoped for.

CAT MAGIC AND SUPERSTITIONS

THE VERY NATURE OF THE CAT LENDS ITSELF TO IDEAS OF SORCERY and divination. Its large eyes dilate fully to see in the dark, causing the mirror-like retinas to give a ghostly reflection, and its nocturnal habits have always been shrouded in mystery. Another puzzle is the cat's ability to live through plights that would be fatal to other animals, making it one of nature's best survivors. Combined, these factors have worked both for and against the cat. In some places it has been celebrated, even treated like a god, while in others it has been feared and treated with contempt. In short, wherever the cat has travelled, stories of magic and superstition have followed.

Although the cat is no longer worshipped and revered, as it was by the ancient Egyptians, many customs and superstitions that built up around it still exist in other cultures today.

FERTILITY ASSOCIATIONS

The domestic cat became associated with the Egyptian goddess Bastet (*see page 26*) about the Twenty-second Dynasty (945–715 BC). But it was not until the Ptolemaic period (332–30 BC) that the cat-headed Bastet's popularity reached its peak, when she was worshipped at a huge annual festival of revelry and sacrifice for bestowing the gifts of life and fertility.

Cat protection
This mummified cat is an unfortunate relic of the old European practice of bricking up a cat into the wall of a building as a charm against rodents or evil spirits.

Portent of nocturnal powers (below)
In early Tarot cards, the Fool is often a vagabond being bitten on his leg by a cat; today a dog is more usual. Symbolically, the cat warns of the powers of darkness.

Centuries later, in many parts of the world, cats were either buried beneath newly sown crops or beaten to death as a symbol of threshing and smoothing of the grain. This was to encourage a stronger crop next spring, and is perhaps testament to Bastet's fecundity. With the advent of Christianity the cat lost its position as a deity (*see page 28*). Instead, cats were often persecuted in the name of religion, which may be why the Church perpetuated ritual cat sacrifices. Based on the original Egyptian superstition that fire symbolized purity and fertility, cat sacrifices by fire became a part of the Halloween and Easter festivals. These feast days are still important to the Christian Church but luckily for the cat, sacrifices are not.

AGENTS OF FORTUNE
Superstitious beliefs persist that cats, and especially black ones, can bring both good and bad fortune – a belief that is often based on geography and ownership. As Moncrif, the first

Witch with magical cat
The notion of magical, witch's cats has died out as a superstition, and now only survives in semi-humorous illustrations like this one.

naturalist to rehabilitate the cat, wrote in his *History of Cats* (1727): "the colour black works very well against cats in unsophisticated minds: it heightens the fire of their eyes, which is enough to make people believe they are witches at the very least". Black cats were said to be in league with the devil and often sacrificed as a result. Later, the unfortunate black cat became a portent of good luck in Britain when it crossed your path: this was based upon the idea that evil had passed you by unharmed. In North America this is reversed on the basis that the black cat is an evil spirit: its mere presence is dangerous.

The cat's legendary introduction to Japan is said to have occurred on the tenth day of the fifth month of the year 999, when a Chinese mandarin presented the emperor, Ichijo, with a white female cat. The birth, shortly after, of a litter of five kittens in the imperial palace of Kyoto was a harbinger of a benign future for the cat in Japan. Today, the image of a cat with one paw raised in a beckoning movement is a

Good luck symbols (above)
The black cat on the right of this card is wearing a swastika, an auspicious sign associated with Buddha. In Britain, the black cat is viewed as a lucky omen.

Shooting stars (right)
In 1805, the cat-loving French astronomer Lalande introduced the cat in the sky map, with a constellation for ailurophiles called Faelis, since lost to us.

Japanese symbol of good luck and prosperity. It is known as the *Maneki-neko*, or Beckoning Cat. The legend began when some samurai followed a beckoning cat to a shrine, where they took shelter from a storm. They spread the fame of the shrine, and people brought their cats to be buried there and prayed for their feline souls and for luck in their own lives. The national cat of Japan, the Japanese Bobtail (*see page 164*) is also associated with fortune. It has been adopted by sailors, because its naturally occurring bobtail resembles the royal family's emblem – the chrysanthemum – and it is used as a talisman to ward off storms at sea.

Talisman for success
Here, the raised left paw of the Japanese Maneki-neko, *or Beckoning Cat, promises* sen ryo, *or great wealth. A raised right paw promises* fuku, *or luck and happiness.*

MAGICAL POWERS

There is a long-standing tradition that cats can affect the weather, which perhaps stems from their earlier association with crop fertility. For example, in Cambodia an ancient custom involves taking a cat from village to village, where it is sprinkled with water to persuade Indra, the compassionate Vedic god, to send the rain necessary for a good harvest.

Throughout Europe and beyond, a cat with its paw behind its ear was a portent of rain. There may be some truth in this; one theory is that changes in air pressure before a storm may affect the cat's ears, causing it to paw at that area. Many other weather predictions have arisen from cat behaviour, such as a cat purring and rubbing its nose for good weather, or yawning as a sign of rain.

The origin of the widely spread superstition that a cat has nine lives, because of its amazing ability to survive, probably started in ancient Egypt, where the number nine held magical values. This idea was later adopted during the 17th century, when accusations of witchcraft were rife (*see page 28*). It was believed that a witch could take the shape of a cat nine times, and so a cat could effectively be reincarnated nine times as well. Maltreatment of cats was never institutionalized in China, as it was in Western Europe. This was perhaps because superstition gave cats the power to avenge themselves on their killers after death.

Medieval engraving
In this earliest surviving Western representation of the ages of man, the 70-year-old is shown with a cat. Both the cat and the dog have been linked with age and wisdom.

SECOND SIGHT

Astrology and alchemy were tolerated by the early Church, and practitioners used the cat in some horrendous rituals. However, it was believed that a harmless way to acquire second sight was to live from early childhood with a tortoiseshell cat. Western astrology ignored the cat, but in the Chinese horoscope it is sometimes a substitute for the Rabbit, which represents the fourth sign of the zodiac. The cat has also appeared in the Tarot – the art of divining the future from the fall of cards. In early sets of Tarot cards the Fool is often shown as a vagabond with a cat biting his leg. In the Tarot the cat is linked with nocturnal powers.

ANCIENT BURIAL RITUALS

One superstition for which there is much material evidence was that if the body of a cat was built into a house wall, it would keep away rodents. This belief survived as late as the 18th century, and the dried, mummified corpses of cats have been found in many buildings in Europe.

In Thailand, the ritual burial of cats is based on religious creeds in the ancient manuscript scrolls, known as the *Cat Book Poems*. They state that when a man of great spiritual advancement dies, his soul enters the body of a cat, and then ascends to heaven on the death of the animal. At one time, this belief was part of the ritual burial of members of the Thai royal family, when a live cat was left in the tomb. There was a small hole in the tomb's roof, and when the cat emerged from it the monks knew that the departed soul had passed into the body of the cat.

This Thai ceremony, like all the other rituals mentioned here, is no longer performed. However, it seems that cat-based superstitions are still a large part of many cultures, and such old prejudices, especially against the black cat, die hard. In truth, cats are harmless animals – unless you happen to be a mouse.

Japanese shapeshifter
This picture is based on the story of a witch cat at Okabe, on the Tokaido road. In this scene, a demon cat with a typical double tail dances with the witch's attendant, watched by a huge spectral cat in the background.

Legendary creature
Sightings of a cat-headed creature with a snake's body, reported to attack livestock, reached legendary status in Italy, probably because it was never caught.

CAT FOLKLORE AND FAIRYTALES

WHEREVER CATS HAVE LIVED, STORIES AND PROVERBS HAVE EVOLVED around their behaviour. Cats carry an air of mystery along with an aloof attitude, and these attributes have captured our imagination. One feature that is peculiar to the cat, more than to any other animal, is an association with magic, which is often manifested through the cat's ability to transform itself into different guises. Across the world, these magical cats, both good and bad, have become immortalized through pantomimes and plays. Cat stories, proverbs, and sayings enrich many languages, but most countries will have their own unique examples as well.

In all cultures, stories have been passed down through the ages by word of mouth, until certain gifted story tellers decided to record these tales in print only a few centuries ago.

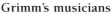

POPULAR FABLES AND PANTOMIMES

Many well-known fables are attributed to the Greek author Aesop (*c.* 6th century BC). Although these tales were rarely written down, they travelled verbally around the world until the French poet La Fontaine (1621–1695) adapted and embellished many of them. From this large collection of stories, the cat is often portrayed as a sly and devious creature, whose predatory greed is always at the forefront.

One fable in particular – *The Cat, the Weasel, and the Rabbit* – was an obvious criticism of human behaviour. The rabbit and the weasel go to the elderly cat Raminagrobis to ask him to settle a dispute. In order to do this, the cat asks them to move closer and stand on some scales so that he can weigh his judgment for their case. When they do this, he lashes out and kills

them, thereby settling the dispute. La Fontaine neatly concludes: "This strongly resembles the squabbles which petty rulers sometimes have among themselves, before they are swallowed up by mighty kings". La Fontaine's Rodilard is a more sprightly version of Raminagrobis. In order to catch mice, he plays dead and also camouflages himself with flour.

Not all stories paint the cat in a bad light. There are tales of reciprocal kindness between humans and cats, as shown in the story of a Tuscan woman who is considerate to cats and is rewarded with gold, whereas her sister, who has no time for them, receives only scratches.

Travelling tales (left)
Many countries illustrate their folk tales on national stamps, such as this Mongolian cat-and-mouse fable. The other stamp features a giant cat from the Faroe Islands; a figure from a traditional ballad.

Grimm's musicians
Around 1812, the Brothers Grimm recorded various stories, including The Bremen Town-Musicians, *who are a troupe of animals, among which is a cat. They combine their vocal s*

Pantomimes are a theatrical amalgamation of the most popular folk stories and fairytales, and two famous ones, *Puss in Boots* by the French author Charles Perrault (1628–1703) and *Dick Whittington*, both feature cats prominently. These cats, through cleverness, even magic, bring good fortune to those who are associated with them. Dick Whittington's cat is usually depicted as black, because in Britain black cats are a good omen (*see page 30*).

Cinderella is another well-loved pantomime. In modern productions the fairy godmother is played as a kindly woman, but in the original Italian version the part was written for a cat.

The world turned upside down (left)
For many centuries, before reading was widespread, moral stories were recorded in pictures so that people could understand them. Man Shoeing a Cat *is one of about 1,500 wall paintings from Taagerup Church in Denmark dating back to the early 1500s, in which the order of nature is reversed so that animals and people engage in strange, unexpected, and unreasonable behaviour.*

Unlikely story (below left)
Little is known about the Egyptian "working classes" who were employed in the construction and decoration of royal tombs. Images such as this tabby cat herding a flock of geese, taken from a practice sketch (c. 1150 BC), may be part of some long-forgotten fable, or simply the artist's attempt to inject some humour into the scene.

Cautionary tale (right)
Aesop's tale of Venus and the Cat *is vividly illustrated here as the cat-maiden spies a mouse and gives chase. This fable is perpetuated in proverbs, such as the Biblical saying* "A leopard cannot change its spots", *and the succinct German proverb,* "The cat cannot leave the mouse".

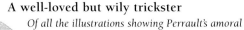

METAMORPHIC CATS

From ancient times, stories have been told of cats that are transformed into humans, often becoming beautiful women. Aesop's tale of *Venus and the Cat* concerns a cat who falls in love with a handsome young prince. Venus, the goddess of love, agrees to transform the cat into a beautiful girl, and when they meet the couple eventually marry. But Venus is not convinced that the cat's transformation is complete, so on their wedding night, she introduces a mouse into the couple's bedroom. Forgetting herself, the girl pounces on the mouse and devours it. The goddess, confirmed in her suspicions, decides she must turn the girl back into a cat.

In Japanese folklore, the shapeshifting cat is of a more sinister nature. In a tale similar to Western vampire myths, a demon cat attacks a fair maiden and sucks out her lifeblood. It then assumes her identity. In its new female form, the cat turns its attentions to the maiden's lover, Prince Nabéshima. By gradually extracting his life-force, during nightly visits, the vampire cat weakens the prince to the point of death. He is saved just in time through the vigilance of his faithful servant, and the impostor is eventually destroyed. A similar fable comes from Spanish-Jewish folklore. Lilith, Adam's first wife, leaves him to become a vampire in the form of a black cat, who then descends upon helpless, sleeping babies and sucks out their blood. This story may have contributed to the widespread belief that a cat may lie on a sleeping baby, causing suffocation.

FAVOURITE PROVERBS AND STORIES

All languages are rich in cat proverbs and cat sayings, many of which can be applied to feline or human behaviour. The following selection cites only a few, out of hundreds of examples: from Spain "A cat with gloves cannot catch mice", from Italy "Old cats mean young mice", from The Netherlands, "The man who does not love cats will never get a pretty woman", from Britain, "Curiosity killed the cat", and finally from Russia "A cat and a woman are for the house, a man and a dog are for the yard".

There is an African folk tale that neatly explains how the cat came to be the favourite companion of women. A cat is seeking the best animal to live with for protection. Each time the cat chooses a new protector, that animal is preyed upon by a stronger one, so the cat keeps moving on until finally it settles with a man. When the man goes home he is confronted by his wife, who in anger chases him from the house. From then on, the cat took shelter under the woman's protection. From Poland comes one of the most poignant feline folk stories. Early one spring, some kittens were thrown in a river to drown. Their mother wept so loudly that the willow trees along the banks of the river consulted about what they could do to help. They bent their branches out over the river so that the desperate drowning kittens could cling to them. Each spring, willows deck themselves in silky velvet buds, soft as kitten fur, in memory of their rescuing the young.

Romantic tale (left)
In the late 17th century, Comtesse D'Aulnoy wrote Le Chatte blanche, *or* The White Cat. *Its roots probably lie in the "Venus" story, although in this version the prince cuts off the cat's head to turn her into a princess, and they then live happily ever after.*

Magical cat (right)
Dick Whittington, one of the most popular British folktales, is also a splendid subject for pantomime. According to the legend, Dick came to London as a poor boy and, with the help of his magical cat, made his fortune.

CATS IN CHILDREN'S LITERATURE

CHILDREN'S STORIES ARE OFTEN REMNANTS OF FOLK TALES, DISTANT myths and legends, and ancient pagan beliefs. In the 18th century, after centuries of persecution in Europe, the cat's reputation began to be ameliorated; as a consequence, it appeared in many a tale. Some children's stories are true to the real nature of cats, teaching young minds to recognize and understand typical feline behaviour, but many tales are fictional, and were created purely for children's entertainment and delight. Stories of mysterious, fantastic, and magical cats have become particularly popular, and the cat is now an enduring figure in children's literature all over the world.

It was probably through fables and folk stories that the domestic cat first made its appearance in literature (*see page 32*). Because many of these tales were of a didactic nature, they came to be regarded as "morality tales" for teaching children, and became the basis for well-known fairytales and children's stories.

PROTOTYPE CATS

In its original form, *The Adventures of Pinocchio* by Carlo Collodi (1826–1890) was a morality tale for adults, which later became classified as children's literature. In the original version, the cat was used to represent human hypocrisy, but first and foremost he was a street cat, surviving on his wits, and scavenging for food.

The real nature of the domestic cat is best described by Rudyard Kipling (1865–1936) in his classic story *The Cat that Walked by Himself*. Kipling tells the tale of how Woman tamed a number of animals to serve her and Man, and how she and Man almost, but not quite, tamed the cat. After a while, the cat went to them of his own accord and they eventually struck up a bargain. In return for milk and shelter, the cat agreed to kill mice and be kind to babies when he was in the house, but essentially "he is the Cat that walks by himself, and all places are alike to him".

A little-known British children's story from the 18th century is *The Life and Adventures of a Cat*, in which the hero is a cat named Tom. Until it was published, male cats were called "boars" or "rams", but ever since, in English-speaking countries, male cats have been known as "toms".

HUMANIZED CAT HEROES

The celebrated children's author Beatrix Potter (1866–1943), in common with the cartoonist and illustrator Louis Wain (*see page 50*), found success by humanizing her animal characters. Her tales began life as a series of letters to the children of a former governess, but are now household favourites across the

Dual nature

This confident feline is from The Cat that Walked by Himself, *written and illustrated by Rudyard Kipling. In the story, the cat is part tame and part independent.*

world. Her characters have faults, just like real people: Mr. McGregor's white cat is someone to be avoided, according to Benjamin Bunny in *The Tale of Peter Rabbit*, but in *The Tale of Tom Kitten* and *The Tale of Samuel Whiskers*, cats are portrayed as likable characters who are naughty but nice.

In the mid-20th century, new feline heroes appeared in children's literature. The American author Kathleen Hale (1898–) based the marmalade cat Orlando, his wife, Grace, and their kittens, Pansy, Blanche, and Tinkle, on her own cats. Orlando's adventures grew into a series of 18 books that, with their humorous

Famous grin (above)
In Alice in Wonderland, as Alice talks to the Cheshire Cat, it slowly disappears until only a disembodied grin remains. Among the many possible origins of this cat is a brand of Cheshire cheese, which featured a cat's grin on its packaging.

Marmalade cat (right)
For over 30 years, the children's author Kathleen Hale wrote and illustrated her Orlando adventures, in which her affable marmalade cat takes a trip abroad, becomes a doctor, celebrates his wedding, and, in 1968, even goes to the moon.

The ultimate survivor (left)
This illustration from The Adventures of Pinocchio *shows the cat escaping from an altercation between Pinocchio and his creator. The story of a puppet that becomes a boy is now a children's classic, but originally it was an Italian morality tale in which the cat represented hypocrisy.*

Lear's seafaring characters (below left)
While living and working in Italy, Edward Lear was inspired by his beloved tabby, Foss, to write the magical verse of The Owl and the Pussycat. *Edward Lear illustrated all his work with simple line drawings that were strikingly modern for their time.*

Cautionary tale (below)
The lessons to be learned in children's literature are often harsh. In the French story La Boîte d'allumettes, *or* The Box of Matches, *a young girl plays with matches, accidentally sets herself alight, and burns to death. Her two faithful cats extinguish the flames with their tears.*

plots and distinctive illustrations, are now classics of children's literature. Since Hale's long-running success, the fashion for the "cat series" appears to have died out. Although modern cat protagonists such as Mog, the witch's cat, and the ghost cat Fred have their fans, the current trend in children's literature seems to be for unusual animals, fantastic creatures, and even teddy bears.

POETIC CATS
Many creative people have gained inspiration from their cats, and poets are no exception. The British-born artist and humorist Edward Lear (1812–1888) based his famous verse, *The Owl and the Pussycat*, on his much-loved tabby, Foss. However, it was possibly through rose-tinted glasses that Lear addressed Foss: "What a beautiful Pussy you are", because Foss had a stocky body and a shortened tail. Despite these unconventional looks, Foss made a perfect model for Lear, and is now also immortalized in cartoon form.

Another devoted cat fan, T.S. Eliot (1888–1965), based his collection of 14 cat poems, *Old Possum's Book of Practical Cats*, on his numerous cats. Each poem is a play on the English language – which can be lost through translation – with characters such as Growltiger, Rumpuscat, and the mystery cat Macavity. Eliot gave his cats human traits, while managing to keep the essence of their feline nature.

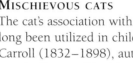

Moral lesson
In the 19th-century morality tale Poor Minette, *good and bad cat spirits fight over the tempted soul of the ingénue cat, Minette.*

MISCHIEVOUS CATS
The cat's association with magic and sorcery has long been utilized in children's literature. Lewis Carroll (1832–1898), author of the *Alice* books, put the famous Cheshire Cat among the characters found in Wonderland. When Alice sees a large cat, grinning from ear to ear, she asks, "Please would you tell me why your cat grins like that?" The answer she receives is, "It's a Cheshire Cat and that's why". This opaque reply may have come from the saying to "grin like a Cheshire caterling": caterlings were deadly swordsmen, who protected the Royal Forests in the reign of Richard III. Over time, caterling became shortened to cat, and anyone seen with a notably evil smile was said to be "grinning like a Cheshire cat".

Writing under the pen name of Dr. Seuss, Theodor Seuss Geisel (1904–1991) created perhaps the most successful and undoubtedly the most mischievous of all modern literary cats in *The Cat in the Hat*. A gawky, subversive cat in a stovepipe hat decides to keep two bored children entertained while their mother is away. Chaos ensues, but this magical feline manages to return everything back to normal just before mother returns. If only reality were as anarchic, and at the same time as controlled, as in these bewitching cat books.

CATS IN LITERATURE

DOMESTIC CATS PLAYED A MINOR ROLE IN THE HISTORY OF EARLY writing, appearing more often in fables, proverbs, and children's stories than in major works of literature. This began to change in the 19th century as the cat's popularity increased and it became an acceptable literary subject, even starring as the central character. Cats, with their associated charm and mystery, have proved to be universally captivating – writers as diverse as Pierre Loti, Anton Chekhov, H.G. Wells, and Don Marquis have all paid tribute to them in stories, essays, or poems. Today, a wealth of publications are available to suit the tastes of every kind of cat lover.

Cats and writers often live together, perhaps because these individuals are temperamentally suited. Unlike dogs, cats never interrupt the creative flow with loud barking or demands for a walk; instead they are quite happy to amuse themselves or sleep quietly nearby.

PROVERBS AND APHORISMS

Cat proverbs are found in all languages; some make reference to the cat's prowess and special instincts, a few to its grace and beauty. From China comes the comment, "A lame cat is better than a swift horse when rats infest the palace". The Irish say, "If a cat had a dowry she would often be kissed". A proverb from the southern states of America has it that "The cat is mighty dignified, until the dog comes by". From Britain comes the saying, "In the cat's eyes, all things belong to cats".

Many cat-loving writers have made memorable comments about their admired felines. Canadian novelist Robertson Davies (1913–1995) wittily described the mutually beneficial relationship in his line, "Authors like cats because they are such quiet, lovable, wise creatures, and cats like authors for the same reason". Davies was not alone in his fond feelings for cats. The novelist Sidonie-Gabrielle Colette (1873–1954), author of *Gigi*, was one of the literary world's great cat lovers. "By associating with a cat", she wrote, "one only risks becoming richer". Another commentator was the British Prime Minister Sir Winston Churchill, who was awarded the Nobel Prize for Literature in 1953.

Commemorative stamp
Russian author Aleksandr Pushkin (1799–1837) almost always kept feline company. He included many in his stories, and sketched them for relaxation. Issued by his homeland, this stamp commemorates his literary works and acknowledges his affection for cats.

Generally known as a commanding orator, he reportedly said, "All dogs look up to you. All cats look down to you. Only a pig looks at you as an equal". Théophile Gautier rather pithily added that "The cat is a dilettante in fur".

The American writer of books for adults as well as children, Samuel Langhorne Clemens (1835–1910), better known by his pen name Mark Twain, was devoted to his cats. He was often photographed with them and liked to give them whimsical names such as Apollinaris, Blatherskite, and Zoroaster. He explained that the names were not given "in an unfriendly spirit, but merely to practice the children in large and difficult styles of pronunciation".

THE CAT AND THE POET

One of the earliest, and perhaps most beautiful, homages to the cat was written in Gaelic by an anonymous Irish monk during the 8th or early 9th century. Famous throughout Ireland as *Pangur Bán*, meaning "whiter than white", the poem describes their life, with the monk hunting for words while the cat hunts mice.

> Oftentimes a mouse will stray
> Into the hero Pangur's way;
> Oftentimes my keen thought set
> Takes a meaning in its net.

The Italian poet Torquato Tasso (1544–1595) succumbed to delusion and was interred in an asylum, as was the British poet Christopher Smart (1722–1771). Strangely, both found the environment conducive to feline verse: Tasso produced the work *Sonnet To My Cats*, whereas Smart penned *Jubilate Agno* – with a section to *My Cat Jeoffrey* – which was turned into an operetta by Benjamin Britten (1913–1976). Other famous poets from Wordsworth to Baudelaire have eulogized their feline friends.

Gruesome story (left)
This illustration, from The Black Cat *by Edgar Allen Poe (1809–1849), shows the gruesome moment when Pluto, the black cat, reveals his master as the murderer of his wife. Oddly, Poe was reputed to be a cat-lover.*

Poetic sphinx (below)
In his epic poem, The Sphinx, *the Irish writer Oscar Wilde (1854–1900) extols the feminine traits of this mysterious, mythical creature.*

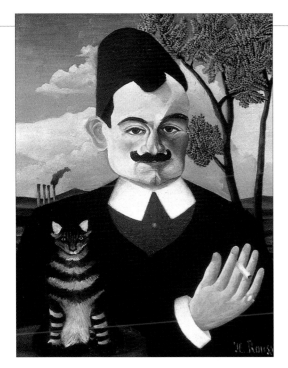

Portrait of an artist (left)
This portrait (1910) of the French writer Pierre Loti (1850–1923) was made shortly after the success of his book Au Maroc, *or* In Morocco, *which probably explains why he is wearing a fez-like hat. He is shown with a cat because Loti was well known as a great cat-lover, and wrote repeatedly about his feline pets.*

Feline sleuth (right)
Like many authors, American-born crime writer Patricia Highsmith (1921–) chooses feline company, and like others in her genre, she chooses a Siamese. This breed has a peculiar attraction for the sleuthing novelist, possibly because of its age-old reputation for almost human intelligence.

Cat-like behaviour
Colette loved cats. She was photographed with them, she included them in her works, and even portrayed them herself in the theatre. As well as making an excellent sphinx, she played a cat on stage in La chatte amoureuse.

Don Marquis (1878–1937), the American writer, is known for *archy and mehitabel*, the verse adventures of the cockroach archy and the cat mehitabel, written entirely in lower case. As the story unfolds, mehitabel emerges as an aspiring lady whose natural "alley cat" inclinations win through in the end.

REVEALING CAT STORIES
Fiction is often based on autobiographical material and many authors have reconstructed their "cat experiences" to great effect. Anton Chekhov (1860–1904) wrote of how his uncle Pyotr tried to train his kitten to kill mice. Each time a mouse was caught in a trap, Uncle Pyotr grabbed the kitten and pushed its face towards the unfortunate rodent. Naturally, when it was released, the terrified kitten ran away. Forever after, on catching sight of a mouse, the cat trembled and took to ignominious flight. Chekhov was taught Latin by the same uncle, and forever after reacted to the Classics much as the cat was reacting to mice.

In her short story *The Cat*, Colette recounts the short-lived marriage of the petulant Camille to the doting Alain. Sadly, Alain's doting affection is for Saha, his Russian Blue cat, and in a jealous rage Camille attempts to kill Saha. Alain guesses Camille's wicked deed, and, faced with an ultimatum, chooses Saha. This tale is not unlike the author's real-life experiences with men.

People have a lasting fascination with the idea of the human in animal form, as seen from folklore and fairytales (*see page 32*). Humanized cats make interesting characters. One of the most outspoken is Tobermory, the feline creation of Saki (1870–1916), whose real name was H.H. Munro.

Tobermory acquired the gift of speech and, with a feline disregard for the consequences, used his gift with devastating effect on members of a high-society house party.

GENRE FICTION
For those with a combined love of cats and mystery, there is a specific genre of books available – the detective novel. At the turn of the 20th century, Guy Boothby, an Australian novelist, based his books on the sleuthing skills of Dr. Nikola and Apollyon, a black cat. A little later, Dorothy L. Sayers (1893–1957) gave her detective Lord Peter Wimsey a crest, featuring a cat with the motto "As my whimsy takes me".

The contemporary American author Lilian Jackson Braun began writing her *The Cat Who...* mysteries when one of her treasured Siamese cats mysteriously fell to its death from an

Farmyard tales
The story of Turi and Kamphau'ern, two cats from adjacent farms, was told and illustrated by Norwegian artist Olaf Gulbrannson (1873–1958). The dramatic tale of love and death reads like an account of human intrigues.

apartment block. To date there are more than a dozen books to choose from, all featuring the highly talented Koko and Yum Yum, Siamese cats with a bent for detection. Another contemporary author, Akif Pirinçci, (1959–) this time from Germany, wrote the compelling novel of cats and murder, *Felidae*, later made into a cartoon film (*see page 46*).

The ultimate genre, for fantastic invention, must be science fiction. In *The Invisible Man*, one of the first writers in this genre, H.G. Wells (1866–1946), made his ill-fated scientist try out his new drug on a cat before taking it himself. Cats find a more comfortable home elsewhere in science fiction. C.J. Cherryh (1942–), for example, based one of her alien races on cats in her *Chanur* series of books. Robert A. Heinlein (1907–1988) was also a fan of felines. He immortalized his own cat, Pixel, by giving that name to the title character in *The Cat Who Walked Through Walls* – a cat who does just that, because he thinks that he can.

During the 1980s, publishers discovered that "the cat book" sells, and in consequence a glut of them appeared on the market. Of these, many are either fictional or anecdotal, but some are prized for originality or comprehensive coverage. Serious tomes, such as *The Literary Companion to Cats* and *The Cat in Ancient Egypt*, now vie for the book buyer's attention with many more recreational titles, including *The Internet for Cats*, *Testing Your Cat's IQ*, and *Yoga for Cats*. All of these books have their own particular appeal. In fact, hundreds of new titles are published worldwide each year, and before too long the "cat book" may well qualify as a separate genre in itself.

CATS IN EASTERN ART

LARGE CATS SUCH AS LIONS AND TIGERS WERE FREQUENTLY PAINTED or sculpted as images of power and leadership, from Persia to India and on to China. In Persia, lions were symbols of royalty, and can often be found on the walls and pillars of ancient palaces such as the magnificent ruins of Isfahan. In China, the tiger is considered the "Lord of the Beasts", and features heavily in myth and symbolism. As the cat spread eastwards, it naturally basked in the glory of its larger relations, and consequently found its own related niche in art. Artists had little difficulty in capturing the essence of these relations of the venerated lion and tiger.

Eastern art forms, although highly structured and stylized, can be vividly natural and much truer to feline behaviour than contemporaneous Western art. In Eastern art, bi-colours and tortie-and-white cats are by far the most commonly portrayed, suggesting that those patterns were preferred by the travellers who took the cat to the East, or by the people who lived there.

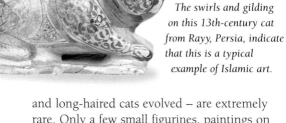

Artistic licence
The swirls and gilding on this 13th-century cat from Rayy, Persia, indicate that this is a typical example of Islamic art.

PERSIAN ART
Muslims regard the cat with favour because it was the pet of Mohammed (*see page 26*), but there is no tradition of representational art in Islam. Therefore, cat images from the Middle East – where cats were plentiful and long-haired cats evolved – are extremely rare. Only a few small figurines, paintings on tiles, and bottles and boxes in the shape of cats have been found. It is in the works of visiting Western painters that we first see cats lazing luxuriously at the feet of Arab scribes or being fed tasty morsels by Arab women.

However, in the 14th century, anonymous Ottoman artists did paint striking narrative scenes of national folk-tales and stories from the Bible, often including domestic cats in an incidental role in their tableaux.

INDIAN AND FAR EASTERN ART
In India, most early interpretations of cats are found in carved or painted religious images: for example, a goddess of fertility may be seen riding a cat. In painted scenes that told a

Breed beginnings
By the 19th century, cats that could be the forebears of the Japanese Bobtail (see page 237) appear in Japanese art. This example, by Hokusai, shows the characteristic short tail, long hindlegs, and the preferred coat pattern.

story, perhaps of a feast day or the arrival of Western visitors, cats signify high status and wealth. Other proof of the cat's exalted status in India exists in local folk-art traditions, including portraits, still lifes, and observations of animals. The house cats in these paintings resemble the exclusive pets in court portraits, implying that there may have been little or no effort to create pedigree-type breeds.

Farther east, in Thailand, cats were highly regarded, and their wide variety of coat colours and postures were recorded in vivid paintings. The many manuscript copies of the *Cat Book Poems* (*see page 94*) depict pointed cats – ancestors of the Siamese – copper-brown "Sopalak" cats and silver-tipped blue "Si-Sawat" cats, as well as tabbies, bi-colours, and patterns that would baffle geneticists and breeders. The most famous manuscript copy is now in the Thai National Library in Bangkok, but others exist in museums around the world.

Cats only became widespread in Chinese art during the Sung Dynasty (AD 960–1279). They appeared in many court portraits, suggesting that here, as in India, cats were status symbols.

Human-looking cat (left)
An anonymous Indian artist in the 19th century created this anthropomorphic cat. Its dark-lidded eyes, showing an expanse of white around the irises, appear far more human than feline.

Indian trading scene
In a vivid "story" painting from the 17th century by an unknown Mughal artist, wealthy traders are portrayed in opulent surroundings. An alert white cat sits contentedly beside the chair of the most important Westerner.

Cats at play
Much Eastern art has been lost because it was worked on fragile paper, or, like this 18th-century painting, on silk. These tabbies are typically Oriental, being striped; Western cats at this time were often blotched tabbies.

Most Chinese artistic renditions of the cat are expressively natural. *Spring Play in a T'ang Garden*, attributed to the Chinese Emperor Hsuan Tsung, who reigned from 1426 to 1435, depicts kittens frolicking among flowers in a garden. On silk and in woodblock, cats stalk bees or climb trees to avoid bug-eyed dogs. By the 1500s, the cat was a popular subject for *cloisonné* figures and, because of its superb nocturnal vision, for lamp designs. These lamps took the form of crouching cats in painted porcelain. The models were open at the top, with hollowed-out eye sockets, and the light from a candle within shone through the eyes, both to illuminate and to ward off evil spirits.

JAPANESE ART
The cat achieved lasting fame in Japanese art, where it was represented in scroll paintings, pottery, bronzes, and ivories for almost a thousand years. From the 1700s onwards, cats became even more widespread in Japanese art. As in China and throughout Europe, the cat was associated with evil spirits (*see page 28*), and Japan's double-tailed witch cat of Okabe was a particularly common subject.

Feline opportunist (right)
Kuniyoshi's cats are strikingly real. The boldness of the tabby as it steals a fish in Girl Chastising a Thieving Cat *is familiar to anyone remotely acquainted with felines.*

Most Japanese paintings and woodblock prints show cats in the typical tortie-and-white Mi-ké patterning, seen in the Japanese Bobtail today, but in earlier works the cats tend to have longer tails. Through the 18th and 19th centuries, bobtails and predominantly white cats appear more and more often in illustration.

The most accessible renditions of cats in Japanese art are those portrayed in woodblock prints. Beginning with Utamaro (1753–1806) and Koryusai (late 18th century), through to the magnificent portrayals by Ando Hiroshige (1797–1858) and Utagawa Kuniyoshi (1797–1861), cats are portrayed with a reality seldom seen in Western art.

Utamaro included cats as the companions of beautiful women, while Hiroshige portrayed cats in scenes, such as the series *Famous Sites of Edo*, in vividly realistic attitudes. But it was Kuniyoshi, more than any other Japanese artist, who depicted cats with supreme accuracy and acute observation. He adored cats, and his studio was overrun by them. As a young boy, Kawanabe Gyosai (1831–1889) was Kuniyoshi's pupil, and he later sketched Kuniyoshi holding a cat in one arm and painting while other cats washed themselves and frolicked at his feet.

Perhaps Kuniyoshi's most famous work is the triptych woodblock print of the *Fifty-three Stations of the Tokaido Road*, with each station represented by a cat. Most of the cats are short-tailed and particoloured, but there are also striped tabbies and solid-coloured cats, as well as the double-tailed witch cat of Okabe.

Untamed hunter (right)
The kitten in this Japanese scroll is happily trawling a fish bowl in search of an extra snack. The pretty bow around its neck reveals that this kitten is well-loved, and presumably well-fed, but despite its comfortable life the feline's natural hunting instinct remains intact.

Tiger netsuke
The Japanese used netsuke toggles, often carved in the form of an animal like this ivory tiger, to fasten containers to their kimono sashes.

Cats also play a significant role in Kuniyoshi's other works. In his theatrical scenes, demonic and spectral cats from *kabuki* plays are invested with recognizable feline characteristics, while his portraits of actors, nobles, and courtesans often include incidental cats. In one of these, he portrays a beautiful young noblewoman painting butterflies while a kitten pounces on her finished paintings. Although it is intended to be a homage to the skill of the young woman, Kuniyoshi's portrait is also an astute comment on the natural predatory instinct of cats. This is typical of the best of Eastern art, revealing acute natural observation within an apparently rigid, stylized medium.

CATS IN WESTERN ART

IMAGES OF DOMESTIC CATS HUNTING, FISHING, SUCKLING THEIR YOUNG, and playing with children are abundant in Egyptian art, but Greek and Roman artists ignored the small cat, preferring to portray the more spectacular lion and tiger. The cat's association with the old religions from antiquity until the Middle Ages brought them into disfavour with the Christian Church, and they rarely feature in art of that period. During the Renaissance, the cat began to appear in allegorical paintings, often as a symbol of perfidy and evil. Later, in American and European art, cats came to symbolize beauty and harmony, often appearing in domestic scenes and family portraits.

With the exception of the illustrators of the Lindisfarne Gospels in AD 700 and the Irish Book of Kells in AD 800, European artists and artisans throughout the Middle Ages disregarded the cat, although cats occasionally appear in carvings in Celtic churches (*see page 28*). These carved cats may owe more to fables and pagan stories known to the carpenters and masons than to Christian influence.

Family psalter
Psalters were prayer books commissioned by wealthy families. Illustrated with both Christian and folklore narratives, they were often used to teach children Latin. A Lincolnshire knight, Sir Geoffrey Luttrell, commissioned his psalter around 1330. It is filled with an unusually large profusion of figures and foliage, including this cat and mouse, all rendered in exquisite detail.

CATS AS SYMBOL AND ALLEGORY
The Christian Church's hostility towards cats is the most likely reason for their infrequent appearance in early Western art. When a cat does appear, it is often as a symbol of treachery or evil. The Italian artist Domenico Ghirlandaio (1449–1494) followed an artistic tradition of portraying a cat sitting at the feet of Judas in his painting *The Last Supper*. The cat is thought to indicate Judas' treachery. Jacopo Tintoretto (1518–1594) also included a cat of rather sinister appearance in his version of the same scene, and in another painting, *The Annunciation*, Tintoretto painted a wicked-looking cat on its guard under a menacing cloud. The Venetian artist Jacopo Bassano (*c.* 1517–1592) was one of the few artists of this period to use animals in a natural manner, and included two black-and-white cats in his *Animals Going into the Ark*.

The German painter and engraver Albrecht Dürer (1471–1528) produced the intricate woodcut *Adam and Eve*, in which a number of animals appear. The mouse is caught by Adam's foot, and may represent the weakness and vulnerability of man, while the cat is linked by its tail to Eve, and is likely to symbolize sexuality and evil. The great master of fantasy, Hieronymus Bosch (*c.* 1450–1516), painted demonic cats in *The Temptations of Saint Anthony*, and in many of his paintings the demons resemble the feline form. As the 18th century approached, the association of cats with the devil waned. In his paintings *The Skate*, *The Thief in Luck*, and *The Dead Hare*,

Jean Baptiste Chardin (1699–1779) showed cats as thieves and gluttons rather than symbols of the devil. During this period, the English artist William Hogarth (1697–1764) produced a series of works named the "Stages of Cruelty", which includes callous acts inflicted on cats. This suggests that the cat was no longer regarded as the perpetrator of evil, but as the victim.

CATS IN DOMESTIC SCENES
The cat rarely appeared in naturalist painting before the 18th century, although Leonardo da Vinci (1452–1519) appears to have found the anatomy of the cat fascinating. Some of his sketches are as accurate as photographs in their portrayal of cats in action and asleep, while others are more fantastical and dragon-like. As European artists moved away from the old subjects of religion, history, and mythology, and turned to illustrating scenes from daily life, cats began to appear in domestic scenes. In England, William Hogarth painted his famous portrait *The Graham Children*, which includes an alert cat, and the great animal portraitist George Stubbs (1724–1806) painted his only known study of a cat, *Miss Anne White's Kitten*. In France, artists Antoine Watteau (1684–1721)

Allegory or realism? (left)
In The Last Supper *by Francesco Bassano the younger (1549–92), son of Jacopo, the cat is positioned facing a dog, a typical symbol of treachery opposed to fidelity. However, this cat and dog appear in several paintings by the Bassano family, and it is likely that they were domestic pets.*

Reluctant plaything (right)
Dressing the Kitten, by Joseph Wright of Derby (1734–1797), shows a kitten suffering the indignity that has befallen many family pets through the ages. This sentimental scene is a model of lighting and composition in portraiture.

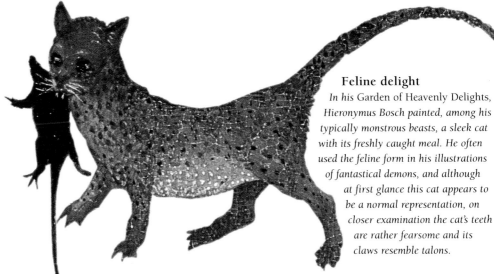

Feline delight
In his Garden of Heavenly Delights, Hieronymus Bosch painted, among his typically monstrous beasts, a sleek cat with its freshly caught meal. He often used the feline form in his illustrations of fantastical demons, and although at first glance this cat appears to be a normal representation, on closer examination the cat's teeth are rather fearsome and its claws resemble talons.

Languid beauty
Perhaps better known for the cat adventures in the comic strip Images sans paroles, *Théophile Steinlen was an accomplished artist. This study of a reclining brown tabby,* L'hiver: chat sur un coussin *(1901), is a testament to his skill at capturing the essence of the feline nature.*

and Jean Fragonard (1732–1806), and in Italy Giovanni Tiepolo (1696–1770), all included contented cats in their paintings, curled on the laps of shepherdesses, in the arms of mistresses, or simply idling in luxury with their owners. A portrait of a boy, *Don Manuel Osorio de Zuniga* by Francisco Goya (1746–1828), shows three plump cats eyeing the young boy's tethered magpie. Few of these representations come close to the anatomical perfection of Leonardo da Vinci's best sketches, and it was not until the influence of the famous French painter and naturalist Jean-Baptiste Oudry (1686–1755) spread that realism was once more achieved.

CAT PORTRAITURE
During the 19th century, the Swiss-born artist Gottfried Mind (1768–1814) produced many carefully observed watercolours and skilfully executed pen studies of his cat Minette and her kittens. The works of Dutch-born Henrietta Ronner (1821–1909) were thought sentimental at the time, but are highly prized today for their accurate observation of cat behaviour, while the Swiss-born artist Théophile Steinlen (1859–1923) produced thousands of drawings of cats that, although intended as comic strips, brilliantly caught cat behaviour (*see page 44*). Other European painters were using cats to symbolize the sexuality of their subjects. For example, Edouard Manet (1832–1883) put a self-contained and serene cat at his subject's foot in his painting *Olympia*, perhaps as a subtle code for sensuousness, a technique used in portraits of geisha in Japan. Auguste Renoir (1841–1919) was far more robust in his affinity with cats.

In many of his paintings, such as *Madame Julie Manet, Girl with a Cat*, and *Sleeping Girl with Cat*, his cats are playful, spirited, and sensual, much like the women he portrays. In Russia, the poet Aleksandr Pushkin (1799–1837) drew subtle pencil sketches of his cats, and these can now be seen at Boldino, his house near Pskov.

AMERICAN FOLK ART
For almost 150 years, beginning in the early 18th century, folk art was the dominant indigenous form of painting in the United States, although it was often underestimated

Companionship (right)
In Girl with a Cat, *Auguste Renoir illustrates the close relationship between cats and humans. The cat appears relaxed and trusting in the girl's arms, while she bends her head towards it in a touching demonstration of affection.*

Domestic bliss (below)
American folk art often shows family groups or single portraits. A typical example is Girl in Red with Her Cat and Dog, *by Ammi Phillips.*

and undervalued until recent times. Portraitists, often self-taught, travelled around the country painting middle-class families. The names of many of these artists are unknown but some, such as Ammi Phillips (*c.* 1788–1865), left what are now treasured bodies of work. Both dogs and cats were regularly portrayed in these family paintings, and they communicate a vivid impression of domestic security and stability in the early years of American settlement. No longer associated with paganism and the devil, the cat had become a modern symbol of beauty and peaceful civilization.

CATS IN MODERN ART

DEVELOPMENTS AWAY FROM REALISM IN THE TWENTIETH CENTURY have produced a rapid succession of styles, and the popularity of the cat as a suitable model for painting and drawing has never waned. Practitioners of every movement, from German Expressionism to Pop Art, have been drawn to the cat's natural beauty and unique presence, and have attempted to capture the cat on canvas, in print, or in three-dimensional form. Cat art now abounds in both traditional and modern styles, and mass reproduction has ensured its popularity worldwide.

Until the advent of World War II, many artists regarded Paris as the Mecca of art, choosing to settle and work there. From this large, creative melting-pot various contemporary art styles arose, and within each genre are innumerable fine works of art that depict the cat.

CAT-LOVING ARTISTS

A tradition for living with feline companions and creating representations of them has been established during the 20th century. One of the first artistic ailurophiles to emerge was the British painter Gwen John (1876–1939), who surrounded herself with cats in her home and painted them on many occasions. She was artistically overshadowed during her lifetime by her more famous brother Augustus, but her paintings are now highly sought after. The Japanese painter Tsuguharu Foujita (1886–1968), the British Surrealist Leonora Carrington (1917–), and the media-aware Pop artist Andy Warhol (c. 1928–1987) have all produced a variety of cat studies simply because they were

Ceramic icon
This 1980s' raku cat portrait by Jill Crowley (1946–) is part of a series based on Crowley's own Burmese and a neighbour's tom that used to pester it.

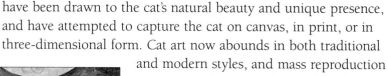

Cubist cat
Pablo Picasso's interest in the wild, untamed qualities of all cats is evident in La Technique (c. 1962), in which a domestic, or possibly feral, cat clearly shows its savage nature as it contemplates an assembly of food.

enamoured of this fascinating animal. In the work of the early German Expressionists there was a declared and deliberate distortion of reality, and cats began to appear in works of art that exaggerated aspects of their form to convey their unique feline qualities. The chief subjects for the Expressionist Franz Marc (1880–1916) were animals and nature, which included cats. The Swiss painter Paul Klee (1879–1940) was extremely fond of cats and painted them frequently in imaginative scenes. Klee was influenced by the Expressionists, although he never affiliated himself fully to any one style.

MODERN CAT STUDIES

The development of Cubism – another great art movement – is often credited to the master of form and volume, Pablo Picasso (1881–1973). Of his works depicting cats, he is quoted as saying: "I want to make a cat like those that I see crossing the road. They don't have anything in common with house pets; they have bristling fur and run like demons". It was clearly the cat as stalking hunter that inspired his portrayals.

Later came Surrealism, which was concerned with exploring and illustrating the workings of the unconscious mind. The Spanish artist Joan Miró (1893–1983), was a Surrealist, and his work *Tête* (*Le chat blanc*, 1972) shows an apparently feline animal, combining eight "spider leg" whiskers on a mottled background. Certain artists are not easily identified with the various movements that proliferated during the early decades of the 20th century. For example, Marc Chagall (1887–1985) preferred to take from each the lessons that were useful to him. In 1910, Chagall left Russia for Paris, where his Jewish folk background fused with modern French techniques to produce highly

Perfect balance
Martin Leman, the most sophisticated of "naive" painters, has earned an international reputation through his widely sold gift books, containing cat images such as this interlocking pair named Winsor and Newton.

Behaviour study (left)
This early 20th-century Franz Marc cat sketch may appear stylized, but it superbly captures the fervid intent of these circling felines. Marc made an intensive study of animal anatomy in order to arrive at a system of rules of form.

Feline high jinks (right)
Working in Europe, Japanese-born Ryozo Kohira (1947–) combined traditional woodblock printing, modern techniques, and vivid colours in this imaginative study of drunken, gluttonous cats, The Cat's Voyage (1964).

Curious portrayal

In Girl with a Kitten (1947) by Lucian Freud (1922–), a kitten is held in an oddly tight grip by Kitty Epstein, a close friend of the painter. The girl's overtly anxious expression and the kitten's direct stare combine to unsettle any viewer of this startling portrait.

Animal observation (right)

The graphic artist Tsuguharu Foujita made numerous, closely observed cat studies, and Le Chat (1939) is a typical example of his meticulous style. He was highly proficient in lithography and used many of his own cats as models for his behavioural observations.

Personal gifts

During the 1950s, Andy Warhol frequently sketched his favourite cats in frivolous poses. The results were originally printed as presents for his close friends, but have now been published posthumously in gift books, as well as appearing on mass-produced mugs and T-shirts.

stylized images filled with nostalgic, fairytale qualities. He utilized this skill in the 1930s with illustrations for the *Fables of La Fontaine* (*see page 32*), many of which involve cats.

COMMERCIAL CAT ART

Following World War II, the art world shifted its centre to New York, triggering the next major wave of art styles. Some of the artists that flourished in this atmosphere found a niche in new mass-market ventures such as posters and gift books, even collectable plates. Andy Warhol was one of these purveyors; perhaps best known for his colourful lifestyle

and Marilyn Monroe screen prints, he was also passionate about his cats, reproducing them in thousands of witty, whimsical drawings. The British painter, Martin Leman (1934–) has built a highly successful career with his many books featuring sleek and sophisticated cats.

Although cats are not the central theme of paintings by the American artist Will Barnet (1911–), Madame Butterfly, the beloved family cat, is often included in his uplifting family portraits. Similarly, the Scottish artist Elizabeth Blackadder (1931–) has been inspired mainly by everyday subjects, such as domestic items, flowers, and her cats. At one time three cats

joined the household and she began to record their unexpected and independent movements. But probably the best-known scene to include a cat is the portrait *Mr. and Mrs. Clarke and Percy* (1971) by the British artist David Hockney (1937–). Seated in designer Ozzie Clarke's lap, Percy was only so-named for the portrait.

The cat has been a celebrated subject in the modern art scene, and with its ever-growing popularity as a domestic companion there is every reason to believe that it will remain so.

Fantastic creatures (right)

Leonora Carrington's mysteriously named painting Tuesday (1964) is peppered with felines and cat-headed females, all presented in a dream-like, Surrealist style.

Watercolour portrait (below)

In this Expressionist portrait, Zu Apollinaire (1935) by Hannah Höch (1889–1978), the form of the sleeping cat is stretched to convey its extreme comfort and relaxation.

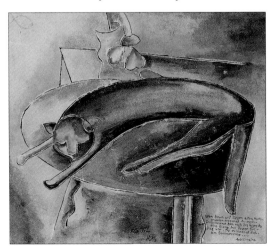

CARTOON CATS

FROM THE EARLIEST FELINE IMAGES PENNED BY ARTISANS FROM ancient Egypt, through to the humanized characters of the early 1900s, and on to the more recent comic strips, the domestic cat has been an ideal medium for caricature, satire, and humour. Some cartoonists have used cats to convey political or social comments, whereas others have drawn attention to the obvious similarities between feline and human behaviour. During the 20th century, a host of imaginative cartoon cats has been unleashed on a delighted public, creating a highly lucrative and internationally successful form of entertainment.

Cartoonists often give cats a valuable dual role in their work, as representations of both feline and human behaviour. The cat's fluid, sinuous movements translate effectively on to paper, allowing cartoonists to capture the essence of the feline, but the cat's facial features are ideal for conveying human expression, making it possible for cartoonists to portray human characteristics in a feline guise.

HISTORICAL CARICATURES
The divinity of the cat in ancient Egypt ensured that it played a prominent role in Egyptian art. Stonemakers, sculptors, and painters were employed by the state to decorate the tombs of the New Kingdom pharaohs (1540–1196 BC). In addition to their work on official monuments, these artisans created the first "cartoon cats": humorous, satirical drawings on papyrus (reed paper) and ostracon (smooth white limestone). At Deir el-Medina, in the region now known as the Valley of the Kings, a wealth of papyri and ostraca have been discovered. Now spread throughout the museums of the world, these informal drawings provide a fascinating window to the irreverent and appealing cynicism of that time (see page 32). Some ostraca are simple cartoon sketches of cats, but others are scathing commentaries on aristocratic, ancient Egyptian life.

In Russia, an unknown cartoonist portrayed Tsar Peter the Great (1672–1725) as a cat in the cartoon, *The Cat of Kazan*. In 1698, the tsar was reviled by the old aristocracy and church hierarchy because he outlawed the wearing of beards. Facial hair in the form of a moustache was allowed, however, mainly because the tsar sported a fine specimen – known as his "whiskers" – which was considerably exaggerated on the cartoon cat.

Political burlesque
The Russian ruler Tsar Peter the Great made himself so unpopular during his reign that, on his death, he was lampooned as a large cat, inelegantly lying on its back with legs and tail in the air, being buried by mice.

In recent years, the cartoon cat has played a less political role in Russia, starring in light-entertainment comic strips such as *Holidays in Prostokvashino* and *The Adventures of Leopold*.

ORIENTAL OBSERVATION
For centuries, Japanese artists (see page 38) have been able to capture the true nature of cats in their woodblock prints. The accuracy of their observation and illustration has been profound. This is also true of many modern Japanese cartoonists, who often employ the fierceness and frailty of cats to symbolize these same aspects of human nature.

Clean and tidy
By nature, cats, as well as humans, are fastidious in their hygiene. Ando Hiroshige has artfully captured this shared attribute in his anthropomorphic drawing.

Early satire
Although the Egyptians duly revered the cat, they were not without a sense of humour. This satirical papyrus (c. 1150 BC), was penned by Theban scribes. It shows a procession of cats acting as household servants, attending to a seated lady mouse or rat and her young offspring.

The popular Japanese artist Ando Hiroshige (1797–1858) created humanized cats that not only conveyed human actions and emotions, but also retained their feline nature. Western influences permeate current Japanese cartoons, with the escapades of the cat-shaped robot *Doraemon*, and the celestial cats Luna and Artemis in the comic book *Sailormoon*.

COMMERCIAL CARTOONS
In the West, humanized cats were often used for satirical commentary until sentimentality became a popular theme in Victorian Britain. The second president of the British National Cat Club was the artist Louis Wain (1860–1939).

Early cartoon hero (left)
Although its heyday was the 1920s, Felix the Cat is still well loved today. Felix Kept on Walking, was his theme song, ideal for a cat that always seemed to attract trouble.

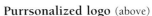

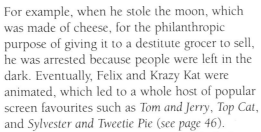

Purrsonalized logo (above)
The French artist, director, and cat-lover Jean Cocteau (1889–1963), artfully combined a few symmetrical ink blots with flowing calligraphy to create this highly distinctive cat emblem for the Club des amis des chats.

Stories without words (above left)
Artist Théophile Steinlen is best known for his picture stories, Images sans paroles. These three sketches conclude an amusing comic strip story, which shows how the cat's inherent curiosity can end in disaster. Steinlen was probably the first artist to successfully show cats in violent motion.

His innumerable caricatures of clothed, semi-human cats were, and still are, highly popular (*see page 50*). His characters have distinctly human features and enjoy many human pastimes, including smoking cigars, taking tea, and playing cards. Later in life, suffering from schizophrenia, his humorous, gentle images evolved into intricate, psychedelic patterns.

Another deeply cat-conscious artist was Théophile Steinlen (*see page 40*), whose Parisian home was known locally as "Cat's Corner", due to its large colony of former strays. These cats were the inspiration for Steinlen's posters, advertising commodities ranging from tea and milk to cafés and veterinary clinics.

FAMILY FAVOURITES
Since the images of Wain and Steinlen, there have been many other cartoon cats to delight us. The prototype was the cheeky rogue *Krazy Kat*, by American cartoonist George Herriman (1880–1944), who first appeared in 1910. Krazy Kat paved the way for the highly successful *Felix the Cat*, penned by Otto Messmer (1892–1983). Felix's character was that of a resilient survivor in a hostile world, conveyed with a subtle but clever comedy.

Feline from the future
Created in 1970 by Japanese cartoonist Fujiko F. Fujio (1936–1996), Doreamon is an atomic-powered robot in the shape of an earless blue cat. A visitor from the future, he helps children who are weak at studies or sports, and his comical exploits are avidly followed worldwide.

For example, when he stole the moon, which was made of cheese, for the philanthropic purpose of giving it to a destitute grocer to sell, he was arrested because people were left in the dark. Eventually, Felix and Krazy Kat were animated, which led to a whole host of popular screen favourites such as *Tom and Jerry, Top Cat,* and *Sylvester and Tweetie Pie* (*see page 46*).

Other best-selling cat cartoons include the cleverly illustrated cat words, such as "cat alog" and "platy puss", of the French visual joker Siné (1928–), the signatory Kliban cat – a strongly striped and round-eyed feline – by American Bernard Kliban (1935–1990), the inimitable cat creations of Englishman Ronald Searle (1920–), and the blatantly sexual exploits of *Fritz the Cat*, created by the artist Robert Crumb (1943–).

But no cartoon cat has been as successful as the slovenly, lazy, cantankerous, self-centred, not forgetting fat, *Garfield* by Jim Davis (1945–). People the world over recognize both their cats and themselves in the character of Garfield, and love him all the more for it.

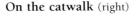

On the catwalk (right)
Louis Wain's highly sentimentalized, anthropomorphic cats can be seen taking part in a variety of typical Victorian pastimes. Here they are bedecked in their Sunday finery, enjoying a stroll around town.

CATS IN ENTERTAINMENT

CATS RARELY MAKE HOLLYWOOD HEROES AND THEY CERTAINLY HAVE not yet climbed the pinnacles reached by dogs such as Beethoven, Benjie, and Lassie. This is simply because cats are a great deal more difficult to train than dogs, and their faces are not usually so expressive. Cats with successful careers in entertainment to their name have usually received many hours of painstaking training, which are only rewarded if the cat has any inclination to learn the tricks it is being taught. Those cats that do adapt to a life within the entertainment industry, as a star or in a supporting role, will always steal the show.

With very few exceptions, cats tend to play supporting roles in film and television, but in advertising, an area that is an important part of entertainment, cats are almost on a par with dogs. As human cultural changes continue to swing in favour of cats, and computer animation becomes more sophisticated, it is likely that their most "entertaining" years lie ahead.

EARLY ENTERTAINMENT

Historically, cats have only briefly appeared in circuses and other entertainment. In 1549, in Brussels, a "cat organ" was played to a visiting king, Phillip II. This cruel instrument prodded or pulled the tails of cats that were trapped in tiny cages in order to elicit mews and squeaks. In the early 19th century, in Italy, a troupe of cats were trained by Pietro Capelli and toured around Europe. These cats balanced on high wires, performed trapeze acts, and juggled with their hindlegs. Yuri Kouklachev of the Moscow State Circus continued the tradition in the 20th century. He said it took three years to teach a cat to do a "handstand" on its forelegs, and he found it easier to train his cats at night when they were more alert. However, the concept of performing animals is declining in popularity, and it is unlikely that cats will be trained for circus acts in the future. There are now more acceptable occupations for well-trained felines.

STAGE AND SCREEN CATS

Cats have long found employment in films, with walk-on roles that rarely demand acting ability. The 1958 film *Bell, Book, and Candle* has Kim Novak in the role of a witch who lives with a Siamese cat called Pyewacket. The cat won a PATSY (Picture Animal Top Star of the Year) award for its performance. Blake Edwards' *Breakfast at Tiffany's* (1961) features Audrey Hepburn's marmalade confidant "Cat", who is temporarily abandoned on New York's rain-swept streets. Sheila Burnford's novel *The Incredible Journey* was filmed in 1963, and this tale of a cat and two dogs who travel hundreds of miles to return to their home was believed by many filmgoers to be a true story. Paul Mazursky's touching film *Harry and Tonto* (1973) is the

Curtain call
The designer Alexander Benois painted cats on witches' broomsticks for the backdrop of the "St. Petersburg at Night" sequence in Igor Stravinsky's 1911 ballet Petrushka.

story of a widower, played by Art Carney, who is evicted from his New York home and travels to Chicago with his cat Tonto. Both stars won awards for their roles. In *The War of the Roses* (1989) the feuding couple, played by Kathleen Turner and Michael Douglas, have a cat and dog respectively. The film implies that "cat people" should not marry "dog people".

Not surprisingly, given their long association with evil, cats are popular in horror films. From Mexico, where an aristocrat keeps man-eating cats in *The Night of the Thousand Cats* (1972), through Hollywood B-movies such as *Cat People* (1942) and *The Shadow of the Cat* (1961), to *The Black Cat* (1985), an Italian film about a killer cat, the medieval image of the cat is perpetuated.

Feline foe (left)
The comic strip Batman *was made into a film in 1966. Batman's sidekick was Robin, and his opponents included The Joker and Catwoman. Despite the rivalry between Catwoman and Batman there were sexual undertones to the relationship that were made explicit in Tim Burton's 1992 film* Batman Returns.

Best friends (right)
When Truman Capote's novel Breakfast at Tiffany's *was filmed in 1961, the character of Holly Golightly was played by Audrey Hepburn. Her confidant was a ginger tom called Cat, who won a PATSY award for his acting prowess.*

La Belle et la bête (left)
In Jean Cocteau's 1945 film version of the fairytale The Beauty and the Beast, *the beast, played by Jean Marais, has strikingly feline features. The mask was actually based on the features of Cocteau's blue Persian cat.*

Curious cat (above right)
The animated film Felidae *is a far cry from Disney. The hero, Francis, investigating the death of a cat in his neighbourhood, uncovers a world of mysterious sects and bestial murder.*

Cartoon cruelty (right)
Despite overtones of cruelty and sadism, the cartoon series Tom and Jerry *has been highly popular since 1939. Its success may lie in the fact that the victim, the mouse, is always the eventual victor over poor Tom.*

Cats fare better in science fiction films. In the 1978 film *Superman*, with Christopher Reeve in the starring role, the hero rescues a Devon Rex from a tree. In the same year, Disney made *The Cat from Outer Space*, featuring a cat called Jake (actually played by a female named Amber), who captains a spaceship, and in 1979 a red tabby accompanied Sigourney Weaver in *Alien*.

On stage and television, cats are more often played by people. The feline character Catwoman made her first screen appearance in the 1950s' television version of *Batman*, and in the 1980s' British space-comedy series *Red Dwarf*, the hip-cat character "Cat" evolves from the spaceship's ordinary moggie. On stage, the relatively minor role played by real cats has been more than made up for by the huge success of the musical *Cats*, based on T.S. Eliot's poems (*see page 34*).

CARTOON CATS
The first animated cat was George Herriman's *Krazy Kat*, transferring from paper to screen in 1916. More successful was Pat Sullivan and Otto Messmer's *Felix*, who in his prime during the 1920s was as popular as Charlie Chaplin. *Felix* was succeeded by *Tom and Jerry*, created in the late 1930s by William Hanna and Joseph Barbera. This team's production company also made *Top Cat*, a popular 1960s' television series.

The first full-length feature cartoon was Warner Brothers' *Gay Purree*, in the 1960s, with Judy Garland as the voice of Mewsette, the feline leading lady. This was followed in 1970 by Walt Disney's *Aristocats*, featuring the elegant Duchess, who was given voice by Eva Gabor, and the stout-hearted drifter O'Malley, spoken and sung by Phil Harris. Robert Crumb's sexually explicit *Fritz the Cat* was transferred from a strip cartoon to screen in 1972.

Since then, with the exception of *Felidae* (1994), the animated film that director Michael Schaak adapted from the novel by Akif Pirinçci, felines have returned to peripheral roles such as the nightmarish urban cats in *Stuart Little* (2000).

CATS IN ADVERTISING
Advertising has brought fame to many cats. In Britain, the all-white Arthur is well known for eating his food from the tin with his paw. In America, Arthur's equivalent was Morris, who won a PATSY in 1973 for advertising cat food. Another famous British cat was Solomon, a Chinchilla Longhair who started his career advertising carpets but went on to feature in the James Bond film *Diamonds Are Forever*.

Cats are featuring increasingly in advertising for two simple reasons. First, cat ownership is rising and now exceeds dog ownership in North America and many parts of Europe. The second reason is that market research has shown that people actually remember advertisements better when cats appear in them. One detergent company studied the effectiveness of their advertisements and discovered that more people remembered their product seven days after viewing a commercial with a cat in the starring role than when a dog, horse, chimpanzee, bird, "woman next door", or even a famous personality was featured.

Cats seem to suggest cleanliness and sensuality. Good advertisements appeal to our emotions and cats are ideal for representing what advertising experts classify as "female" values – warmth, beauty, capriciousness, and elegance. With these associations, the future is bright for cats in entertainment.

Glamour puss (below)
One of the most successful stage musicals of all time is Andrew Lloyd Webber's Cats, *based on a book of poems by T.S. Eliot. The musical includes the character Grizabella, excluded from the book because her life was considered too louche for children's ears.*

CAT EPHEMERA

HUMAN ATTITUDES TOWARDS THE CAT ARE CONSTANTLY CHANGING. Ideas and images that appeal to one generation pass rapidly out of fashion as tastes change and we learn more about feline behaviour. In many early civilizations, big cats, especially the lion, were used to symbolize power and royalty in official sculptures and royal crests, much as they still are today. Meanwhile, the domestic cat has been modelled in softer images such as religious amulets and collectable ornaments. Examples of ancient, rare, and priceless artefacts are now housed in museums and galleries throughout the world, so that the cat's place in aesthetics is easily traced.

Since records began, people have used artistic representations as part of folklores, customs, and religious beliefs. Early civilizations, such as those of the Egyptians, Aztecs, and Babylonians, produced a range of feline-inspired artefacts, some of which still survive. Many of these primitive images represent members of the big cat family – lions, tigers, leopards, and cheetahs – that were prevalent at the time. But a greater selection of cat images were produced by later societies, when the domestic cat became a part of everyday lives.

EARLY FELINE ICONS
The most dramatic and enduring legacy of the Egyptian and Greek use of feline symbolism is the sphinx – a mythical beast with the head of a human and the body of a lion. Some of the sphinxes that flanked the entrances to important funerary monuments can still be viewed in their original settings. However, many were removed by early archaeologists to various museums, including the fine set now adorning a staircase at the Louvre in Paris.

The lion, another enduring symbol, has been associated with royalty since time immemorial (*see page 26*). One of the most impressive finds from the tomb of King Tutankhamum was his stately, gold, lion throne (*c.* 1327 BC). Much later, the great jewellers Cartier used the lion, and other big cats, to create a selection of royal gems for the Duchess of Windsor.

If big cats evoke masculine images of power, strength, and royalty, then the domestic cat embodies beauty, poise, and cleanliness, which are all stereotypical feminine traits. For example, statuettes of the cat goddess Bastet (*see page 26*) wearing a long, tight dress were commonplace in early Egypt. A few of these are now on permanent display in collections housed at the British Museum in London, the Egyptian Museum in Cairo, and the Metropolitan Museum of Art in New York.

THE POPULARIZED CAT
Ornamental representations of the cat have always been popular. Fertility amulets in feline form, dating from *c.* 2300 BC, along with cosmetic pots and gold jewellery, are all evidence of a strong intimacy between cats and the Egyptians. Relationships of a similar type can be traced to many parts of the world through artistic relics. From Persia comes a well-preserved bottle in the shape of a cat, from Crete a fresco of a cat hunting scene, and from Italy an intricately decorated vase (*see page 26*). Much later, the Toby jug – originally a beer mug, shaped like a man – was produced in feline form in many European countries.

From the 17th century onwards, cat figures became fashionable ornaments in both the East and the West, as they represented home values and domestication. These figurines, sized for the mantelpiece, were made in abundance, but only a few are worthy of museum space. Cats produced by Gallé and Fabergé inspired a host of cheap imitations. Other fine examples of rare and expensive collectors' items include German Meissen cats, Dutch Delftware cats, Scottish Wemyss cats, and Ch'ing cats from China.

Some cat forms are prized for sentimental or superstitious reasons. The former is the case for a bronze cat made by the French Art Deco sculptor Chassagne. Since the mid-1920s the cat has been a permanent fixture at Heal's, the famous London department store, and is now considered the company mascot. Superstitious beliefs have kept Kaspar, a black cat, employed by the Savoy Hotel in London. He is placed in the 14th seat, with a napkin tied around his neck, at dinner parties for groups of "unlucky 13", and the waiters are instructed to change his place-setting for each course of the meal.

Model cat
Entrepreneur and artist Emile Gallé (1846–1904) was well known for decorative art of high quality. This brightly coloured, green-eyed cat is typical of the faience (tin-glazed earthenware) figurines that were, and still are, eagerly sought by collectors of feline artefacts.

Feline adornment
Cats are central to the design of these well-preserved pieces of Egyptian jewellery. The gold spacer-bar was part of a bracelet from the tomb of Queen Sobekemsaf, dated around 1650 BC, and the gold finger-ring with seated carnelian cat is of a similar date. Sadly, it is now impossible to know whether these cats were employed for decorative or religious reasons.

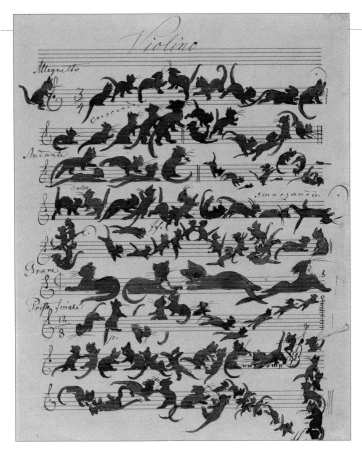

Cats chorus (left)
This mid-19th century painting, Die Katzensymphonie *or* Cat Symphony *by Moritz von Schwind (1804–1871) is more of an artistic diversion than pure music. His cats get bolder in the crescendo and scamper off at the finale.*

Venerable playing card
This forgotten Rheinish card game "Bird's Play" illustrates that games – like art, literature, and advertising – reflect the fashions and social values of their time.

"Concatonated Order of Hoo-Hoo" adopted the black cat to promote the lumber and forest industry. There are now over 8,000 members, all of whom advertise using the cat.

CAT CURIOSITIES AND TRENDS

Although music cannot be put on display, it is nonetheless a part of cat history, as numerous composers have been stimulated by the beauty and movement of cats. Domenico Scarlatti (1685–1757) claimed that his cat, Pulcinella, helped in his compositions. Pulcinella had a habit of walking along the keys of Scarlatti's harpsichord, and on one occasion this inspired him to compose *Fugue in G Minor, L.499,* now generally known as *The Cat's Fugue.*

This musical association was taken to the extreme by Japanese makers of the samisen, an instrument similar to a guitar – cat skin was stretched across its soundbox, and its strings were made from cat intestines. In 17th-century Italy, violin-makers advertised the use of catgut in their instruments, although in reality it was sheepgut. They did this to protect their design, but the term has stuck due to the din produced by inexpert violinists, which is often compared to the caterwauling of mating cats.

The pre-neutering days generated many unwanted cats, which led to a fashion during the 19th century for preserving these animals in the name of art. For example, taxidermist Walter Potter produced a number of works comprising stuffed kittens posed in human scenarios. Such a practice seems distasteful today, although the success of the 1980s cartoon book, *101 Uses for a Dead Cat,* suggests that modern tastes have barely changed.

HERALDRY AND EMBLEMS

The lion, known as "king of beasts", is the only feline to hold a prominent position in heraldry. Early records show that cats were first used as heraldic emblems during the Roman Empire. A famous legion, the *Felices seniores,* bore the cat on its coat of arms, an early pun on the words *felis* (cat) and *felix* (fortunate or happy). This military pun was echoed by the popular cartoon character Felix the Cat (*see page 44*) that appeared on badges of U.S. Marine Corps pilots in World War II. Tomcat replaced Felix in the 1970s to celebrate the night-flying abilities of the Grumman F–14 Tomcat carrier-borne interceptor aircraft – comparisons were made with the cat's night vision.

In the 5th century AD, the king of Burgundy, Gundiracus, adopted the cat as a symbol of liberty and independence. In due course the cat came to be featured on more than a hundred European coats of arms, including the German Katzen family shield, with a silver cat holding a silver mouse on an azure background, and the Italian Della Gattas' cat *couchant.* In Scotland, a home of the wildcat (*see page 18*), various branches of the great Clan Chattan displayed the cat on their family crest with a variation on the motto "Touch not the cat but [without] a glove". These families were fierce warriors and the untameable, aggressive wildcat was an appropriate image for their coat of arms.

Images of the cat are not restricted to formal emblems. For example, in Germany the word "katz" is often found in the names of towns and villages, whereas in neighbouring Austria it is incorporated in family names, such as Katzler.

In British, pubs feline images are adopted in a similar way to heraldic signs. The Red Lion is a popular name, but establishments such as The Cat and Fiddle, or the Cat and Custard Pot are more memorable pub names. In France, shop signs incorporating the cat are common, as a result of the banning of coats of arms after the French Revolution. In fact, inns, clubs, cafes, and restaurants have all appropriated the cat. Across the Atlantic in North America, a fraternal organization founded in 1892 as the

Heraldic cats (left)
Cat emblems abound. This 1970s' badge features Tomcat, *an American pilots' mascot, and the kilt pin is taken from a Scottish family's coat of arms.*

Anthropomorphic ceremony (right)
The Jamaica Inn in Cornwall houses the exhibits of the British taxidermist Walter Potter. The Kittens' Wedding consists of the usual wedding party attended by a full congregation, and all are dressed in their wedding finery. This may have been popular with its original Victorian audience, but many people today find it macabre.

CAT COLLECTABLES

MOST CAT LOVERS ARE TEMPTED AT ONE TIME OR ANOTHER TO BUY some picture or keepsake that reminds them of their feline friend, but for some ailurophiles collecting can become a major pastime. Due to the international surge in the cat's popularity during the mid-19th century, artists, manufacturers, and later advertisers recognized potential in exploiting the cat. This led to a variety of cat merchandise, ranging from postcards and stamps to toys and games, most of which is easily accessible and reasonably priced for the collector. If you would prefer to view, rather than collect, then there are now many private collections open to the public.

Cultural artefacts that exemplify the role of cats in history are usually found in museums, art galleries, or in the sales of international auction houses. Whatever the case, they are inaccessible to most potential collectors. Perhaps this is why the "cat collectable" has become such big business. Modern manufacturing techniques have produced a plethora of collectable feline images that sell successfully worldwide.

MASS-PRODUCED CATS

Since the 17th century in Europe, and before that in the East, ladies' fans have been variously decorated to make them more appealing to the user. Animals were always a popular choice of adornment, and the cat featured frequently in the early 1900s in the work of the great French fan-makers Duvelleroy, among other artists, on cotton, silk, or paper fans.

It is likely that figurines in the shape of a cat were manufactured specifically with the collector in mind, perhaps as a direct result of the popularity of the cat fans.

Spanish stamp
Lindbergh's cat is featured on this commemorative stamp (1930), although she remained behind.

During the 18th century, many manufacturers made cat figurines (*see page 48*). From that era only the mass-produced ones are still available, although there are plenty of cat figurines being produced today for the collector. However, a cat figurine may be pleasing to the eye, but it is uninspiring as a plaything. Luckily, German soft-toy makers Steiff, renowned for developing the Teddy Bear, rectified this with a line of black cats, which ran until the 1980s. The early ones, especially those that still display the company name with a button on the ear, are the most sought-after today. The toy world went on to exploit the adulation surrounding cats with other items that now fetch high prices, including Victorian metal cats that nod their heads, key-wound plush cats from Europe, Japan, and China, and early American cat-based board games.

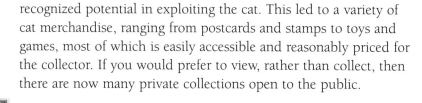

Cat postcards are the ideal starting point for any novice collector. Most countries joined the international postal system in the late 1800s, and postcards became fashionable as a method for sending short communications. During this time, manufacturers produced a seemingly infinite variety of cat postcards, many of which included the caricature cats of German artist Arthur Thiele (1841–1919) and the delicate illustrations of Helena Maguire (1860–1909). Yet it is probably the anthropomorphic subjects of Louis Wain (*see page 44*) that are well known to collectors and non-collectors alike. This trend in the early postcards to illustrate clothed cats participating in human events, was replaced later by black-and-white pictures of real, clothed cats. Often their eyes were artificially coloured, sometimes with tinted glass, to emphasize this very appealing attribute. In the 1930s, the fashion was for cat breeds, which were replaced during the 1950s by brightly coloured photographs of kittens, often held in the arms of young girls.

Popular ornament
For those who cannot afford the fine porcelain cats of Meissen or Gallé, a 19th-century, mass-produced earthenware cat, such as this ginger and white, is an affordable alternative.

French fan
(above right)
With technical developments in the 1920s it was possible for fan-makers Duvelleroy to produce this popular fan shaped like a cat's head.

Folk art (left)
Rugs similar to this one were once a common sight in 18th-century American homes. Many artists prefer everyday subjects, which is why the family cat often appears on rugs.

POSTAGE STAMPS

Naturally associated with the postcard is the postage stamp, which has proved to be another useful medium for displaying cats. The first stamp to feature a cat was issued in 1930 by Spain to commemorate the first, non-stop, transatlantic flight of Charles Lindbergh (1902–1974). Lindbergh's pet cat, Patsy, accompanied him on many flights, but not on this one, which is why she is portrayed gazing wistfully at the departing plane.

As with the postcard, stamps that feature cat breeds are immensely popular, especially complete cat stamp sets. The first "all-cats" set of stamps was issued in Poland in 1964, possibly because then, as now, it had one of Europe's highest cat populations per household.

In 1995, the West Indian island of St. Vincent issued a sheet of nine different $1 stamps depicting cats of the world. The sheet was designed in such a way that it formed a composite picture of a verandah, facing the ocean with mountains in the background, and a coterie of cats engaged in various activities – a real find for the dedicated ailurophile. In fact, most countries have capitalized on the collectable cat stamp, including Britain with a set of Elizabeth Blackadder's cat paintings (see page 42), Sweden with the Nordic goddess Freya, and France with a centenary stamp to celebrate the birth of their cat-loving novelist, Colette (see page 36). The popularity of the cat stamp is now so great that a specialist organization in North America is devoted purely to this subject.

Interestingly, the production of cat stamps is not exclusive to the country of the cat, so that the Russian Blue may be featured on a Danish stamp, or the Scottish Fold may be found on a stamp in China. However, the Isle of Man has capitalized on its unique feline, the Manx. This very important emblem appears not only on the island's stamps, but on its coins as well.

HALLOWEEN MEMORABILIA

Another large range of cat collectables can be found in the merchandise for Halloween. The reason lies in the age-old association made between cats and witches (see page 30), as well as the naturally mysterious image of the black cat itself. Today, most superstitions about black cats and evil have been discarded, although their association with Halloween is as strong as ever. Sending special cards and decorating the house is still part of the fun: the difference now is that

Playing cards
These 20th-century playing cards penned by Austrian cartoonist Manfred Deix are now collectors' items. His humorous illustrations are very appealing, especially the chain-smoking King of Diamonds and cross-eyed Ace.

Halloween symbol
For many years, black cats have been associated with the Halloween festival. In some places, and especially in North America, people decorate their homes, and these cats are typical of Halloween decor.

designers depict the black cat as a benign creature and the Halloween cat is more decorative than frightening. For sure, Halloween merchandise will brighten up any cat collection.

CAT PROMOTIONS
Cats have figured prominently in advertising campaigns for over a century. Originally, the black cat was used, not because it was popular, but because the colour lent itself to inexpensive mass production. One brand in particular became very successful, and advertisements for *Black Cat* label cigarettes, matches, stockings, shoe polish, whisky, and even apples can now be bought at reasonable prices. Other manufacturers have seen the potential in collectable cat advertising and have designed different labels for the same product, such as the German wine *Zeller Schwarze Katz*. However, do not think that advertising remained the prerogative of the black cat. Advanced reproductive technology and fluctuations in fashion for different breeds has left a huge range of collectable items.

It is easy to start collecting any of these cat curios as a hobby, but for some enthusiasts the collection takes over their lives, as well as their homes. Special museums for private collections can now be found in Britain, the United States, The Netherlands, Switzerland, and in Malaysia.

Advertising alcohol
Cats have been associated with gin for centuries in Europe, where Old Tom *was once a well-known Victorian brand. In Germany a black cat has featured on the wine label for Zeller Schwarze Katz, or Black Cat wine, for many years. One such cat was once seen standing on their winning barrel at a local wine-making competition.*

Feline greetings
Postcards were a popular form of correspondence throughout Europe during the early decades of the 20th century. The German card (above) shows an exaggerated, arch-backed tomcat ready to defend his turf, while the French postcard (left) depicts feline natural grace and beauty enhanced by Parisian haute couture.

WORKING AND COMPANION CATS

ORIGINALLY USED AS RATTERS AND MOUSERS AT HOME AND WORK, cats have proved to be highly effective pest-controllers all over the world. But it is the cat's endearing personality that has secured its continued usage now that there are other forms of rodent control. From Egyptian fowler and ship's mouser to parliamentary confidant, the domestic cat has fulfilled many roles in both work and play. In North America and many parts of Europe, cats are now the most popular pet, partly due to their predatory skills but more because they are easy to care for and excellent company.

From the time that people first stored grain, the cat has been a safe and most effective method of vermin control. Now, with the shift to urban living, there is less grain to protect, but there are just as many rats and mice to destroy.

INVETERATE HUNTERS

Theban tomb frescoes from the Twelfth Dynasty (c. 1980–1801 BC) indicate that cats may have assisted Egyptians during fishing and fowling expeditions. Whether or not these cats were domesticated is unclear, but they show that the family cat is an inveterate hunter.

Domestication was probably a symbiotic process; humans encouraged the cat to stay by providing meals and shelter, while in turn the cat purged the household of vermin. Over the years, cats have found work in all manner of places, from offices to museums. Many of these cats have killed a prodigious quota of pests. For example, Towser (1963–1987), a Scottish distillery cat, was credited with

Egyptian hunting fresco

A few hunting scenes that include cats have been found in Theban tombs. In this one (c. 1450 BC), the sculptor Nebamun, and family, are employing the services of a cat to flush fowl out of their secluded, marshy hiding place.

catching a record-breaking 28,899 mice. There is even a convent in Cyprus – St. Nicholas of the Cats – named in honour of the cats that protected it, and the local population, from an infestation of venomous snakes. These cats were introduced to Cyprus from Egypt about AD 328, and they easily adapted to island life.

Cats have worked equally well indoors by protecting sacred scrolls in temples throughout Asia, rare books in libraries across Europe, and food stores in homes and farms the world over.

More recently in the United States, librarians have formed The Library Cat Society, in order to preserve the work of library cats that prevent rodents from damaging book bindings.

UNUSUAL MILITARY COMMISSIONS

Compared with dogs, cats have not only played a minor role in human warfare but also an unhappy one. Stories of their misuse start as early as 525 BC, when Cambyses of Persia took possession of Egypt. It is said that the Persian troops managed to take the city of Pelasium without a blow being struck. They went into battle with live cats secured to their shields, which rendered the Egyptians powerless to resist in case they harmed their sacred animals.

Other military cats have not been so lucky. Two thousand years later, in Germany, vessels were allegedly strapped to the backs of terrified cats, which were forced to run about in enemy territory, spreading poisonous fumes.

A very recent tale concerns the US Army's attempt to employ the cat's excellent nocturnal vision by training them as night scouts for their troops in Vietnam. As plausible as the story sounds, its origins lie with a writer for the satirical magazine *National Lampoon*.

Friends in high places (above)
Very few pictures exist of the president John F. Kennedy (1917–1963) with any of the family cats, although his wife Jacqueline and daughter Caroline look happy and relaxed here, photographed in 1960 with Tom Kitten. Past associations of untrustworthy, sly, or evil cats may have persuaded Kennedy not to appear with cats, even though he was known for liking them.

Public relations opportunity (below left)
Cats are loved regardless of the political beliefs of their owners. Vladimir Lenin (1870–1924), leader of the Communist Party and founder of the modern Russian state, is pictured here stroking a cat. It does appear to be an official pose rather than a casual moment, as Lenin pays more attention to the camera than to the cat.

Work, rest, and play (right)
The cats at the Nga Phe Cahung Monastery in Burma have shown a natural inclination for jumping through hoops. This was first discovered by the Buddhist monks when, during many long hours in meditation, the playful cats jumped through their outstretched arms which were statically positioned in prayer.

We do know, however, that the British Army employed cats during World War I to act as gas detectors. When they were introduced to front-line trenches (much like canaries in coal mines), their sensitive olfactory system could detect unfamiliar smells well before humans could, therefore acting as an excellent alarm system.

At one time, before the introduction of quarantine laws, all ships carried a rodent-catching cat. Most are unsung heroes, but a few are remembered with honours. There is a bronze statue in Sydney, Australia, of Captain Flinders (1774–1814) and his feline companion, Trim. During the late 1700s they were among the first to chart the Australian coastline. Another ship's cat, Oscar, has been celebrated for his amazing survival rate after an extraordinary number of shipwrecks during World War II.

FAMOUS CAT LOVERS

A cat is the perfect confidant, which may be why so many leaders – American presidents in particular – have kept one. Abraham Lincoln (1809–1865) took back to Washington three bedraggled cats from the Civil War. Slippers, a cat of President Roosevelt (1858–1919), often attended glittering White House dinners. President Clinton (1946–) made "first cat" Socks off-limits to overeager photographers.

Grinding corn (left)
Attempts to train cats to work are usually unsuccessful. This crude 18th-century model of a mill driven by a cat, trying to catch a mouse, is fun but impractical. A horse or a donkey is certainly better suited to this type of work.

Pampered kittens (right)
The prominent French church leader Cardinal Richelieu (1585–1642) delighted in having kittens around him. Unfortunately, even though he made provision in his will for their continued care, Richelieu was an unpopular leader and his cats suffered when he died.

The official home of the British Prime Minister, 10 Downing Street, has had a succession of cats. Sir Winston Churchill (1874–1965) was known for his love of cats, and his last one, Jock, was provided for in his great owner's will.

The assembly of prestigious ailurophiles includes many other people: from royalty and religious leaders to writers, poets, and artists. When a reporter asked the Swiss sculptor Alberto Giacometti (1901–1966) which of his masterpieces he would save from a fire, he replied, "It depends on what is in my house. If there was a cat, and my works, I would save the cat. A cat's life is more important than art". Remarks like this by influential people help to promote the cat's image.

Gainfully employed (left)
In an attempt to reduce the risk of potentially dangerous microbes entering into Britain through smuggled food, customs officials commissioned a poster in 1935 showing cats sniffing at baggage, on the prowl for unusual scents. The sum of 3d was even promised for these feline services.

Artistic calling (right)
There are many artists that paint cats but are there any cats that can paint? According to a recent publication, Why Cats Paint, *there are many artistically gifted cats, and Max – shown painting* Birdies *(1991) – is one of them. It is a brilliantly executed book that almost convinces us of innate feline artistic talent.*

IMPROVING HUMAN LIVES

In recent years, scientists have investigated our relationship with cats. An Australian study confirmed that cat owners visited the doctor less often than non-cat owners. In the United States, fewer cat owners died during the year following a major heart attack than those in similar circumstances who did not own a cat.

What remains to be determined is whether an improved quality of life is created by living with cats, or whether living with cats is but a manifestation of a good quality of life.

CHAPTER THREE

FELINE
DESIGN

The cat is an almost perfectly designed predator. Its superb balance and flexibility enable it to catch small prey and escape from larger predators. Its brain, nerves, and hormones harmonize to avoid wasting energy, while still being capable of explosions of activity. The domestic cat's anatomy is almost identical to that of its close wild relations; most medical problems are caused by injury or illness rather than poor design. Its internal organs and body functions have adapted for survival; its digestive and excretory systems allow it to cope without food for longer than other domesticated animals; and its unique reproductive system is adapted to ensure successful matings each year.

ABOVE MALE FAWN SIAMESE

LEFT ADULT CAT LEAPING THROUGH THE AIR

THE SKELETON

LIGHT YET MARVELLOUSLY STURDY, THE FELINE SKELETON EVOLVED for a lifestyle that called for sudden bursts of speed combined with dextrous agility. This substructure is the basis of the extraordinary grace of the cat's movement. Cats have more bones than humans have, mostly because of their tails. All parts of the feline skeleton, from the skull to the tail vertebrae, are in proportion: the legs are slender but robust, supporting a narrow ribcage and a highly supple spine. The cat's shoulder blades are unattached to the main skeleton, permitting superb flexibility at any speed. The entire structure is held together by strong but elastic ligaments.

The hard, rigid structure of the cat's skeleton protects the internal organs, provides points of attachment for muscles, and acts as a system of levers and joints necessary for fluid movement. It is held together by fibrous ligaments, elastic tendons, and rapid-acting, powerful muscles.

THE STRUCTURE AND GROWTH OF BONE

Internally, a bone is a latticed structure of hard, calcified struts, called trabecullae, filled in its core with bone marrow. Bone is nourished by its own blood supply that enters each bone through a nutrient membrane, or foramen.

Bones grow continuously during kittenhood. The skull begins as separate bones, to permit birth, then fuses along suture lines. In some young kittens a small soft spot can be felt on top of the skull; this disappears rapidly.

The long bones of the limbs and ribcage begin as hollow tubes of cartilage, and calcify in infancy, replacing cartilage with bone. The outside of the tube, the periosteum, produces new cells, thickening the bone. Inside, the old bone is constantly remodelled to maintain uniform thickness.

Bones increase in length by constant production of new bone at the growth plates, or epiphyses, at their ends, which are nourished by a rich supply of tiny arteries. The growth plates are prone to injury during skeletal development, and both growth and sex hormones influence growth. Curiously, sex hormones inhibit activity: cats neutered very early tend to grow slightly longer leg bones.

If a bone breaks, cells on the periosteum activate, producing new bone to bridge the gap. Usually, too much new bone is made, but the excess is gradually remodelled from the inside.

JOINTS AND LIGAMENTS

Cats have three very different kinds of joints: fibrous, cartilaginous, and synovial. Each of these has a different level of flexibility, and each of them performs a different function.

The joints at the suture lines between the fused bones that form the skull are made of hard fibre. These joints have no flexibility at all. The mandible or jawbone, for example, is really made up of two bones with a fibrous joint at the midline between the incisor teeth. If a cat lands heavily on its jawbone in a fall, this hard fibrous joint may split: technically, the cat has not actually broken its jaw, although that is the apparent effect, but has torn this fibrous joint.

Other joints, like the thick discs between the spinal vertebrae, are made from tough cartilage. These cartilaginous joints are looser and more supple than similar joints in other species, such as humans and dogs, providing the cat with a far greater degree of flexibility in its torso.

During growth, the growth plates at the ends of the long bones are temporarily cartilaginous joints, and as such they are less sturdy and more prone to damage than in adulthood.

Synovial joints are found where the greatest degree of movement is needed. They are hinged or ball-and-socket joints, with pearly, smooth, articulating cartilage on their contact surfaces, and surrounded by a joint capsule filled with synovial fluid. This construction is found in the highly flexible joints of the legs and the jaw.

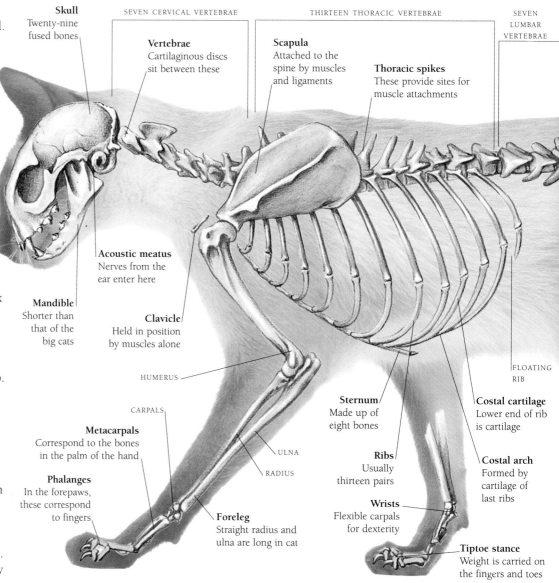

Skull
Twenty-nine fused bones

SEVEN CERVICAL VERTEBRAE

THIRTEEN THORACIC VERTEBRAE

SEVEN LUMBAR VERTEBRAE

Vertebrae
Cartilaginous discs sit between these

Scapula
Attached to the spine by muscles and ligaments

Thoracic spikes
These provide sites for muscle attachments

Acoustic meatus
Nerves from the ear enter here

Mandible
Shorter than that of the big cats

Clavicle
Held in position by muscles alone

HUMERUS

CARPALS

Metacarpals
Correspond to the bones in the palm of the hand

Phalanges
In the forepaws, these correspond to fingers

Foreleg
Straight radius and ulna are long in cat

ULNA

RADIUS

Sternum
Made up of eight bones

Ribs
Usually thirteen pairs

Wrists
Flexible carpals for dexterity

FLOATING RIB

Costal cartilage
Lower end of rib is cartilage

Costal arch
Formed by cartilage of last ribs

Tiptoe stance
Weight is carried on the fingers and toes

Synovial joints sometimes suffer from excess production of the lubricating synovial fluid or inflammation due to arthritis or synovitis through injury, disease, or allergic reaction.

Ligaments, the tough bands that hold bones together, are important in all joints, but vital in synovial joints, which inherently lack stability. The feline hip joint, in particular, is prone to dislocation; this is often, although not always, prevented by the ligament that holds the ball of the femur firmly in the socket of the pelvis.

SKELETAL VARIATIONS AND PROBLEMS

The cat's skeleton is one of evolution's superb achievements. In the natural cat, problems such as arthritis and decalcification are caused by disease or hormonal or dietary upsets, not poor design. Until recently, there was little skeletal variation in cat breeds: unlike the dog, which has been radically changed through selective breeding, the cat escaped extreme modification.

Environmental pressures do create natural variations. In hot climates, cats are naturally small; a small animal has a higher surface-area-to-weight ratio, which helps it to keep cool. Free-breeding cats in cold climates, such as Scandinavia, Russia, Canada, and the northern United States, have larger, heavier skeletons.

In natural circumstances, severe skeletal anomalies disappear, most often because lethal problems or survival drawbacks are associated with them. Some, such as taillessness or extra toes (polydactyly), may be perpetuated by their occurrence in a limited gene pool. The natural tailless breeds all developed on islands, and isolated gene pools also perpetuated the high proportion of seven-toed cats around Boston, Massachusetts and Halifax, Nova Scotia.

Active selection for a breed standard has led to more dramatic changes recently. Siamese and Orientals today have much longer and thinner legs bones than their predecessors, while the British Shorthair's skeleton has become more compact, with heavier long bones. This kind of active intervention has perpetuated the most considerable, and worrying, skeletal problems. Exaggerated shapes increase the risk of painful inherited arthritic problems.

EXTREME SKULL SHAPES

The cat has a short face compared with other domestic animals. Selective breeding has not compromised this natural shape as much as in the dog, but changes have occurred, and different breeds have more or less round or angular faces. The most dramatic alterations are those in the length of the head or muzzle.

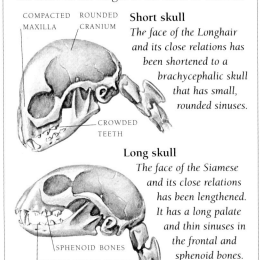

COMPACTED MAXILLA · ROUNDED CRANIUM · CROWDED TEETH

Short skull
The face of the Longhair and its close relations has been shortened to a brachycephalic skull that has small, rounded sinuses.

Long skull
The face of the Siamese and its close relations has been lengthened. It has a long palate and thin sinuses in the frontal and sphenoid bones.

SPHENOID BONES · WIDELY SPACED TEETH

RETRACTABLE CLAWS

Claws grow from the last bone of the toe and are anchored by tendons. They are modified skin: an outer cuticle of hard protein, keratin, protects the dermis, or quick. Cats' claws are kept sheathed for protection on the move.

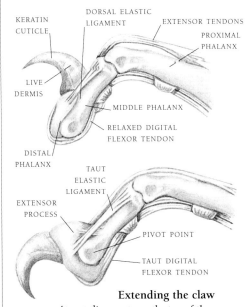

KERATIN CUTICLE · DORSAL ELASTIC LIGAMENT · EXTENSOR TENDONS · PROXIMAL PHALANX · LIVE DERMIS · MIDDLE PHALANX · RELAXED DIGITAL FLEXOR TENDON · DISTAL PHALANX

TAUT ELASTIC LIGAMENT · EXTENSOR PROCESS · PIVOT POINT · TAUT DIGITAL FLEXOR TENDON

Extending the claw
At rest, ligaments on the top of the paw naturally sheath the claws. A cat exposes its claws by contracting digital flexor muscles in its legs, pulling taut the flexor tendons under the paw.

ILIUM

THREE FUSED SACRAL VERTEBRAE

Pelvis
Ring made up of fused ilium, ischium, and pubis

PUBIS

HIP joint
Flexible ball-and-socket joint

ISCHIUM

Structure of the skeleton
The domestic cat's skeleton is almost identical to that of the big cats in both shape and proportion; the only real difference is in size. In all cats, the many vertebrae give superb mobility, but the forelegs also provide highly visible flexibility; when a cat walks, its shoulder blades rise above the spine, just as in the lion or tiger. Less apparent to the eye is the intricate bony structure of the wrists, which allows exceptional dexterity in actions such as retrieving prey from its hiding place or walking along narrow purchases.

Tail
Each vertebra of the supple tail articulates

EIGHTEEN TO TWENTY CAUDAL VERTEBRAE

FEMUR

TIBIA

FIBULA

Tarsus
Cat's foot starts here

PHALANGES

METATARSALS

BIOFEEDBACK AND THE BRAIN

ALL THE SENSES AND THE BODY'S HORMONE-PRODUCING GLANDS send information to the brain. This interprets the chemical signals and instructs the body on how to respond via the nervous system. The brain also sends instructions to the master gland of the cat's hormonal system, the pituitary, which lies in the base of the brain.

Information from the brain stimulates the pituitary to produce the hormones that control functions and activities such as metabolic rate and sexual behaviour. This demands a great deal of energy and, although the brain accounts for less than one per cent of body weight, it receives 20 per cent of the blood pumped by the heart.

All of the hormones that manage and maintain body functions or regulate feline behaviour are under the direct or indirect control of the brain. In both cases, chemical messages are sent from the brain to the hypothalamus, which controls the pituitary. In turn, the pituitary produces hormones that stimulate events elsewhere.

BODY-FUNCTION CONTROLS
Hormones produced in the brain control most of the body's daily functions. Deep in the brain, the hypothalamus gland produces antidiuretic hormone (ADH), which controls the concentration of urine (see page 68), and oxytocin, which stimulates labour and milk release in female cats (see page 70). The pituitary, the "master gland" at the base of the brain, produces a number of hormones. Growth hormone releasing hormone (GHRH) and its opposite, growth hormone inhibiting hormone (GHIH), regulate the production of growth hormone (GH) in the cat's brain. GH production continues after kittenhood, although its role in adult cats is not clear.

Thyroid stimulating hormone (TSH) initiates activity in the thyroid, which controls metabolic rate (see page 68). Adrenocorticotrophic hormone (ACTH) stimulates the adrenal gland to produce cortisol in response to stress or danger. Production of sexual hormones, eggs, and sperm (see page 70) is controlled by follicle stimulating hormone (FSH) in females and luteinizing hormone (LH) in males.

One of the more recently isolated hormonal controls is melatonin. The pituitary produces melanocyte stimulating hormone (MSH), stimulating melatonin synthesis in the pineal gland. Melatonin is involved in triggering sleep cycles and controls the body's internal clock, maintaining the day-and-night pattern known as circadian rhythm.

THE ADRENAL GLANDS
The adrenal glands lie adjacent to each kidney, and consist of an encapsulating cortex and a central core, or medulla. The adrenal cortex produces cortisol and other hormones, which are instrumental in controlling metabolic rate and determining the body's response to injury.

Early learning
A kitten's brain is almost fully developed at birth. By seven weeks of age, most areas have matured. Skills are acquired fastest during this development stage; learning before seven weeks of age is more important in cats than in dogs, whose brains continue to develop for another month, or humans, whose brain development takes years.

The adrenal medulla produces epinephrine and norepinephrine, better known as adrenalin and noradrenalin. These hormones control the heart rate and blood-vessel dilation.

The adrenals are a vital component of the biofeedback circuit. This system controls the fight-or-flight response, and has the most direct effect on feline behaviour. The biofeedback mechanism dictates the innate disposition, sociability, and tameability of the cat.

THE BIOLOGICAL COMPUTER
Sitting on top of all this activity is the brain. The brain is an almost incalculably intricate information-storage system, which consists of billions of specialized cells called neurons, each with up to 10,000 connections with other cells.

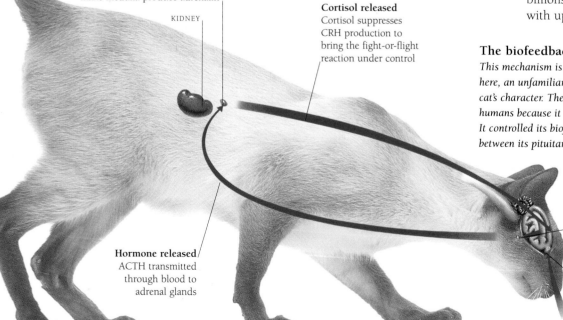

The biofeedback mechanism
This mechanism is the cycle of response to any external stimulus: here, an unfamiliar scent. The balance of the cycle determines a cat's character. The African wildcat was willing to live close to humans because it had an innate ability to keep calm around them. It controlled its biofeedback mechanism, fine-tuning the interplay between its pituitary and adrenal hormones and resisting panic.

Adrenal gland
ACTH stimulates cortex to produce cortisol, while nervous stimuli make medulla produce adrenalin

KIDNEY

Cortisol released
Cortisol suppresses CRH production to bring the fight-or-flight reaction under control

Brain activity
CRH stimulates pituitary to release adrenocorticotrophic hormone (ACTH)

Hormone released
ACTH transmitted through blood to adrenal glands

Scent stimulus
Unknown scent is picked up and stimulates hypothalamus to produce corticotrophin releasing hormone (CRH)

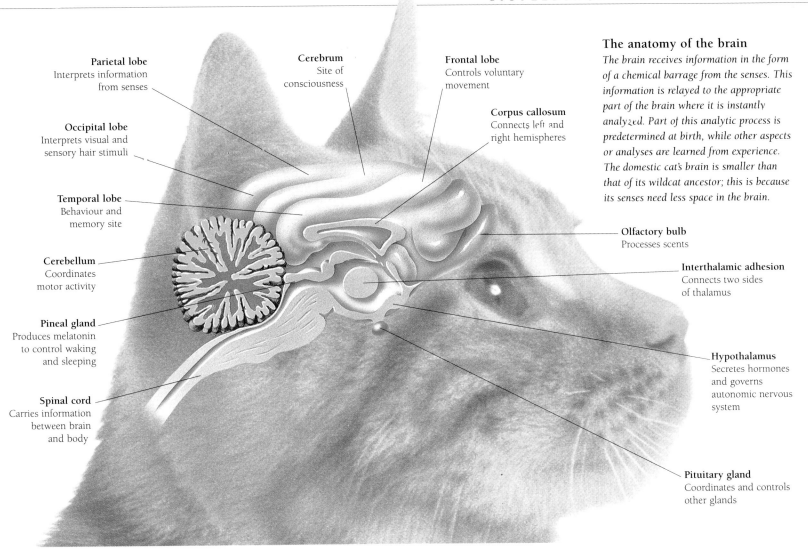

Parietal lobe
Interprets information
from senses

Occipital lobe
Interprets visual and
sensory hair stimuli

Temporal lobe
Behaviour and
memory site

Cerebellum
Coordinates
motor activity

Pineal gland
Produces melatonin
to control waking
and sleeping

Spinal cord
Carries information
between brain
and body

Cerebrum
Site of
consciousness

Frontal lobe
Controls voluntary
movement

Corpus callosum
Connects left and
right hemispheres

Olfactory bulb
Processes scents

Interthalamic adhesion
Connects two sides
of thalamus

Hypothalamus
Secretes hormones
and governs
autonomic nervous
system

Pituitary gland
Coordinates and controls
other glands

The anatomy of the brain
*The brain receives information in the form
of a chemical barrage from the senses. This
information is relayed to the appropriate
part of the brain where it is instantly
analyzed. Part of this analytic process is
predetermined at birth, while other aspects
or analyses are learned from experience.
The domestic cat's brain is smaller than
that of its wildcat ancestor; this is because
its senses need less space in the brain.*

Neurons communicate with each other through chemical substances called neurotransmitters. By seven weeks of age, messages move through the cat's brain at almost 386 kph (240 mph). This rate of transmission slows down with age, but recent studies have shown that this can be reversed. Elderly cats offered mental stimulation can actually grow heavier brains. The increased weight is due not to new brain cells, but to more connections between existing ones.

The brain's capacity to store information is genetically predetermined. Some stored data is instinctive, or present at birth, and controls sexual behaviour, territorial marking, aggression, and attachments made during kittenhood. But the feline brain can also absorb new data, such as how to use a cat flap or obtain food.

Anatomically, the cat's brain is similar to that of any other mammal. The cerebellum controls the muscles (*see page 64*),

the cerebrum governs learning, emotion, and behaviour, and the brain stem connects to the nervous system (*see page 62*). A network of cells called the limbic system is believed to integrate instinct and learning. Conflict between what a cat instinctively wants to do and what learning impels it to do probably takes place here.

FELINE INTELLIGENCE
Circuitry and hormones form the framework on which personality develops. This structure varies from cat to cat, and is genetically influenced. It is also only part of the picture; development also depends on learning. Just as humans have an instinct to communicate vocally at birth, but must learn how to use a language, cats are born with instincts to mark and defend territory, or hunt for sustenance, but must learn how to achieve these things. Cats that need not hunt to survive simply never

Not idle dreams
*Sleep is hormonally controlled. During sleep cats
dream much as we do, probably sorting out
memories of the day's events.*

develop the "hunting" parts of their brains. Many still hunt, but do not succeed like those that have learned and practised (*see page 84*). Feral cats develop a valid suspicion of other animals and people, because cats are small and relatively defenceless against such adversaries.

By raising kittens in our homes, we actively intervene in the development of their brains, biofeedback mechanisms, and behaviour. A cat raised in a human home before seven weeks of age has modified biofeedback. It learns that humans are safe, knowledge that is fixed into its brain and hormonal responses to us.

Some people believe that cats cannot learn; they may appear unable to, because their brains are designed for solitary living, and respond poorly to the social learning that we, as pack animals, naturally use. Cats do not usually obey for praise, because esteem holds no survival advantage to a solo hunter, but they may comply with human wishes for a food reward. The clearest example of feline thought is the cat that fights against entering its carrier to go to the veterinarian, but walks straight in for the return journey. Faced with two evils, the cat is quite capable of deciding which is the lesser.

THE SENSES

FROM THE MOMENT IT IS BORN, A CAT NEEDS SENSES TO SURVIVE. At birth, heat and smell receptors in its nose help it to seek out its mother. Over the next three months, all of the senses develop to maturity. Information from the cat's environment is picked up by sensory organs, which convert the information into chemical or electrical signals that are transmitted through specialized nerves to the brain. Feline senses are similar to ours, but they have developed along different lines. In addition to the classic five senses, cats have refined balance, the sex-scenting sense of the vomeronasal organ, and possibly even, an electromagnetic "homing" sense.

The cat is the most sophisticated and perhaps the most successful of all the land-based predators. Its small size, however, means that it is also prey to larger animals. The cat's senses have evolved to equip it for hunting prey, but also for sensing and avoiding danger.

TOUCH-SENSITIVE WHISKERS

Touch receptors are found all over a cat's body, but mostly in the paw pads and whiskers. Kittens grow their touch-sensitive whiskers, or vibrissae, while still in the womb, before they grow any other hair, and newborn kittens, although blind and virtually deaf, have fully functional whiskers. Later in life, the specialized sensory ability of the whiskers helps the cat in its nocturnal hunting, providing information about the environment and movement of prey. The mandibular and mystacial vibrissae, sited on the chin and upper lip, are the longest and most abundant. These whiskers can be angled forwards in greeting or backwards out of harm's way when fighting or feeding. The top rows can move independently of the bottom rows. Superciliary facial whiskers above the eyes and genal whiskers on the cheeks warn of dangers to the eyes as a cat explores, and a sixth set of whiskers extend from the back of each foreleg.

HEARING AND BALANCE

Evolution has equipped the cat with excellent hearing, well adapted for hunting small rodents. It can detect the faintest high-pitched squeaks of a mouse or the rustle of tiny movements. Noise consists of pressure waves travelling in the air. These are channelled through the ear, converted to electrical impulses, and carried to the brain, which interprets them. Pressure in the inner ear is regulated by its connection to the back of the throat via the Eustachian tube.

The other function of the ear is to help the cat to balance. While most domestic animals live on the ground, the cat is a natural climber, and has a sensitive sense of balance. In the inner ear is the organ of balance, the vestibular apparatus. Changes in direction or velocity register instantly in this organ, allowing the cat to compensate by changing its orientation.

Ears
More than a dozen muscles are used to control ear movement precisely, rotating the ears, independently if necessary, to listen for prey or danger. A cat can hear frequencies as high as 65,000 cycles per second, a full octave and a half higher than our maximum of 20,000 cycles per second. Like humans, however, cats lose their ability to hear the higher frequencies within their range with advancing years.

Nose and mouth
Cats have twice as many smell-sensitive cells, or olfactory receptors, in their noses as we do, allowing them to register scents we are not even aware of. Their taste buds are specialized to detect the amino acids of meat, and are less able than ours to detect the carbohydrate constituents of vegetable matter.

Eyes
Protruding eyes give cats a wider angle of vision than ours, and they also have superior peripheral vision; both are important to an animal that is both prey and predator. A cat can see in one-sixth of the amount of light we need; its pupils can dilate to 90 per cent of the eye area, but in sunlight they must close almost entirely to protect the eye.

Whiskers
Whiskers are highly mobile and incredibly sensitive, capable of detecting anything that displaces them by a distance 2,000 times less than the width of a human hair. Tip-to-tip, the facial whiskers indicate the smallest gap a cat can comfortably pass though.

Senses for survival
Although many pet cats hunt nothing more than fluffy toys or their owner's ankles, they retain virtually all of the sensory abilities of a wild hunter. Selective breeding and a protected life have not diminished the cat's sensory capacity, although this could happen in the future.

Breeding certain cats perpetuates sensory defects that would be limited under natural circumstances. Deafness, for example, is not life-threatening for the indoor cat, and some breeders deliberately breed deaf blue-eyed cats, knowing they will live safely indoors. Rexed cats may have whiskers more brittle than those of other breeds.

FELINE VISION

Most experts believe that cats are colour-blind. Tests have shown that the colour-sensitive cone cells on their retinas are sensitive to blue and green, but not to red. In trials, cats differentiate between green, blue, and yellow, but do not recognize red. In normal life, colour appears meaningless to cats. Cats' eyes are more sensitive to movement than ours, with more motion-detecting rod cells in the retina. The cat's world is seen in soft-focus, however: it cannot resolve detail sharply because the lens in its eye is large, to gather as much light as possible.

But by far the most spectacular modification in the cat's eye is the tapetum lucidum, a layer of reflective cells behind the retina. This bounces light back through the retina to give rods and cones a heightened ability to extract information.

THE CHEMICAL SENSES

Smell and taste are chemical senses. Cats have a good sense of smell, better than ours but not as refined as a dog's. A newborn kitten's nose is sensitive enough to smell differences between its mother's individual nipples. Later, the nose is used to scent prey or food, to detect potential danger, and to identify friend from foe and read the chemical messages left in urine or faeces. Unlike us, but like many other mammals, cats have a vomeronasal, or Jacobson's, organ in the hard palate that forms the roof of the mouth. When a cat uses this organ, its mouth opens in something between grimacing and gaping, called flehming. Assisted by the tongue, odours are "lapped" into the vomeronasal organ, and sent to the hypothalamus area of the brain (*see page 58*) by a route separate to odours from the nose.

The cat's mouth is exquisitely equipped for discriminating eating. The tip, sides, and back of the tongue, and part of the throat, contain mounds of tissue called papillae. An adult cat has about 250 mushroom-shaped fungiform papillae, each holding anything from 40 to 40,000 taste buds. Cone-shaped vallate papillae at the back of the tongue contain more taste buds. Taste tests indicate that cats can detect sour, bitter, and salty tastes, but not sweet ones.

A SENSE OF PLACE

Scientific studies suggest that cats are equipped to find their way home. Research from Germany and the United States implies that older cats in particular show a homing ability if released within 12 km (7½ miles) of home. Researchers conclude that cats are probably aware of electromagnetic fields: magnets attached to cats disrupted the homing ability. This could explain why there are minute traces of iron in the cat's brain cells.

THE ANATOMY OF THE SENSES

The cat's senses work in a similar way to ours. Sights, sounds, physical contact, or chemicals in the cat's environment affect one or more of its sensory organs. These external stimuli are captured by the sensory organ, which then converts the information into either electrical or neurochemical messages. These travel at lightning speed to the brain. Analysis of some messages is instantaneous; others may take a fraction longer to interpret. The brain produces a physical, mental, or hormonal response. For example: a cat's left ear captures the squeak of a mouse and passes it to the brain. At the same time, its nose picks up scent and passes it to the pituitary gland. The cat instantly turns its head to capture sounds in both ears and look for any movement, and its pituitary secretes hormones as the cat prepares to pounce.

Organic camera
A protective, clear cornea covers the eye's anterior fluid-filled chamber. Behind this is the three-part uvea: the coloured iris, stabilizing ciliary body, and focusing lens. Behind the lens is the fluid-filled posterior chamber. The retina lining the back of the eye "reads" the light, sending information down the optic nerve; behind it is the tapetum lucidum.

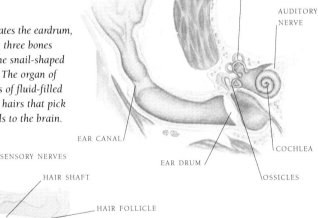

Funnel for sound
Sound travels down the ear canal and vibrates the eardrum, or tympanic membrane. In the middle ear, three bones called the ossicles transfer vibrations to the snail-shaped inner ear, or cochlea, and auditory nerve. The organ of balance, the vestibular apparatus, consists of fluid-filled chambers and canals, lined with sensitive hairs that pick up movements in the fluid and send signals to the brain.

Sensitive roots
Beneath the skin, a whisker's follicle is about three times deeper than that of a normal hair. A special blood-filled capsule, the sinus, acts like hydraulic fluid and amplifies the signal sent to the web of sensory nerves around it.

Chemical receptors
Taste buds, nests of microscopic taste-sensitive cells, cover the tongue. Each has a taste hair that detects the chemicals in foods. Inside the nasal chambers, odour molecules adhere to the sticky membranes that cover the curved bones, called turbinates. Other chemical odour molecules are captured in the vomeronasal organ in the roof of the mouth.

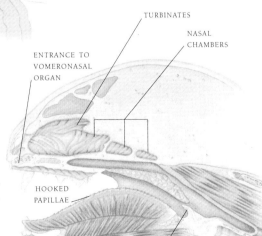

THE NERVOUS SYSTEM

THE NERVOUS SYSTEM COORDINATES EVERY ASPECT OF A CAT'S LIFE. It works in close conjunction with the hormonal system, to which it is linked through the brain and pituitary gland, to coordinate all of the cat's natural functions second by second, day by day, even season by season. While hormones have slow but lasting effects, the nervous system, with its blindingly intricate circuitry, responds swiftly, accurately, and directly to both internal and external events. Some areas of the nervous system are under the cat's voluntary control; others function apparently on their own, but in reality they are controlled at deeper, unconscious levels.

Information travels through the nervous system in two directions: sensory nerves inform the brain about how a cat feels, and motor nerves carry information away from the brain, telling the cat's body what to do about how it feels.

CENTRAL AND PERIPHERAL

The nervous system is classified as two parts, central and peripheral. The central nervous system consists of the brain (*see page 58*) and spinal cord – the body's command centre and the superhighway for two-way transmission of nerve impulses. The peripheral nervous system receives information about temperature, touch, pressure, and pain, and delivers instructions to muscles (*see page 64*). It consists of the cranial nerves and the spinal or peripheral nerves.

The cranial nerves are responsible for facial muscles and the transmission of information from the senses (*see page 60*). They spread out from under the brain in a region called the brain stem. The spinal nerves leave the spinal cord along its length, linking the extremities of the body to the central nervous system.

This division of the nervous system is for ease of understanding; it is not a physical split. Many nerve cells are partly in the central nervous system and partly in the peripheral nervous system.

CHEMICAL MESSENGERS

The nervous system consists of nerve cells, called neurons, and support cells that provide structure and produce myelin (*see page 63*).

A neuron's body is covered with branch-like structures, called dendrites, which receive messages from other cells. Each cell also has a long, tail-like structure, called an axon, which sends messages to other nerve cells or directly to organs. All of these messages are carried by neurotransmitters, chemicals produced in axons, which navigate the tiny gap between nerve cells, known as the synaptic cleft. Neurons never touch each other.

At any moment, even in deep sleep, a cat's nervous system is sending and receiving vast numbers of messages. Any cell commonly sends messages to thousands of other cells.

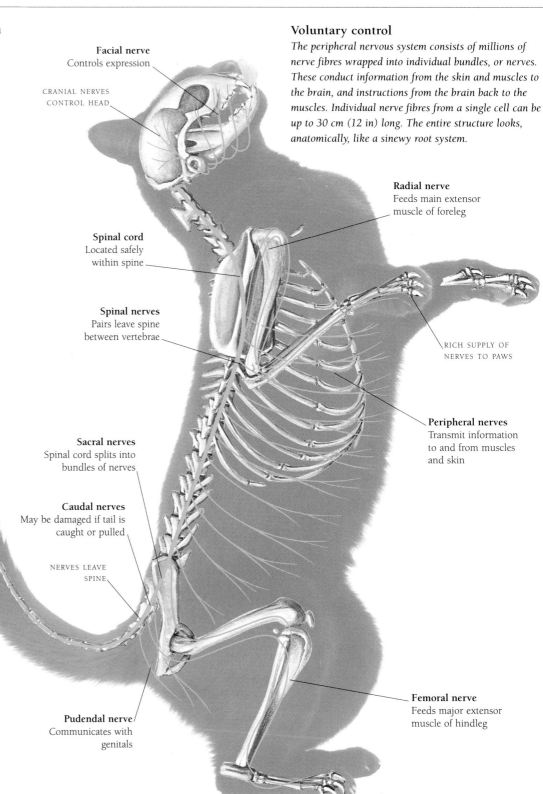

Facial nerve
Controls expression

CRANIAL NERVES
CONTROL HEAD

Spinal cord
Located safely
within spine

Spinal nerves
Pairs leave spine
between vertebrae

Sacral nerves
Spinal cord splits into
bundles of nerves

Caudal nerves
May be damaged if tail is
caught or pulled

NERVES LEAVE
SPINE

Pudendal nerve
Communicates with
genitals

Voluntary control
The peripheral nervous system consists of millions of nerve fibres wrapped into individual bundles, or nerves. These conduct information from the skin and muscles to the brain, and instructions from the brain back to the muscles. Individual nerve fibres from a single cell can be up to 30 cm (12 in) long. The entire structure looks, anatomically, like a sinewy root system.

Radial nerve
Feeds main extensor
muscle of foreleg

RICH SUPPLY OF
NERVES TO PAWS

Peripheral nerves
Transmit information
to and from muscles
and skin

Femoral nerve
Feeds major extensor
muscle of hindleg

THE MYELIN SHEATH

Myelin is a fatty, protective membrane found around the largest axons, which increases the speed of communication along the nerves. Technically, a nerve fibre consists of an axon, its myelin sheathing, and the cell that makes the myelin. Myelin is produced by cells called oligodendrocytes inside the central nervous system, and neurolemmocytes in the peripheral system. Few nerves have myelin sheathing at birth, but the cat's nerves are myelinated quickly, and very effectively.

Protective layers
Myelin layers, seen here as coloured lines, sheathe axons. The cat's well-myelinated system is one of the reasons why its physical responses are so good.

UNCONSCIOUS CONTROL

Many functions of the nervous system are under conscious, or voluntary, control. When a cat sees prey, it voluntarily controls its muscles so that it can pounce on its prey. Sensory nerves carry messages to the brain, and motor nerves carry messages back to the muscles, stimulating them to work in the controlled way necessary to pounce accurately.

But other activities are involuntary. These usually involve the internal organs, regulating heartbeat and breathing (*see page 66*) and the many digestive processes (*see page 68*). Such involuntary activities are controlled by the autonomic nervous system.

The autonomic system consists of two parts: sympathetic and parasympathetic. The former stimulates activity, the latter damps it down. When a cat is relaxed, the parasympathetic part controls involuntary activity: the pupils are relaxed, and the heart rate and breathing are slow and regular. When a cat becomes stressed, the sympathetic system takes over, triggering the hypothalamus and the pituitary in the brain to stimulate the adrenal glands (*see page 58*) into the "fight-or-flight" response. This chain of response, although complex, is instantaneous. Blood flows from internal organs into muscles, tiny subcutaneous muscles cause body hair to stand on end (*see page 72*), the heart speeds up, and pupils dilate for better vision (*see page 60*).

Inherent problems
Manx cats often suffer from a congenital neurological disorder called spina bifida, which interferes with normal nerve operation. Completely tailless Manx cats may have malformed sacral and caudal nerves and be incontinent as a result.

Even a perceived threat will trigger this cascade of activities in the autonomic nervous system. If the alarm is false, the sympathetic system "stands down", and the parasympathetic system takes charge of involuntary activity once more.

NEUROLOGICAL PROBLEMS

Neurological disorders are remarkably rare in cats, a tribute to their evolutionary credentials. Viral infections like rabies, panleucopenia virus, and feline infectious peritonitis (FIP), as well as poisons and a variety of parasites, such as toxoplasma, damage nerves. But road traffic accidents are the most common causes of nerve damage: almost half of all cats hit by cars have head injuries. Also broken backs, almost inevitably with severed spinal cords, are not uncommon. Unlike other bodily cells, nerve cells cannot regenerate; there is no repair possible.

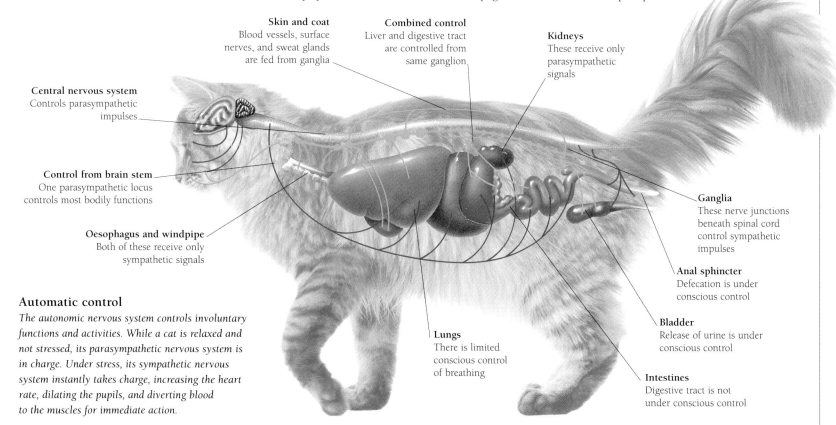

Skin and coat
Blood vessels, surface nerves, and sweat glands are fed from ganglia

Combined control
Liver and digestive tract are controlled from same ganglion

Kidneys
These receive only parasympathetic signals

Central nervous system
Controls parasympathetic impulses

Control from brain stem
One parasympathetic locus controls most bodily functions

Oesophagus and windpipe
Both of these receive only sympathetic signals

Ganglia
These nerve junctions beneath spinal cord control sympathetic impulses

Anal sphincter
Defecation is under conscious control

Bladder
Release of urine is under conscious control

Lungs
There is limited conscious control of breathing

Intestines
Digestive tract is not under conscious control

Automatic control
The autonomic nervous system controls involuntary functions and activities. While a cat is relaxed and not stressed, its parasympathetic nervous system is in charge. Under stress, its sympathetic nervous system instantly takes charge, increasing the heart rate, dilating the pupils, and diverting blood to the muscles for immediate action.

MUSCLES AND MOVEMENT

THE FELINE BODY RESPONDS SWIFTLY AND SUPERBLY TO MESSAGES sent by the brain via the nervous system. The cat's extraordinary balletic grace is partly attributable to its highly specialized skeleton, but its fast-acting muscles are also responsible. The strong, flexible musculo-skeletal framework has evolved to fulfil two purposes: it makes the cat an efficient hunter, and enables it to escape from dangerous situations. This combination of power, flexibility, and balance has enabled the cat to survive in its ecological niche of small predator. Its speed and grace in movement depend upon its muscles or, more accurately, the types of cells within its muscles.

All muscles are divided into three basic types. One is cardiac muscle, found only in the heart (*see page 66*). The muscle controlling the other internal organs is "involuntary", or outside the cat's control, and is called smooth or non-striated, because this is how it appears under the microscope. The rest of the body's muscles are called striated or striped. They are controlled at will and used in all conscious or instinctive movement. Cats have fewer striped muscles than humans, in part because they use less facial expression for communication.

Each muscle has a Latin name according to its shape, attachments, function, location, or number of parts. Feline muscles are inherently more flexible than those of other mammals, giving a great range of movement. Most muscles are attached to bones (*see page 56*) on either side of a joint, and act to operate the joint itself.

Muscles will only contract, they cannot extend, so opposing muscles perform opposing actions: the gastrocnemius and quadriceps extend the hindleg, and the biceps femoris and tibialis bend or flex it. If the extensors, or their tendons or ligaments are torn in an accident, the leg will be permanently retracted; if the retractors are torn, it will be permanently extended.

MUSCLE CELLS

Each individual muscle consists of many muscle fibres, which are held together as a group by connective tissue. The muscle tissue consists of three different types of muscle cells, all named after their actions. "Fast-twitch fatiguing" cells work quickly, but they tire equally quickly. "Fast-twitch fatigue-resistant" cells work quickly but tire more slowly. "Slow-twitch" cells work comparatively slowly and also tire slowly.

A FLOATING SHOULDER

The cat's shoulder is a feat of muscle, in that the forelimb is connected to the rest of the body only by muscle. Unlike our collarbone, which connects the shoulder and breastbone, the cat's vestigial clavicle "floats", anchored in place by muscle. The shoulder's freedom of movement effectively lengthens the cat's stride and enhances its range of motion.

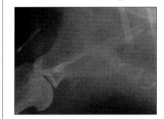

Floating bone
The collarbone, seen in the top left corner of this x-ray, is a mere sliver of bone held in place by the muscles.

The striped muscles
The striped muscles are symmetrical across the body and under the control of the central and peripheral nervous system (see page 62). These muscles act on joints, rather than individual bones. Generally, they are arranged in opposing groups with opposite actions.

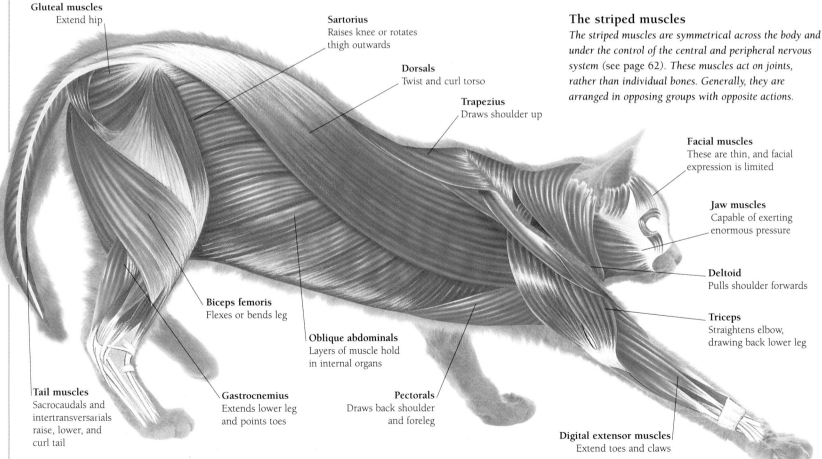

Gluteal muscles
Extend hip

Sartorius
Raises knee or rotates thigh outwards

Dorsals
Twist and curl torso

Trapezius
Draws shoulder up

Facial muscles
These are thin, and facial expression is limited

Jaw muscles
Capable of exerting enormous pressure

Deltoid
Pulls shoulder forwards

Triceps
Straightens elbow, drawing back lower leg

Biceps femoris
Flexes or bends leg

Oblique abdominals
Layers of muscle hold in internal organs

Tail muscles
Sacrocaudals and intertransversarials raise, lower, and curl tail

Gastrocnemius
Extends lower leg and points toes

Pectorals
Draws back shoulder and foreleg

Digital extensor muscles
Extend toes and claws

Muscles in motion
The powerful hindquarters are used for leaps. In a well-timed jump, a cat might spring six times its own length.

The cat's muscles mostly consist of fast-twitch fatiguing cells: they give the cat its speed and permit it to leap several times its own length in a single bound, but they use up all their energy in an instant. They make cats superb sprinters, capable of reaching 48 kph (30 mph) in a few seconds, but this speed has a price. Because it has relatively few fast-twitch fatigue-resistant cells, the cat is a poor endurance athlete. Even when just trotting, a cat uses more energy than a dog of similar weight and size. If this exertion continues, a cat's muscles will generate so much power that it overheats. Its body temperature rises after less than a minute of sprinting, and it must stop and pant.

The third type of cell, slow-twitch, produces slow, sustainable contractions. These cells are called into play during hunting activities: they enable the cat to move almost imperceptibly slowly and stealthily, or wait for prolonged periods in a crouched, ready-to-spring position.

THE FELINE GAIT
When a cat walks, almost all of its forward movement is generated by the hindlegs; the forelegs act like brakes when they hit the ground, almost negating the slight forwards push they give on leaving it again. The same is true when a cat trots. In this gait the legs move in what is called contralateral fashion: the right front moves forward together with the left hind and vice versa. When a cat bounds or gallops, its body is propelled forward by the power of both the hindlegs pushing off at the same time, and the whole body becomes airborne until the forelegs touch down.

POUNCES AND JUMPS
The cat's pliant muscles and flexible spine allow it to curl in a circle to sleep, or to rotate its body through 180° in mid-air. This snake-like flexibility also gives the cat a wide repertoire of graceful leaping movements suitable for different circumstances. When pouncing, a cat springs with its hindlegs, arches its back, and lands with its forepaws on its prize. A refined arrangement of wrist muscles allows it to rotate its wrists, just as humans can, to grasp prey and to climb efficiently.

The vertical jump is carefully executed. A cat judges the distance to be covered and calculates how much propulsive power from its hindleg muscles is needed. This movement is different from the impromptu jumps made when chasing or being chased, and these are different again from the startle jump. In this, extensor muscles in all four legs activate simultaneously, and all the feet leave the ground at once, as if on springs.

HIGH-RISE SYNDROME
The excellent righting reflex, which allows the cat to flip right-side-up, and the soft paws and flexible joints that absorb the impact of landing, give the cat its legendary ability to survive falls. As ever-more cats live in high-rise buildings, vets have observed a strange phenomenon. Cats that fall fewer than four floors usually survive their mishaps, but falls from five to ten floors are usually fatal. Computer modelling trials show that cats reach 97 kph (60 mph) after falling five floors, and the force of impact is too great to absorb. Curiously, falls from even greater heights sometimes cause little injury. This is because, once the righting reflex has served its purpose, the cat assumes the skydiver's free-fall position: limbs outstretched and relaxed, and head held high. This muscular relaxation, taken together with the deceleration effected by outstretched limbs, lessens the impact and injury. These reflexes did not evolve to save cats in falls from windows, but natural abilities often have unexpected new uses.

Vertical climbing
Powerful leg muscles, flexible joints, extendable claws, and a sophisticated sense of balance allow the cat to live in a vertical as well as horizontal world.

THE HEART AND LUNGS

THE CAT'S LUNGS, HEART, AND CIRCULATORY SYSTEM ARE DESIGNED for an animal whose actions are usually controlled and planned, but occasionally demand sudden, rapid bursts of great energy. Instructions from the nervous and hormonal systems can rapidly increase the pumping action of the heart and divert circulating blood from the brain and internal organs to muscles. Blood carries oxygen from the lungs and nutrients from the intestines through the body. It collects waste to be expelled from the lungs, detoxified in the liver, or excreted through the kidneys. It is made primarily in the bone marrow; an emergency reservoir is held in the spleen.

Each time a cat takes a breath, it draws in vital oxygen and exhales waste carbon dioxide. Every beat of its heart pumps these and other substances through the circulatory system to and from the farthest reaches of the body.

THE RESPIRATORY SYSTEM

Inhaled air first passes through the scenting apparatus of the nose (*see page 60*), surrounded by the frontal sinuses, where it is is warmed, humidified, and filtered. It is then drawn down the trachea, or windpipe, and into the lungs through two bronchi. Each bronchus splits into many smaller bronchioles that end in tiny pockets called alveoli.

THE CIRCULATORY SYSTEM

The body of a typical 5 kg (12 lb) cat contains about 330 ml (12 fl oz) of blood, about the same amount as in a soft-drink can. Pressure is highest in the arteries, which carry blood from the heart, and lowest in the veins returning it to the heart. The muscular, elastic walls of arteries expand and contract as the heart pushes blood through them: this is the pulse. The thinner walls of the veins are more easily damaged; they have no pulse, and contain valves to ensure that the blood in them moves only one way.

Different parts of the body need different amounts of blood. The brain is a small part of the body's weight, but takes 15 to 20 per cent of its blood. Resting muscles get about twice this, but during pursuits or in escape situations, up to 90 per cent of the blood can be diverted into the muscles from the internal organs and even from the brain. The amount of blood that each part of the body receives is controlled by nerves and hormones that cause arterioles to dilate in response to local activity, dramatically increasing the blood supply to that region.

THE COMPOSITION OF BLOOD

The bulk of blood volume is pale-yellow plasma. Another 30 to 45 per cent is red blood cells, and the remainder is white cells and platelets. In kittens, the liver and spleen make blood cells; in adults bone marrow produces them.

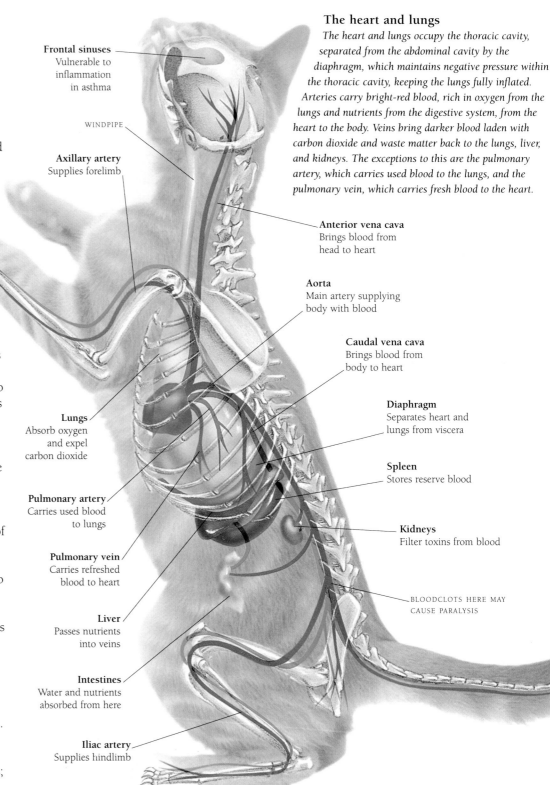

The heart and lungs

The heart and lungs occupy the thoracic cavity, separated from the abdominal cavity by the diaphragm, which maintains negative pressure within the thoracic cavity, keeping the lungs fully inflated. Arteries carry bright-red blood, rich in oxygen from the lungs and nutrients from the digestive system, from the heart to the body. Veins bring darker blood laden with carbon dioxide and waste matter back to the lungs, liver, and kidneys. The exceptions to this are the pulmonary artery, which carries used blood to the lungs, and the pulmonary vein, which carries fresh blood to the heart.

Frontal sinuses
Vulnerable to inflammation in asthma

WINDPIPE

Axillary artery
Supplies forelimb

Anterior vena cava
Brings blood from head to heart

Aorta
Main artery supplying body with blood

Caudal vena cava
Brings blood from body to heart

Diaphragm
Separates heart and lungs from viscera

Spleen
Stores reserve blood

Lungs
Absorb oxygen and expel carbon dioxide

Kidneys
Filter toxins from blood

Pulmonary artery
Carries used blood to lungs

BLOODCLOTS HERE MAY CAUSE PARALYSIS

Pulmonary vein
Carries refreshed blood to heart

Liver
Passes nutrients into veins

Intestines
Water and nutrients absorbed from here

Iliac artery
Supplies hindlimb

Plasma is the transport portion of the blood, carrying nutrients from the digestive system (*see page 68*), and transporting waste. Levels of plasma are maintained by liquid absorbed along the length of the large intestine.

Red cells carry oxygen through the arteries to the cells of the body. Oxygen attaches to a red-pigmented protein called haemoglobin in the cells to create oxyhaemoglobin, which is bright red in colour. The oxygen diffuses through capillary walls into cell tissue. In return, carbon dioxide diffuses into the veins and attaches to haemoglobin to make blue-red methaemoglobin.

White blood cells defend the body against microbes and parasites, clear waste from injuries, detoxify substances released in allergic reactions, and help the body to acquire immunity against infections by producing antibodies. Disc-shaped platelets help to clot blood around wounds.

BLOOD TYPES

Cats have three blood types: A, B, and AB. Most cats are type A, but the incidence of types varies geographically. Virtually all random-bred cats in Switzerland are type A, but this drops to 97 per cent in Britain, Austria, and Holland, 94 per cent in Germany, 89 per cent in Italy, and 85 per cent in France. In the United States, almost 100 per cent of cats are type A in New England, but this falls to 95 per cent on the West Coast.

Many pedigree breeds are almost exclusively type A, but others show varying levels of type B, most notably the Exotic Shorthair (*see page 108*) and Cornish Rex (*see page 166*) with up to 25 per cent, and the British Shorthair (*see page 114*) and Devon Rex (*see page 168*) at nearly 50 per cent. Type AB is extremely rare and not linked to breed.

BLOOD PROBLEMS

Some problems with circulation have external causes. If a cat does not eat, there is little water in the large intestines to be absorbed to maintain plasma levels, so water is drawn from elsewhere in the body, causing dehydration.

In anaemia, the level of red blood cells falls, to be corrected by the bone marrow producing new cells. Heavy flea infestation, injury, stomach ulcers, or tumours can cause temporary anaemia. Illness can inhibit the production of new cells.

The cat's miaow
The ability to miaow that sets the small cats aside from the big cats is due to a difference in the throat structure. Less certain is how a cat purrs. Some anatomists feel that it is the sound of turbulence in the anterior vena cava, creating vibrations that are transferred up the windpipe.

Such non-regenerative anaemia is often caused by feline leukemia virus (FeLV), kidney failure, nutritional deficiency, or poisoning. Haemolytic anaemias, resulting from the destruction of red blood cells, may be caused by the blood parasite *Haemobartonella felis*, or by paracetamol, the active ingredient in many non-aspirin painkillers. Aspirin in excess can cause fatal bleeding. The bleeding disorder, haemophilia, is rare in cats.

Cats do not develop hardening of the arteries but may develop diseases of the heart muscle (cardiomyopathy) or valves (valvular disease), or disruption of the heart's rate (arrhythmias). They can suffer blood clots (thromboembolisms) causing pain and paralysis to the hindlimbs.

INSIDE THE HEART AND LUNGS

Pulmonary arteries and their capillaries carry oxygen-depleted blood through the lungs to the alveoli, where oxygen is absorbed from inhaled air. Pulmonary veins return the refreshed blood to the heart, which pumps it out through arteries to the body. The oxygen diffuses into cells in exchange for carbon dioxide, and veins bring the depleted blood to the heart to be pumped to the lungs again.

In the lungs
Carbon dioxide from used blood is exchanged for fresh oxygen inside the microscopic alveoli of the lungs. In a typical domestic cat, the total surface area of the alveoli is about 20 sq m (215 sq ft).

The heart's chambers
Blood from the body enters the right atrium, passes into the right ventricle, and is pumped to the lungs via the pulmonary artery. Blood from the lungs enters the left atrium, passes into the left ventricle, and is pumped to the body via the aorta.

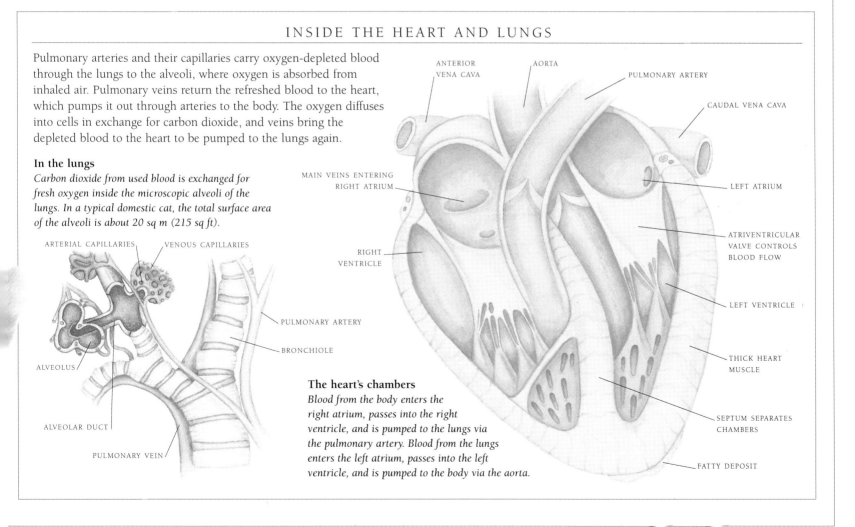

ARTERIAL CAPILLARIES
VENOUS CAPILLARIES
ALVEOLUS
ALVEOLAR DUCT
PULMONARY VEIN
PULMONARY ARTERY
BRONCHIOLE

ANTERIOR VENA CAVA
AORTA
PULMONARY ARTERY
CAUDAL VENA CAVA
MAIN VEINS ENTERING RIGHT ATRIUM
LEFT ATRIUM
RIGHT VENTRICLE
ATRIVENTRICULAR VALVE CONTROLS BLOOD FLOW
LEFT VENTRICLE
THICK HEART MUSCLE
SEPTUM SEPARATES CHAMBERS
FATTY DEPOSIT

DIGESTION AND EXCRETION

THE DIGESTIVE SYSTEM IS RESPONSIBLE FOR BREAKING DOWN FOOD from its complex forms of proteins, carbohydrates, and fats into simple molecules that can easily be absorbed into the bloodstream through the wall of the small intestine. Its other equally important role is to act as a protective barrier against any harmful bacteria or other disease-causing agents that a cat may inadvertently ingest. Food is usually consumed, digested, utilized, dehydrated, and excreted within 24 hours. The cat's digestive system is similar to the dog's or ours, but because it is an obligate carnivore, which must eat meat to survive, there are some important adaptations.

The cat's digestive system is in essence like our own, but simpler. As a carnivore, it does not need a cecum to digest fibre, and its intestines are short in relation to those of herbivores, such as sheep, or omnivores, such as humans.

INGESTION AND DIGESTION

The cat's teeth tear meat like serrated blades, and the barbed tongue scrapes it from bones. Saliva lubricates and binds food for swallowing. Food passes down the oesophagus and into the stomach. The top of the stomach, the fundus, produces acid to break down fibres, and an enzyme to break down protein. The secretion of these digestive "juices" is hormonally controlled. The stomach also secretes mucus to protect it and the intestines against damage from the juices. Muscle contractions mix food, which then passes into the duodenum. The duodenum receives fat-dissolving bile from the gall bladder in the liver and enzymes from the pancreas. Along the length of the small intestine, which is made up of the duodenum and ileum, digestion continues and nutrients are absorbed through the intestinal wall (*see page 66*).

The blood carries these products to the liver, the largest internal organ. The liver processes them into essential fatty acids and amino acids, the "building blocks of life". Unlike the human or canine liver, the feline liver needs animal protein to manufacture the full complement of acids, so a cat will die if it does not eat meat. The liver also breaks down toxic substances.

Calculated bite

The cat's canine teeth sit in beds of sensitive tissue. A cat adjusts its grip on prey until it feels a tiny depression in its neck, just behind its skull. The shape of a cat's canine teeth has evolved to fit perfectly into this depression and deliver a spine-severing bite and instant death.

WASTE DISPOSAL

After the nutrients are absorbed, waste enters the large intestine, or colon. Benign bacteria here break down the waste and neutralize any hostile bacteria in it. Water is absorbed through the colon wall, and mucus is secreted to lubricate the dry waste. When waste accumulates in the rectum, nerves signal that discharge is needed.

The blood carries waste from the liver to the kidneys, where tubules, called nephrons, filter it out and excrete it in urine, of which cats daily produce 60 ml (2 fl oz). Nephrons die with age and disease, but only when 75 per cent are lost does a cat obviously drink and urinate more. Failing kidneys have difficulty clearing urea and creatinine, by-products of protein metabolism and muscle exertion. The blood levels of these chemicals indicate the state of the kidneys, and treatment includes reducing both protein in the diet and the amount of exercise. The kidneys also regulate the blood pressure, maintain the chemical balance of the blood, activate vitamin D, and produce erythropoietin, a hormone that stimulates the production of red blood cells.

DIGESTIVE HORMONES

Behind the mechanics of digestion lie hormones produced by the thyroids and parathyroids and the pancreas. The thyroids, two glands on either side of the windpipe, control the metabolic rate. Pedigree cats tend to have higher thyroid activity than random-bred cats. Overactive thyroids, almost unknown until 1978, but now common in older cats, cause dramatically increased heart rate, a voracious appetite, and weight loss. Underactive thyroids are extremely rare.

THE CAT'S DENTITION

When kittens are born, 26 needle-sharp milk teeth are already in place. The milk teeth are replaced by 30 permanent teeth during the first six months of life. The upper and lower incisors are used to grasp the prey or food, the canines grip and kill, and the premolars and molars shear, cut, and chew meat. The cat has few molars, and the upper ones are almost vestigial because they are not vital in its mainly carnivorous natural diet.

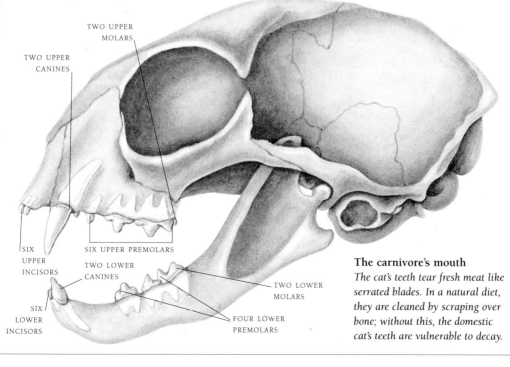

TWO UPPER MOLARS

TWO UPPER CANINES

SIX UPPER INCISORS

SIX UPPER PREMOLARS

TWO LOWER CANINES

SIX LOWER INCISORS

FOUR LOWER PREMOLARS

TWO LOWER MOLARS

The carnivore's mouth
The cat's teeth tear fresh meat like serrated blades. In a natural diet, they are cleaned by scraping over bone; without this, the domestic cat's teeth are vulnerable to decay.

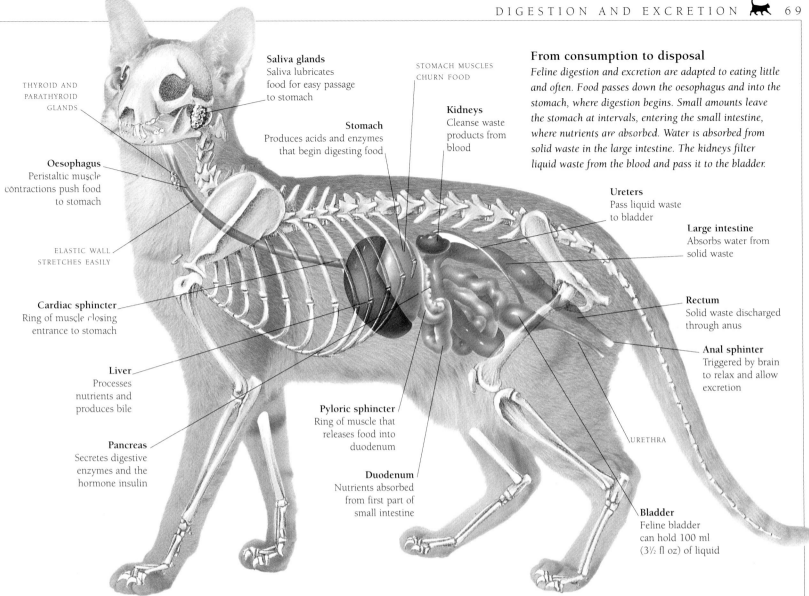

THYROID AND
PARATHYROID
GLANDS

Saliva glands
Saliva lubricates
food for easy passage
to stomach

STOMACH MUSCLES
CHURN FOOD

Kidneys
Cleanse waste
products from
blood

Stomach
Produces acids and enzymes
that begin digesting food

Oesophagus
Peristaltic muscle
contractions push food
to stomach

ELASTIC WALL
STRETCHES EASILY

Cardiac sphincter
Ring of muscle closing
entrance to stomach

Liver
Processes
nutrients and
produces bile

Pancreas
Secretes digestive
enzymes and the
hormone insulin

Pyloric sphincter
Ring of muscle that
releases food into
duodenum

Duodenum
Nutrients absorbed
from first part of
small intestine

Ureters
Pass liquid waste
to bladder

Large intestine
Absorbs water from
solid waste

Rectum
Solid waste discharged
through anus

Anal sphinter
Triggered by brain
to relax and allow
excretion

URETHRA

Bladder
Feline bladder
can hold 100 ml
(3½ fl oz) of liquid

From consumption to disposal
*Feline digestion and excretion are adapted to eating little
and often. Food passes down the oesophagus and into the
stomach, where digestion begins. Small amounts leave
the stomach at intervals, entering the small intestine,
where nutrients are absorbed. Water is absorbed from
solid waste in the large intestine. The kidneys filter
liquid waste from the blood and pass it to the bladder.*

Next to each thyroid is a tiny parathyroid gland,
which produces a hormone needed to extract
calcium, important in muscle contraction, from
bone. Insulin, secreted in the pancreas, allows
cells to absorb vital glucose from the blood.
Many hormones counteract insulin: diabetes in
cats can be caused by overactive pituitary,
thyroid, or adrenal glands (*see page 58*).

DIGESTIVE DISORDERS
Cats regurgitate for a variety of reasons, some
not at all serious: eating too quickly, swallowing
hair when grooming, a tight cardiac sphincter,
or a dilated oesophagus. In vomiting, stomach
and abdominal wall contractions force back up
the stomach contents, which have a distinctive
acidic smell. This is usually more serious than
regurgitation. Vomiting and diarrhoea are fairly
common in cats, and may result from changes
in diet, food intolerances or allergies, poisons,
ingested foreign bodies, infections, or parasites.
Older cats may develop liver, kidney, or pancreas
conditions, hyperthyroidism, inflammatory
bowel disease or colitis, and intestinal cancers.

One common feline food intolerance is to milk.
Kittens produce the enzyme lactase to digest
milk's sugar, lactose, but adult cats often lack
this enzyme, and milk may cause diarrhoea.
Other intolerances are less easily isolated.

The cat's liver is very slow at processing some
drugs. For example, it takes about 72 hours to
break down aspirin, so even small amounts are
dangerous if given more than twice weekly. Other
medications are lethal in even smaller amounts.
The kidneys can be irreversibly damaged by
some ingested poisons, most notably antifreeze.

Ingested objects, such as string, can loop
around the tongue or anchor in the intestines.
Never pull out such ingested objects, as you
can cause damage: seek veterinary help.

Feline immunodeficiency virus (FIV), feline
leukemia virus (FeLV), and the preventable feline
panleukopenia virus (FIP) all cause vomiting
and diarrhoea. The parasite giardia is difficult
to confirm, and may be more common than is
realized, and vomiting is often the only sign of
heartworm. Vomiting caused by inflammation to
the stomach may be stress-induced in some cats.

URINARY PROBLEMS

The kidneys filter the blood of unwanted or
toxic substances and pass them to the bladder.
The lower urinary tract, comprising the bladder
and urethra, may suffer feline lower urinary
tract disorder (FLUTD), caused by diet or
infection. Another problem is cystitis, typified
by inflammation and a burning sensation that
causes the cat to urinate frequently. In some
cats, this can be caused by emotional upsets.

Feline lower urinary tract disorder
*In FLUTD, crystals or stones form in the bladder.
These can be so large that they block the urethra;
they usually cause the cat pain during urination.*

THE REPRODUCTIVE SYSTEM

THE CAT HAS EVOLVED A VERY EFFICIENT REPRODUCTION STRATEGY, ideally suited to the lone hunter. Males are eternal opportunists, always ready to mate. Females come into season as daylight hours increase in spring, ensuring births during the spring and summer, when food is most plentiful. She signals her sexual receptivity to males by leaving hormone-laden urine markings and by "calling", emitting a unique, sonorous tone, which can be heard by any males nearby. Because each cat has its own hunting territory, there may be no males in the vicinity at the time. Rather than waste her eggs, females release them from the ovaries only after mating.

Puberty in cats usually occurs between six to nine months, although some breeds, such as the Oriental types, are sexually precocious, reaching puberty as early as four months.

THE MALE SYSTEM

As with many other mammals, the male cat's reproductive system is on permanent stand-by from puberty, ready for use if the chance arises. Leutenizing hormone (LH) from the pituitary gland (*see page 58*) stimulates the testes to make both sperm and the male hormone testosterone. Sperm production begins in coiled tubes in the testes, and it continues throughout life. Because sperm production is best at slightly less than body heat, testes are held in the scrotal sac outside the body.

Sperm is stored in the epididymis at the base of the testes until needed, when it travels through the two spermatic cords to the prostate and bulbourethral glands. Here a sugar-rich transport medium, semen, is added.

The male shows sex-related behaviour, such as roaming, fighting, and spraying, all year round. His reproductive system is placed on high alert by scent. If his vomeronasal organ (*see page 60*) captures female scent, he will disregard all else to seek out the female giving off the scent, and fight over her if necessary.

FEMALE HORMONES

Like most other domestic animals, the female cat is polyoestrous, meaning she has many oestrous periods, or seasons, throughout the year. But, unlike many other animal species that have oestrous cycles all year round, cats are seasonally polyoestrous: the female's reproductive cycle is most active as daylight hours increase, and then winds down as daylight hours decrease.

At the end of winter, increasing daylight hours stimulate her pituitary gland to produce follicle stimulating hormone (FSH). The FSH induces her ovaries to manufacture eggs and the female hormone oestrogen. The oestrogen is released in the female's urine, acting as a calling card to any available males. Her two ovaries are suspended from the roof of the abdomen, just behind the kidneys. At puberty, which occurs at the same time as male puberty, the necessary eggs are waiting in the ovaries. In contrast to the reproductive systems of virtually all other mammals, however, the ovaries will not release these eggs until after mating has taken place.

Male reproductive system

The reproductive organs of the male cat are kept on permanent stand-by from the onset of puberty at a few months of age. When a male picks up the scent or hears the distinctive call of a female in heat, the brain analyzes the sensory information and sends a hormonal signal to the reproductive organs. This stimulates him into a state of immediate readiness to mate.

SPERMATIC CORDS

BLADDER

SPERM IS STORED IN EPIDIDYMIS

Testes
Respond to LH by producing sperm

PENIS

Penile barbs
Abrade the vulva on withdrawal

Bulbourethral gland
Absolute function is poorly understood in cats

Prostate gland
Produces semen to carry sperm

LH CARRIED IN BLOOD TO TESTES

Scent stimulus
Male picks up female's scent

NEUTERING

Neutering a male is a minor operation. Under general anaesthesia, the testicles are removed through a small incision in the scrotum. The spermatic cords and associated blood vessels are tied. This is usually done at about six months. Neutering a female, or spaying, is more major abdominal surgery. The ovaries and uterus down to the cervix are removed. Spaying can be done before sexual maturity.

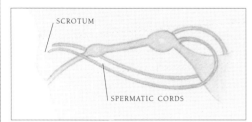

Male neutering

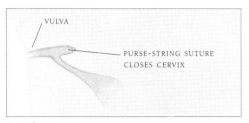

Female neutering

FELINE MATING

Feline sexual activity is flagrant and noisy. It is also curiously dignified. In the wild, cats are lone hunters. The female overcomes her natural solitude by vocally advertising her sexual state. She urinates more frequently, leaving as many scent markers as possible, to attract male cats from a wide area. Other behaviour also changes: she rolls on the ground, stretches out her body, raises her hindquarters, or drags herself on the ground by her forepaws, as if her hind legs no longer work. She may utter plaintive noises, which some owners misread as pain. She does not permit a male to mate with her until she is completely ready. When she is finally receptive, the male is permitted to mount her. He grasps the skin on her neck and mates immediately.

The male's penis is covered in hook-like barbs. As he withdraws, these barbs abrade the vulva. The female "screams" and turns to bite, but an experienced male will maintain his hold on the skin on her neck. He releases her and draws away when he feels it is safe to do so.

The abrasion of the vulva stimulates the release of eggs. Cats are induced ovulators, only releasing eggs under the stimulation of mating. One mating is often not enough to induce this egg release: two or more are usually necessary.

The pair mate repetitively, the male tiring first. The female retains her receptivity: other males are often then permitted to mate. When enough matings have occurred, production of FSH in the female's brain stimulates eggs to leave the ovaries and travel through the fallopian tubes to each horn of the uterus. If no matings occur when a female is in heat, no eggs are released. A period of calm ensues, lasting from two days to two weeks, followed by another heat cycle. Under the influence of artificial indoor light, unspayed females can eventually come to be in permanent, yowling season.

PREGNANCY AND BIRTH

Fertilization takes place in the pencil-like uterine horns, and the foetuses are positioned in rows in each horn of the uterus. Pregnancy lasts about 63 days, during which progesterone, the hormone of pregnancy, brings behaviour changes and the swelling of the mammary glands.

Birth is usually uncomplicated, although some kittens may not live. Milk let-down occurs shortly after the birth, stimulated by suckling. The mother often leaves the birth nest a few days after the delivery, carrying her kittens one by one to a new, more odour-free and thus, to her mind, much safer den.

Female reproductive system

Females conserve their eggs, concentrating their reproductive activity into intense and well-advertised oestrous "heat" periods, during which they seek out potential mates almost tirelessly. Sensory information from the vagina during mating triggers off the production of FSH in the brain and the subsequent release of eggs, which occurs between 24 and 48 hours after mating has finished.

Multiple fathers

Although females frequently mate with several male cats, veterinarians and geneticists still do not know how many fathers can be involved in a litter: the possible combinations of coat colours in cats that are not bred specifically for colour are too diverse. Only by intentionally permitting controlled group matings within a variety of colour-restricted purebreds could information be gathered.

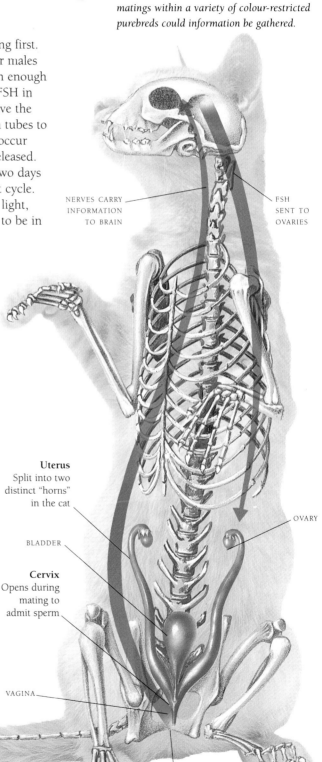

NERVES CARRY INFORMATION TO BRAIN

FSH SENT TO OVARIES

Uterus
Split into two distinct "horns" in the cat

BLADDER

OVARY

Cervix
Opens during mating to admit sperm

VAGINA

Vulva
Stimulated by male's barbs

THE SKIN AND COAT

THE SOPHISTICATION AND IMPORTANCE OF THE FELINE'S SKIN AND coat is sometimes underestimated, because it seems inert to the naked eye. The skin is the body's largest organ. The cells within it constantly carry out surveillance duty, warning of potential dangers, acting as waste managers, and even producing nutrients to maintain a variety of beneficial skin microbes and keep them healthy. Regions of the skin are modified for specific purposes: paw pads for durability, claws for traction, and glands in the chin, lips, and anal area for scent marking. Hair growth is under a variety of external and internal controls.

The skin and hair have prosaic but significant, even life-preserving, responsibilities. The skin contains cells vital to the immune system; it is the cat's first and most important line of defence, keeping harmful micro-organisms or chemicals from entering the body. Sensory functions are carried out by millions of nerve endings that detect heat, cold, pain, itches, and pressure. A profusion of microscopic blood vessels, a far greater blood supply than is necessary simply to nourish the skin, forms a sophisticated thermodynamic system that helps a cat to regulate its body temperature when moving from a warm area to a cold one. Skin secretes glandular substances that help rid the body of waste and help to protect against the harsh world outside, and also produces vitamin D, necessary for building sturdy bones.

COAT STRUCTURE

Hair consists mostly of keratin, the same tough protein that makes up the skin's outer layer (epidermis) and the claws (*see page 57*), which are technically skin. Because hair is made mostly of protein, normal hair growth consumes a good percentage of daily protein intake.

The surface of a hair consists of overlapping cuticle cells, which point away from the body. A hair stroked from root to tip feels smooth, while one stroked from tip to root feels rough because of the "lay" of the cells. The cuticle reflects light and helps to give the coat its sheen; a dull coat may signify cuticle damage. Hair grows in cycles that are controlled by hormones, length of daylight hours, and the surrounding temperature.

Cold-weather coat

The longhaired cat's coat has a protracted growth stage. Hair grows for a longer time, rather than faster, than in shorthairs.

Curled cat

In rexed cats, long- or shorthaired, the growth of hair is genetically retarded. The Cornish Rex has no guard hairs, while the Devon Rex, seen here, has soft guard hairs virtually identical to down hairs.

Humans have simple follicles: a single, large hair grows from each hair follicle. Cats have more efficient insulation, in the form of compound hair follicles: up to six primary (guard) hairs grow from each follicle, each surrounded by finer secondary (down and awn) hairs. Each follicle has its own arrector muscle, which can make the primary hairs stand on end. Cats "raise their hackles" like this when alarmed or angry, but also to increase insulation and cut heat loss.

Cats have two types of hair specialized for sensation. The whiskers, or vibrissae, are thick, stiff hairs found on the head, the throat, and the forelegs (*see page 60*). Other large, single hairs, called tylotrichs, each surrounded by a complex net of blood vessels and nerves, are scattered over the skin and act like short whiskers.

COATS FOR CLIMATES

Evolution has equipped the cat with excellent defences against heat and cold. Breeds that have adapted to northern climates, such as the British Shorthair, Maine Coon, and Norwegian Forest Cat, are densely coated with insulating down. In cold weather, the hair erects, trapping a layer of air, much as double glazing traps a wall of air between the cold outside and the warmth indoors. A layer of fat under the skin insulates the body; heat is lost through fat one third as fast as it is through muscle. In extreme cold, a cat curls up and covers its face with its tail to protect it from freezing air.

Breeds adapted to hot climates radiate heat more efficiently. Siamese lack down hair: cats that have this insulating coat shed it in hot climates. If still too warm, they shed primary hairs. Blood vessels in the skin dilate, speeding the loss of body heat. Cats do not sweat to lose heat as we do. Instead, they lick their fur, and the evaporating saliva carries away body heat.

THE FELINE HAIR GROWTH CYCLE

Feline hair grows to a genetically predetermined length during a 60 to 90 day growth period (anagen). After a short transitional stage (catagen) and 40 to 60 days of rest (telogen), it falls out, giving way to a new hair. Because hairs are at different stages at any time, shedding is gradual. The main external influence on growth is light: increased light stimulates a heavy spring moult. Under artificial light, indoor cats often shed all year. There is no such thing as a non-shedding cat.

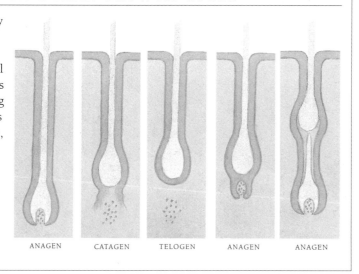

ANAGEN   CATAGEN   TELOGEN   ANAGEN   ANAGEN

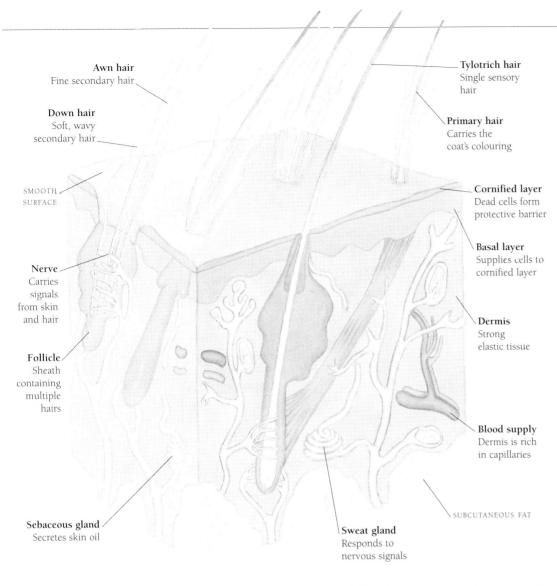

Awn hair
Fine secondary hair

Down hair
Soft, wavy
secondary hair

SMOOTH
SURFACE

Nerve
Carries
signals
from skin
and hair

Follicle
Sheath
containing
multiple
hairs

Sebaceous gland
Secretes skin oil

Tylotrich hair
Single sensory
hair

Primary hair
Carries the
coat's colouring

Cornified layer
Dead cells form
protective barrier

Basal layer
Supplies cells to
cornified layer

Dermis
Strong
elastic tissue

Blood supply
Dermis is rich
in capillaries

SUBCUTANEOUS FAT

Sweat gland
Responds to
nervous signals

The structure of the skin

Hairs grouped on the skin's surface provide insulation and sensory input. The cornified and basal layers of the epidermis provide protection. Beneath them lies the dermis, rich in nerves and blood vessels, containing the hair follicles and their glands, and held together by tough elastic tissue.

NATURAL FLORA AND FAUNA

As on all animals, the surface of the cat's skin is naturally colonized by a variety of microbes. These flora and fauna survive and proliferate without causing skin disease – in fact, they are vital to the health of the skin. Different regions of the skin have different "microclimates", some of them more attractive to microbes than others. Regions that are frequently moist or wet, and areas with dense hair and concentrated sweat glands, are most attractive to microbes. Relatively hairless regions are least attractive.

SKIN AND HAIR PROBLEMS

As a cat's first line of defence, coat and skin are prone to a variety of problems, from allergic or hormonal disorders to parasites, infections, or tumours. Coat problems are obvious because the hair becomes brittle, dry, lustreless, or falls out. Hormonal influences on the skin and coat are not well understood, but hormone-related disorders are usually not itchy.

Many cats carry the spores of the fungal infection ringworm without suffering infections. When it occurs in a cattery, other cats may be symptomless carriers. Ringworm can also infect people: in Paris it is estimated that 90 per cent of human ringworm infections are from cats.

A more common problem is a crusty, itchy condition called miliary dermatitis. This has many possible causes, but an allergy to fleas is overwhelmingly the most common one.

While these control mechanisms are efficient, cats are still prone to both heat prostration and frostbite. Sensible cats avoid high temperatures and humidity but, like dogs, they die quickly if trapped inside a hot car in sunshine. Cats are prone to frostbite, because their fur loses its insulating capacity when wet; cats are most at risk from hypothermia in freezing, wet weather.

SKIN STRUCTURE

Cat's skin has two distinct strata. The surface, or epidermis, consists of about 40 layers of dead, flattened cells, embedded in fat-rich sebum, an oily skin secretion. Beneath this "cornified" layer is the living epidermis, or "basal" area, about four cells thick. The epidermis is not very strong, but plays an important role in the body's immune system. In normal circumstances, it takes about three weeks for cells to migrate from the basal layer to the cornified layer, but this process speeds up when the skin is injured, reducing the replacement time to two weeks. Beneath the epidermis is the dermis, the major structural component of the skin. It consists of strong elastic connective tissue, and contains glands, blood vessels, nerves, and receptors.

Each hair follicle has an associated sweat gland and sebaceous gland, buried in the dermis. The sebaceous glands produce fatty sebum, which controls bacteria and helps to give hair its sheen. They are concentrated on the neck and rump, and especially around the lips and on the chin. Specialized sebaceous glands line the eyelids and produce an oily protective film for the eyes. Other specialized sebaceous glands around the anus and between the toes may produce pheromones, or sex scents.

Unlike humans, cats do not sweat to control their body temperature: they pant inefficiently or lick themselves. Sweat-gland secretions maintain the skin's pliability, excrete waste, provide nutrients for the skin's microflora, and contain substances that protect the body from chemicals and dangerous microbes. They probably also secrete pheromones. Sweat glands associated with hair follicles are called epitrichial (formerly apocrine). Atrichial (formerly eccrine) glands, such as those on the footpads, open onto the skin, rather than into hair follicles.

Sweat glands

Scent-producing glands in cats are concentrated in the areas they use to mark territory: the face, the neck and shoulders, the rump, and between the toes, where they scent pawprints. Specialized anal sacs under the tail also produce scent.

FELINE GENETICS

ALTHOUGH THE DETAILS OF GENETICS ARE COMPLEX, THE BASIS OF these inherited characteristics is quite simple. All the information needed for life is carried in the genes contained in each body cell. Most of that information is common to all mammals. The rest makes cats what they are, and us what we are. The science of genetics began with Gregor Mendel (1822–1884), who showed that some traits are not diluted when crossed, but masked. In 1953, Crick and Watson's modelling of DNA, the genetic storage system that is the basis of life, led to a new understanding of the influence of genetics on physical characteristics, emotional behaviour, and disease.

Genetics follows simple, mathematical laws. Every cell in a cat's body contains a nucleus at some stage in its development. Each nucleus contains 38 chromosomes, arranged in 19 pairs, which are just large enough to be seen with a powerful optical microscope. Each chromosome is formed from a tightly wound double helix of deoxyribonucleic acid (DNA), which in turn is made up of thousands of units called genes, strung together like beads. Each gene is made from four different proteins – A, T, C, and G – and the combinations of these proteins provide the information for all aspects of a cat's life.

COPYING AND MUTATIONS

Each time a cell, such as a skin cell, is replaced, its chromosomes are copied. The DNA unwinds and unzips down the middle. Ribonucleic acid (RNA) is generated to match each half-strand, and then used as a template on which new DNA forms from proteins within the cell. The copying process is so accurate that there may be only one mistake, or mutation, in one gene, for every million copies: this would be like retyping these two pages five million times before typing one wrong letter. These rare mistakes can create cancers, and may contribute to ageing.

Information is passed down the generations in a different way. Egg and sperm cells contain only 19 chromosomes, each one half of a pair. At conception, the 19 chromosomes in the egg unite with the 19 in the sperm, creating a new, unique set of 19 pairs. Each kitten inherits half of its genetic material from each parent. When chromosomes pair up, the genes for each characteristic also pair up, side by side. Mistakes, or mutations, sometimes occur in egg or sperm cells, creating new traits.

DOMINANT AND RECESSIVE TRAITS

Genetic variations in characteristics, for example coat length, are called dominant if one copy of them is needed to show its effect, and recessive if two copies, one on each chromosome in a pair, are needed. Original traits tend to be dominant, and new mutations recessive: cats originally had short coats, and the gene for it is notated *L*, but a mutation occurred long ago, producing a recessive gene for long hair, which is notated *l*.

A cat showing a dominant trait may be heterozygous, carrying the recessive alternative "masked" beneath the dominant one: a cat with a recessive trait must be homozygous for it, carrying no alternatives. Two heterozygous shorthaired cats – both *Ll*, carrying the recessive gene for long hair –

produce, if mated, an average of two *Ll* kittens, one *LL* kitten, and one longhaired *ll* kitten. Appearance gives no clue as to which of the three shorthaired kittens carry the *l* gene that enables them to produce longhaired kittens.

Unfortunately, not all genetics is this simple. Many traits are polygenetic, governed by an unknown combination of genes. Also, not all genes are simply dominant or recessive; some mutations show partial dominance over others, or even over the original form.

CHROMOSOME

X-shape
All chromosomes except for one have this shape

Gene
DNA can be "snipped" into individual genes

DNA
Tightly coiled DNA forms chromosomes

NUCLEUS

Chromosomes
Half come from each parent

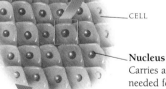

CELL

Nucleus
Carries all information needed for replicating cell

The basis of life

Every living cell contains a nucleus, in which 19 pairs of chromosomes are stored. All chromosomes are X-shaped except the male gender chromosome, which is Y-shaped. In turn, each chromosome, when unfurled, is a complex helix of four proteins. This is called DNA, and it forms the basis of life and all inherited characteristics.

ALLELES

Specific information about a trait is always carried at the same site on each chromosome: in a pair of chromosomes, this pair of sites is called an allele. The information may vary in an allele. If the information is the same at both sites, the instructions are homozygous; if it is different, the instructions are heterozygous.

Matching pair
A genetic trait is determined by the genes from both parents, located at matching sites on the paired chromosomes.

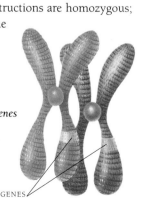

ALLELE OF TWO GENES

THE FOUNDER EFFECT

For hundreds of millennia, the African ancestors of today's domestic cats were almost uniformly shorthaired, striped tabbies. Yet after only a few millennia of migration out of Africa, hundreds of coat colours, patterns, and lengths exist.

In the large population of cats in North Africa, any random gene mutation had little chance of spreading widely unless it offered a substantial advantage to the cat. Most genetic mutations

Spine
Sugar-phosphate spine holds genes in sequence

Protein pairs
These split, unzipping DNA, when cells are renewed

Protein
Each gene combines four proteins in a unique pattern

simply vanished within a few generations. In an isolated feline population, mutations have more chance of surviving. The orange-and-white cats of Scandinavia (*see page 94*) or the polydactyls of Boston and Halifax (*see page 99*) are both mutations from the genetic "norm"; when these cats were taken to regions with few or no cats, they represented a large percentage of a small gene pool. The long-term genetic influence of early members of a cat population is called the "founder effect". The founders have a potent influence on a new population: this is why certain patterns or colours are prevalent in some countries; it is also the basis for the development of new breeds. This leads to the question of just what makes a breed? From a purely genetic perspective, there is no such thing, because the potential genetic differences within a breed, particularly those with large numbers, outweigh the average genetic differences between two different breeds. For example, the DNA profiles of two Siamese can differ far more than those of a Siamese and a Longhair. The definition of a breed is decided on a few obvious effects, such as coat colour or length and body type: from a genetic perspective, these differences are slight.

BREEDS AND GENETIC DISEASES

Breeders use the laws of genetics to select for specific features such as colour or body type. Unfortunately, they may also unwittingly select for other hidden, dangerous genes. This is how genetic diseases gain ground in pure-bred lines. To name only a few of these:

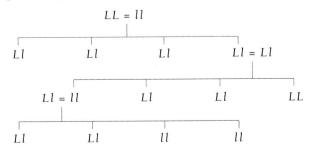

Polydactylous cat
At least two genes are associated with polydactyly, or extra toes. Polydactyls were among the first cats to be brought to Boston and Halifax; this trait is now more common in these places than anywhere else in the world.

several breeds are prone to the lethal disease amyloidosis; certain lines of Longhairs suffer from polycystic kidney disease; Colourpoint Longhairs have a marked incidence of cataracts; Abyssinians can develop retinal degeneration; the Devon Rex carries a spastic muscular disorder; and some Burmese suffer from flat-chested syndrome, episodic weakness syndrome, and a lethal head malformation.

In natural selection for survival of the fittest, such dangerous genes are removed from the gene pool or persist at a very low level: selective breeding has allowed them to survive and be passed on. This is the greatest genetic problem facing cats; responsible breeders now investigate genetics as part of their breeding programmes. Scientists are learning more about cats' genetics. The National Cancer Institute in the United States has a gene-mapping project investigating feline genetics in detail. In England, Alex Jeffreys developed "genetic fingerprinting", which can identify an individual from a sample of DNA from any body tissue. This can be used in cats, as in humans, to investigate paternity. Because pedigree cats often have similar DNA, however, especially in the rarer breeds, genetic profiling can only exclude a sire, not identify one.

In reality, the vast majority of cat matings remain random, and selection for survival of the fittest remains the most powerful influence on the genetic future of the domestic cat. Increasingly, however, advances in genetics will be used to monitor exactly what is happening.

MENDELIAN INHERITANCE PATTERNS

Several major traits in feline appearance have been identified (*see list, right*). Dominant traits are notated in capitals, recessive traits in lower case. A shorthaired cat is notated *L*; unless test matings show it to be *LL*, because *L* needs to appear only once in an allele to show an effect, and the nature of the second gene in the pair usually remains unknown.

Inheritance patterns
This diagram shows how long- and shorthaired traits are passed on, giving the average results over many matings.

A	agouti, or tabby
a	non-agouti, or self
B	black
b	brown, or chocolate
b^l	light brown, or cinnamon
C	full colour, or solid
c^b	Burmese pattern, or sepia
c^s	Siamese pattern, or pointed
D	dense, dark colour
d	dilute, light colour
I	inhibitor, or silvering
i	pigmentation sound to roots
L	short hair
l	long hair
O	orange, or sex-linked red
o	melanistic, non-red colour
S	white spotting, or bi-colour
s	solid colour over whole body
T	striped, or mackerel, tabby
T^a	Abyssinian, or ticked, tabby
t^b	blotched, or classic, tabby
W	white, masking all other colours
w	normal colour

$LL = ll$

$Ll \quad Ll \quad Ll \quad Ll = Ll$

$Ll = ll \quad Ll \quad Ll \quad LL$

$Ll \quad Ll \quad ll \quad ll$

THE AGING CAT

NATURE DESIGNED THE CAT TO PERFECTION BUT EVEN PERFECTION is susceptible to aging. When a cat is in its prime, its anatomy and physiology work in harmony. With time, this superb synthesis of activities falters, and living with us means that cats have more time than ever before. Insidiously, age-related changes develop.

The most important changes occur at the chemical level, with the messenger chemicals of repair losing their immediacy and efficiency. Although this aging is inevitable, increasing scientific understanding of the process means aspects of aging can be delayed, allowing a cat to achieve the full potential of its design.

The cat's unique anatomy and physiology evolved for survival in its original habitat, North Africa. This design was elastic enough for successful life elsewhere throughout the world, in the most varied of habitats and even in close proximity to another species, humans. Everything about the cat's superb design helps it to find food and shelter, to avoid injury, illness, and predators, and to successfully seek a mate and perpetuate its genes.

There is a time limit on all this, however. Aging is controlled by a cat's genetic biological clock, located in the part of the brain called the hypothalamus (*see page 58*), which in turn controls the hormonal biofeedback system.

NATURAL AGING
Studying the physical aspects of aging, Professor Jacob Mosier at Kansas State University found significant changes in the brain. In its prime, a cat depends on its swift mental and physical reflexes for survival. Messages travel along its nerves at approximately 6,000 m (6,560 yd) per second. With aging this slows down to as little as 1,300 m (1,420 yd) per second.

Signs of wear
As a cat ages, its digestive system changes. The intestines and the kidneys are less efficient than they were, and as a result cats often lose weight, and their coats become dull. Changes to a cat's diet, including more easily digested protein and higher levels of vitamins and minerals, may help to counter these effects.

At the same time blood vessels, including those in the lungs and the brain, lose their elasticity. Oxygen exchange in the lungs becomes less efficient, and as a result the brain receives less oxygen, affecting memory and learning.

At the University of California, Professor Ben Hart has studied the behavioural consequences of aging changes to feline design. He found that cats experience changes similar to those that we undergo, including the changes that in humans produce senile dementia. He found that while cats may live 20 years or more, loss of some brain function is natural by 16 years of age. Age-related behaviour changes may include increased irritability, increased hissing and spitting, sleeping pattern changes, loss of toilet habits, and disorientation. He found that by 16 years of age, 70 per cent of pet cats become disorientated, forgetting how to use a cat-flap or simply staring into space; 60 per cent are more irritable with members of their human family, with strangers, and even with family dogs, hissing or spitting with little or no provocation. Seemingly pointless, plaintive meowing tends to increase.

Nocturnal patterns
Although a cat's activity level diminishes with age, some elderly cats change their sleeping patterns and become more wakeful and active during the night. There is little or nothing that can be done to counteract this change.

About 25 per cent of elderly cats change their cycles of sleeping and waking, sleeping more in the day and less at night when they are more restless and demanding of their owners. An additional 20 per cent develop toileting problems for no apparent medical reasons.

AN EXTENDED LIFESPAN
The inevitability of such aging exists in all cats, although some breeds, such as the Siamese and Oriental Shorthair, have aging clocks set to run longer than others. The more important effect we have on the lifespan of cats is through our intervention in caring for them: pet cats are more likely to live out their full potential lifespan than are feral individuals.

Generally speaking, pet cats live more than twice as long as cats in the wild. Evolution did not provide for the security of living with us, with guaranteed food and shelter into old age.

DISTINGUISHING AGING AND ILLNESS

All aspects of feline design change with time. Brain cells die and are not replaced; the delicate balance of biofeedback becomes a little clumsy; the senses become less acute. Physically, bones become brittle and elasticity diminishes, muscles shrink, and the lungs are less efficient picking up oxygen and releasing carbon dioxide. Nutritional needs change: the intestines do not absorb nutrients as well as they did, and the kidneys become wasteful, allowing nutrients to be lost from the system. These are all natural changes, encoded in the genes, as are the alterations in behaviour, appearance, and abilities that result.

Illness occurs when a pathological change occurs to some part of feline design: infection, a growth, or an unnatural change, such as crystals developing in the bladder, fat depositing in the liver, or the muscular heart wall thickening. It is important, but not always easy without proper investigation, to distinguish aging from illness.

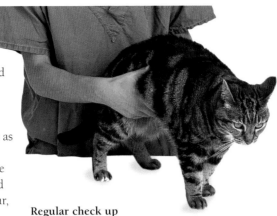

Regular check up
Never assume that changes in an elderly cat's behaviour are simply the natural result of growing older, and not a sign of illness. Any change is potentially significant. Elderly cats will benefit from undergoing a thorough medical examination at least once, or even twice each year. Even the natural effects of aging will mean some changes to the care of your cat, and your veterinarian can advise you on these.

The consequence of this "unnatural" lifespan is the appearance of unexpected, age-related design problems not seen in the wild. For example, a menopause never evolved in female feline design, simply because cats in the wild naturally die relatively young. In the security of our homes, elderly unneutered female cats will continue to have cyclical heat periods. Because feline evolution did not "design" for this long continuation of seasons, problems arise: while young female cats have a low incidence of womb infection, the incidence is quite high in unneutered elderly cats.

The cat is the most fortunate of all the species that humans have domesticated, in that we have interfered least in its natural, evolutionary design. Even so, in the cat's relatively new ecological niche of living with us, we have put unforeseen pressures on its superb anatomy and senses. Although we are often unaware of it, the lives of many of our pet cats now involve chronic low-level stress.

Slowing down
Cats, like humans, can suffer apparent memory loss and "senility", staring into space and becoming disoriented. Like humans, their brains will also respond to continued stimulation and new challenges from their environment. These can help to maintain brain function and slow down the effects of aging.

It is this chronic stress, more than any other factor, that compromises their natural design. Elderly cats are less able to turn off their stress response than young cats: even when they are relaxed, elderly cats will secrete more stress-related hormone than their younger counterparts. Simultaneously, as natural aging progresses, the cat's brain and nervous system also produce fewer of the chemicals called neuroendocrines; specifically, production of the brain chemical dopamine drops. Dopamine is the brain's "master" chemical: if dopamine production can be maintained, it may be that all aspects of the cat's design can continue to function for longer.

INTERVENING TO DELAY AGING
Routine mental and physical stimulation may help to maintain dopamine production and so slow down a cat's natural aging processes. Massaging a cat is a good way to loosen stiff joints and improve circulation, and in a cat that has bonded to you has the additional advantage of stimulating the secretion of beneficial neuroendocrine hormones.

In cats (and also in people, for that matter) the brain shrinks with natural aging until it is 25 per cent lighter than at the peak of its efficiency. Although part of this loss is in the number of cells, most is simply in the loss of connections between the cells. Mental stimulation will not replace the lost cells, but it does maintain or even rebuild the connections between the cells. Scientific studies carried out almost 20 years ago showed that a single brain cell may have up to 10,000 connections with other cells: with time these diminish, but with mental stimulation up to 2,000 lost connections may redevelop.

Cats can age gracefully. Mental and physical activity, appropriate nutrition, warmth, comfort and well-designed toys may help your cat to age comfortably. Aging brings disorientation: some cats pace the room, while others may forget where the cat-flap is or how to use it. Some cats become more vocal, and some, like some elderly people, become snappy. All of these changes may be age-related, but may also be signs of treatable illnesses.

POOR EYESIGHT

Increasing reflection from a cat's eyes is a normal aging process. Diabetic cats develop cataracts to their eyes, but this cat has more typical and natural connective tissue changes in the lenses, called sclerosis. Near vision diminishes with age, just as in humans, but distance vision remains excellent.

CAT WITH SCLEROSIS

CHAPTER FOUR

CAT BEHAVIOUR

ALTHOUGH THE CAT IS A SOLITARY HUNTER, content to live on its own in the absence of other cats or people, it is now in the process of evolving from total independence to more willing dependence. Its behaviour towards other cats and people reflects this evolutionary development. Our intervention in the domestic cat's living environment has modified its natural behaviour. We have encouraged the kitten in cats, making adults more reliant on us, and more sociable, in the process. However, regardless of how much we intervene, the cat's inherent need to hunt, and its ways of conceiving and mothering its kittens, are unlikely to change.

ABOVE A CAT FOLDS ITS EARS BACK AS A SIGN OF FEAR

LEFT FERAL MOTHER AND KITTENS

THE FELINE MIND

PROGRESSIVELY INCREASING IN POPULARITY, CATS OUTNUMBER DOGS across the United States, Canada, Britain, The Netherlands, Austria, and Switzerland. At the beginning of this century, the number of pet cats has just surpassed pet dogs in Sweden, Denmark, and Norway.

As cats replace dogs as our favoured companions, it becomes more important to realize that they are not dogs in disguise. The cat is unique among all the animals that have become domesticated, the only species that first, foremost, and forever follows its own mind.

Like all species, including humans, the cat instinctively strives to survive and leave descendents. Unlike people or dogs, it does this in an independent manner: the cat is not naturally a pack animal.

ALONE IN THE CROWD

Although we are countering its natural independence through selective breeding and early learning, the cat remains a far more "freelance" creature than any other domestic animal. The ways in which cats socialize, hunt, court, and mate differ from those of more communal species.

To us, the cat's behaviour, caring only for its own comfort and security, can appear selfish. Such moral judgments ignore biological reality: the cat behaves like number one because it is number one. In their natural state, cats see other cats only as competition for territory. Even mating (see page 86) can create conflicts: it may be the only time after kittenhood that a cat makes physical contact with another cat.

Reliable food supply
Although their coats differ widely, a sign that they are not closely related, these cats are showing no aggression; this is because food from fishing boats is plentiful.

The cat's mind is plastic enough to permit it to change its behaviour, in part according to how much food is available. The most sociable time in a cat's life is kittenhood, when, together with its mother and siblings, it is a part of a group. This pattern can be extended into adulthood, with cats living in groups usually comprising mothers and their female kittens: males are tolerated rather than welcomed. Although they do not hunt together, their group behaviour, including looking after each other's young, is not unlike that of the all-female pride of lions.

ABILITY TO LEARN

The cat's adaptability has allowed it to enter our lives successfully. Through selective breeding, we increase our cats' sociability, and by raising kittens in our homes and exposing them to a variety of experiences during the period from three to seven weeks of age we perpetuate their kitten-like dependency on us for food, warmth, comfort and security.

Early experiences radically influence the adult cat. Kittens that are raised without human contact will always be fearful of strange humans, but kittens raised with people or other animals will accept and enjoy their company throughout life.

LEAVING A MARK

Cats communicate face to face through voice and body language (see page 80). They also leave signs on their territories that can be read in their absence. Boundaries are marked with urine, faeces, and scented chemicals, rubbed against upright objects. These are not warnings, but simply mark the territory as occupied. When a cat rubs against its owner's leg, we interpret it as affection. In part it is, but it is also a territorial marking.

Visual markers, such as scratches on trees, are also important. This visual aspect of cat's language means that urban cats often use the same markers that we recognize – such as hedges or walls – as their boundaries.

DOMESTIC PREDATOR

While all cats show hunting behaviour, each cat's hunting skills are learned (see page 84). Learning by observation is most effective: ethologist Paul Leyhausen found that the most important factor in a cat's preferences and success as a hunter is its mother. Some mothers teach their kittens to hunt birds, while others teach them to hunt rodents.

The cat's natural diet consists of small rodents and birds, usually only big enough to feed one small stomach. This one factor, more than any other, is the most potent reason for the cat to be independent, to be a lone hunter. In spite of this evolution as a solitary hunter, cats can live in close proximity if food is plentiful. This is clearly seen in some city parks, where cat lovers provide daily food for feral cats, in fishing ports, where cats will gather to eat the unwanted part of the catch, and in our homes (see page 82). Free of the need to defend a hunting territory, cats are able to socialize.

Scent marking
Glands on the cat's lips, face, ears, feet, and body all contain distinct chemicals, which act as scent markers. This cat is leaving its scent on a tree trunk – not to frighten other cats but simply to proclaim its presence.

Instinctive hunter
All young kittens naturally play in ways that mimic an adult cat's predatory behaviour. This kitten uses instinctive stealth to creep silently closer to its victim, which may be a small rodent but is much more likely to be a leaf or a litter mate. Its ears are turned forwards to pick up the faintest sounds most efficiently.

GROUP ACTIVITY

Lions live and hunt as a group, preying on large mammals. A pride, made up of blood-related females, is similar in many ways to a feral cat colony. They eat and sleep together, groom each other, and care for each other's kittens, sometimes even suckling them when the natural mother is otherwise occupied. These rulers of the African savannah are also capable of forming emotional bonds with people, if raised with humans during early kittenhood. Big cats, however, lack the innate tractability of the domestic cat, and remain dangerous.

The family that naps together
Lions can sleep as a group, knowing that other predators would not risk attacking an entire pride. Mothers with young cubs remain vigilant.

THE NEED FOR SLEEP
A unique and curiously appealing aspect of the cat's behaviour is its ability to relax into deep sleep apparently at will. As in all other animals, the cat's natural rhythms are controlled by an internal biological clock, a complex interaction between the body and the mind.

The most common rhythm in animals is a 24-hour cycle called circadian rhythm. This dictates when animals are most and least active. The cat is most alert at dawn and dusk, when hunting is best, and sleepiest around midday and the middle of the night. The cat's need to be active when humans want to sleep, and to sleep when humans are most active, is difficult to change, because its origins are so deeply embedded in feline behaviour.

French research has shown that sleep is induced by a chemical, melatonin, released in the brain (*see page 58*), and that this release is probably under the control of the gene that determines circadian rhythm. Melatonin is thought to be responsible for the cat's habit of

Performing cat (right)
Natural dexterity develops when a kitten plays with its litter mates and practises hunting. Even adult cats can still be actively trained, using food rewards, to complete "assault courses" that include jumps and even tunnels.

Catnapping
Sleep seems to be the natural state for a cat with a full stomach. No one knows why cats sleep more than other species. The benefit for us is that the sight of a sleeping cat relaxes us.

sleeping for 16 hours a day on average. Just why cats sleep so much is not yet properly understood, although in kittenhood, growth hormone is released only during sleep. Sleep is certainly necessary for all animals to maintain general health, and feline dreams, like ours, may act as periods during which the cat's mind can reorganize and reclassify information.

LEARNED BEHAVIOUR
Because cats cannot be trained like dogs, it is sometimes thought that they cannot learn at all. It is possible, however, to train cats to press a button for a food reward. The key to all cat training is the use of reward; discipline, which works in pack species, such as dogs, does not work with cats, who either hide or leave.

The cat's mind is at its most receptive when it is very young, as it loses its ability to absorb new experience earlier than most other species. Experiments from over a century ago show that

Unaware of danger (right)
This kitten (safe on a sheet of glass) has blithely crawled over the apparent edge of the step although it could see it; in a few weeks it will stop and investigate first.

cats can learn by watching what other cats do, and learn best of all by watching their mothers. Other work from the same time showed that cats are not good at solving puzzles, and escaping from mazes was due more to luck than to logic.

There are some things, however, that a cat may never learn or unlearn. Pet cats must adapt to a sometimes alien lifestyle, especially if kept indoors. Some use their hunting instincts to stalk and destroy household objects. Others try to remain nocturnal and demand nighttime activity from their diurnal human companions.

Less certain are the cat's reputed "psychic" abilities. Most of the evidence for these is anecdotal, but there is scientific confirmation that cats have a sense of direction and can foresee earthquakes. Tests in Germany and the United States suggest that cats may have a superior sense of direction (*see page 60*), and Professor Benjamin Hart in California recorded increased unusual behaviour before earthquakes.

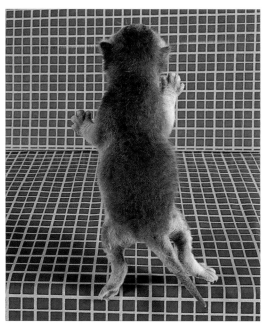

COMMUNICATION

CATS COMMUNICATE SUBTLY AND SUCCESSFULLY WITH OTHER CATS, but less so with us. Social relationships are of little importance to them, so the nuances of facial expression and "distance-reducing" body language are not highly defined: the human smile or a dog's wagging tail do not have overt equivalents in cat communication.

Other methods by which cats exchange information are refined. These include the voice (our favoured medium), touch, scent, and visual displays or markers. With our tendency to rely on speech, we often fail to understand what is meant by scratches on the sofa or faeces left unburied.

Our cats greet us, demand food, complain, and protest vocally, and with experience we know exactly what they are saying. Different breeds have different verbal habits: some "chirp" more than others, some breeds are almost silent, and Oriental breeds are notoriously loud. In the 1930s, psychologist Dr. Mildred Moelk defined 16 different cat sounds; Dr. Patricia McKinley more recently split sounds into 23 categories. In general, however, feline sounds fall into three types: murmurs, vowels, and high-intensity sounds.

Vocal request
Cats' questions and commands all include vowels. The associated body language is alert and confident, as with this relaxed and plaintive young Longhair.

Murmurs include purring and the soft, happy chirp of greeting. The source of the cat's purr remains a mystery (*see page 67*). However it is done, it is not always a sign of happiness. Stressed or traumatized cats also purr, a phenomenon familiar to vets who treat injured or ill cats. The chirp, like the purr, does not involve vowels. It sounds like a "brrrp" or "brrrm", and is uniquely used to show pleasure.

Vowels, the classic "miaows", vary according to what is meant. Domestic cats talk with miaows because we have extended their kittenhood (*see* Family communication, *opposite*). Requests, demands, commands, complaints, and bewilderment have subtle variations that are easy to understand, and with which owners become familiar.

High-intensity sounds are natural adult cat language, and as such they are usually reserved for communicating with other cats, although some are familiar to owners and veterinarians.

Assertiveness
It is easy to interpret shows of confidence or aggression. This cat's fur bristles to enhance its size, and it stares down its opponent, leaning forwards and rotating its ears.

Changing the shape of the mouth creates many sounds, from a grumbling, irritated growl to a threatening snarl, a hiss, or a defensive spit. Shrieks of fear or pain and the wail of females in heat are made in the same way.

BODY LANGUAGE

Vocalizing conveys information over distance, while body language remains the logical and favoured communication method face to face. Feline body language is most refined in sending "go away" signals, and aggressive and defensive displays are overt and simple. The whole body is used to send "loud" signals, while the ears and tail are more subtly expressive.

A safe and secure cat moves with fluid ease and a raised tail. It settles in total relaxation, and on waking it loosens its muscles with a whole body stretch. Nervous cats quietly avert their eyes from others and tend to yawn.

An offensive display involves direct eye contact, often with constricted pupils, and a posture permitting instant attack. Head and whiskers lean forward.

Communicating contentment
Movements are relaxed when a cat is content and secure. Safe in its territory, this feral cat wakes on a warm, sunny windowsill and stretches to prepare for activity – or further inactivity. A yawn and a stretch communicate an absence of tension to other cats, and may also be used to defuse tense situations.

Ears are held up and out so that the opponent sees the backs of them. Fearful aggression is seen in the typical Halloween cat, with its arched back, dilated pupils, and bristling hair.

Apprehension is seen in the crouch assumed by a cat when it feels insecure. As apprehension increases, the crouch becomes lower, the pupils dilate, and the ears flatten against the head. When a cat feels threatened, it rolls over into a defensive position that reveals its claws and teeth. Hair may bristle all over the body.

FELINE GRAFFITI

Cats leave visual markers on their territory to communicate either their presence or their ownership of the domain. While scratching is a method of maintaining sharp claws, it is also used to leave a visible sign of a cat's presence. This is why domestic cats prefer to scratch something highly visible, like the arm of a sofa, and why scratching posts should be placed in a prominent position in a room, rather than tucked away in a corner (see page 276).

Faeces are also used by a dominant cat to mark its territorial boundaries visibly. This is why some cats leave their faeces unburied in neighbours' gardens, but not in their own: the cat feels that its territory extends beyond its own garden and its faeces are a visible sign of the extent of its jurisdiction.

SCENT MARKING

Faeces also serve as a territorial marker through their powerful scent, an important part of the feline communication repertoire (see page 82). Each time a cat empties its bowels, it squeezes a smelly secretion out of the adjacent anal sacs, anointing its deposit with an "identity tag". French research has succeeded in identifying at least 17 different chemicals in anal discharge, which suggests that this malodorous substance transmits a wide variety of information.

The same purpose is fulfilled by feline urine, which is either deposited on the earth as body waste, or used solely as a scent marker when spraying upright objects. Both male and female cats are able to scent mark in this way, and both may sometimes continue to do so even after they have been neutered (see page 70).

Marking the territory
This Siamese scratches a tree in the heart of its territory to leave a high and visible marker, not as a threat, but simply to indicate that the territory is occupied.

Males, in particular, leave a sour, pungent odour, which is sufficiently potent to inform the relatively insensitive human nose that a male has claimed the territory.

Scent from the sweat glands on the paws also leaves an aromatic trail anywhere that a cat has visited. Further glands on the face produce substances that cats use as scent markers. Cats rub their heads against objects, leaving a scent marker that is often sniffed when the cat returns to the area. This most subtle form of scent marking is used by many domestic cats as part of their basic greeting ritual when their human family returns home.

TOUCH-SENSITIVE

While a cat announces its presence by leaving scent markers, its ability to communicate its needs or demands using touch is exquisite. Cats use a nose-to-nose touch, in part to scent each other but also to make physical contact.

Many cat owners know the feather-light touch of their cat's paw on the face early in the morning, a gentle reminder of a new day and impending mealtime. Those who disregard the light touch know that it becomes an urgent tap. If this is ignored, some cats will turn the tap into a veritable punch. They need no words.

FAMILY COMMUNICATION

All of the domestic cat's wild kin use similar methods of communication. In adults, vocal noises are mostly murmurs and high-pitched sounds; vowels are restricted to kittenhood. Territorial body language is highly refined, as is the use of both scent markers and touch. Because territory is so important, most large cats leave obvious scratch marks: in some regions where few trees grow, specimens are densely and deeply marked with generations of scratching. Markers may also be sprayed, to act as scent posts as well as visible signs.

Traditional practice
The heavy scratch-marks on this fallen tree indicate that this ocelot and many others before it have used the tree as a scratching post for a considerable time.

While feral cats may abhor physical contact with people, cats raised from kittenhood with humans usually enjoy physical comfort given by people. Stroking is analogous to the comfort of being groomed as a kitten, and the contact communicates security. Socialized cats seek out people and demand to be touched and stroked.

Taboos, however, remain in the gentlest cat. Touching the abdomen often provokes a mock, or even genuine, aggressive response, because the abdomen is the least defended part of the body. Ordinary stroking may also provoke a sudden aggressive response. This is because, while cats enjoy stroking, it remains a learned rather than a natural adult behaviour. A mental conflict arises, and the cat instinctively bites, then often apologizes, coming back to ask for more "maternal" care.

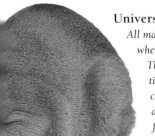

Universal submission
All mammals try to "shrink" when indicating submission. This fearful Abyssinian tightens itself into a low crouch, both to appease another cat and to prepare for explosive activity if it should be needed.

Show of weapons
Subtle body language usually averts conflict, but sometimes it is not enough. This kitten is in full defensive posture, baring all its weapons in a last-second attempt to either dissuade attack or protect itself.

SOCIAL BEHAVIOUR

WE OFTEN EQUATE FELINE SOCIAL BEHAVIOUR WITH OUR OWN OR with that of our other most popular companion animal, the dog: this is misguided. Cats live together on their own terms and for their own reasons, which are quite unlike ours. The cat's social behaviour is governed by territory and food; relationships with other cats are possible, but less important than control of property. Almost any cat feels overwhelmingly threatened when another cat, even a kitten, appears on its territory. From a human perspective, such behaviour appears anti-social, but this is not strictly true. Cats are far less solitary and more sociable than was once thought.

Feline social behaviour is enormously varied. In some circumstances, cats show great intimacy: females may groom each other and even feed each other's kittens. In other circumstances, females fiercely protect their dens and feeding territory from others who happen into them. Some females accept males on their home turf, others drive them off. Males are equally elastic. Young cats may be tolerated or viciously attacked, sometimes even by their mothers.

POPULATION DENSITY

This vastly adaptable range of social behaviours depends upon a variety of environmental factors, but by far the most important of these are the size and proximity of the food source. Sociable behaviour increases when food is plentiful, and decreases when food is scarce. Studies that have been carried out on feral cats to learn how many cats will occupy an area of 100 ha (250 acre) in different circumstances clearly illustrate this. When the only food available to the cats is prey, such an area will not support more than five cats. When both prey and rubbish can be exploited for food, the same area may support as many as 50 cats. But when there is unlimited food, as there is from the cat-lovers who feed the feline colonies in Rome's

Peaceful co-existence
Kittens that routinely meet dogs before seven weeks of age seldom develop fear of this potential adversary.

The family that eats together
If food is plentiful, as in this Greek fishing village, cats live in familial groups, often identifiable by their similar appearance. An outsider, such as the ginger tabby here, may be met with reserve or hostility.

public gardens, or around fish-processing plants in Japan, the area can contain over 2,000 felines. In these circumstances there will be an overlapping of many smaller cat colonies, each of them comprising up to 50 related feline members.

Social interactions naturally increase with population density. Thinly scattered hunting cats will restrict their social gestures mostly to defensive postures. Well-fed cats in colonies are spontaneously and surprisingly sociable, resting together, greeting each other, and sometimes grooming each other. When they sleep, half of the time they are in contact with one or more other cats, although males make only half as many bodily contacts as do females. In one extensive study of a feral cat colony, 64 per cent of interactions involved licking, 29 per cent involved rubbing, and only seven per cent involved forms of aggression. Contacts were initiated by kittens and females, seldom by males.

Feral cats are also flexible enough to change their social behaviour regularly according to the prevailing conditions. On the Hebridean Islands, off Scotland,

each cat has its own rabbit-hunting territory in summer, and defends it from others. But when the winter sets in, these cats move close to the security of local cottages, live on handouts, and accept the presence of other cats.

THE CO-HABITING CAT

Pet cats modify their social activities according to circumstances, just like ferals. A pet cat still needs its own territory, but this may be simply a favourite sleeping place. It may be friendly and sociable to other house cats in all other areas, but defend its chosen spot with fervour.

The level of sociability of pet cats depends to a large extent on their early experiences, but it is affected even more by their sexual status. Most pet cats are neutered, often before reaching sexual maturity, influencing their social activities in a marked manner. While feral tom cats make fewer social bodily contacts than females,

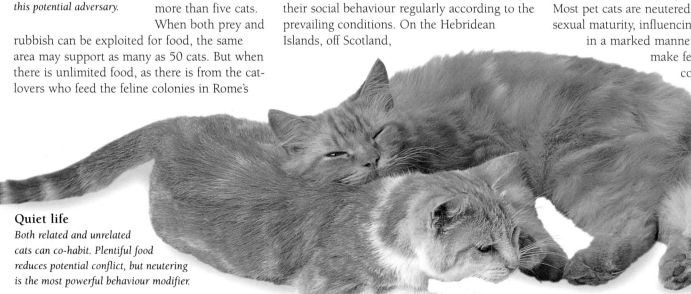

Quiet life
Both related and unrelated cats can co-habit. Plentiful food reduces potential conflict, but neutering is the most powerful behaviour modifier.

neutered toms make as many body contacts with other cats as do neutered females, who make more body contacts than entire females. Although we neuter our cats as a method of population control, the effect on their sociability is both dramatic and very positive.

In a household of several mature pet cats, disputes seldom occur because rank has been established. If two cats do meet unexpectedly, whoever arrives first or holds the higher ground is the social superior, but this ranking is fluid. Rank is of social importance to cats, but it is more variable than in other species, and is often forgotten when all gather for their meal.

In some instances, however, a group of cats treats one in particular as a "pariah", viciously attacking it when it nears, and denying it any social contact. Paul Leyhausen has observed one colony in which the pariah cat was forced to live trapped on water pipes near the ceiling, even toileting from its perch.

ESTABLISHING SOCIAL RELATIONSHIPS

Observations of cats in catteries show that it takes a cat from two to five weeks to adapt to its new territory. It is likely that it takes as long or longer for a cat to adapt to a new cat on its home territory. Early learning has great impact: kittens under seven weeks old bond to new companions better than do older individuals. Forging social bonds is more difficult later in life, and depends in part upon natural characteristics: one study reported that Longhairs adjusted to a new cat in their home more easily than Siamese.

Even a short break in social bonds, such as when one cat is hospitalized for a few days, may affect social structures, with the returning cat being viciously attacked by its fellow cats.

Group dynamics

Social relationships are dynamic, responding to changing conditions. During this two-month study of favoured resting places, Kaspar, the dominant cat, died. This did not alter Maurice and Tom's patterns, but Willow ventured downstairs, and Si, previously confined to the top of the refrigerator, soon became a dominant cat. The kitten Am, at first allowed everywhere, found his territory restricted as he matured. Disputes were settled by eye contact.

Start of study

End of study

SI FEMALE
SIAMESE ADULT

KASPAR MALE
RANDOM-BRED ADULT

AM MALE
SIAMESE KITTEN

MAURICE MALE
RANDOM-BRED ADULT

TOM MALE
RUSSIAN BLUE ADULT

WILLOW FEMALE
INVALID ADULT

In most circumstances, however, social bonds remain strong once established, and anxiety occurs if they are broken. An American study of pet cats living together in indoor households found that when one died, its survivor showed "mourning" behaviour: over half the survivors ate less than before, they became more vocal, and they demanded more affection from people in the household. Over 40 per cent of surviving cats sought out the deceased cat's favourite spot and spent more time there for a period of up to six months after the death of the other cat.

Conversely, although it has not been formally studied, veterinarians are frequently told by cat owners how much "happier" the survivor is now that it has the house to itself. When these cats are medically examined, they appear to be both more robust and more relaxed than they were previously. Author Rudyard Kipling was lucid and eloquently observant in his creation of "the cat who walks by himself" (*see page 36*). While feline social behaviour may sometimes mimic ours, cats often remain social loners through to the depths of their being.

Walking the bounds
Cats often use human territorial markers, such as fences, as their demarcation lines. Social groups remain within the boundaries, making no social contact with outsiders.

DEMARCATION DISPUTE

All the domestic cat's wild relations create and defend their own distinct hunting and breeding territories. When cats from adjacent territories meet at a demarcation point, potential injuries are avoided through displays of intent. During the ritualized stage of a demarcation dispute, cats will arch their backs, bare their teeth, and bristle their fur. They also stare each other down, until one cat eventually turns its head and breaks eye contact. It signals that it will withdraw, but any abrupt movement now provokes a lashing response from the victor. In domestic cat disputes, this is when the vanquished cat is bitten on its tail as it leaves the scene.

Facing off
These jaguars are using body language and voice to intimidate each other, just as domestic cats hiss, spit, and stare each other down, before actually fighting.

HUNTING AND PLAY

HUNTING IS AN INSTINCTIVE ACTIVITY, CENTRAL TO FELINE NATURE. Indeed, the cat is one of the most successful land-based predators in the world. All cats hunt. Rural feral cats refine their techniques to suit the available prey, while urban feral cats scavenge more than they hunt. Pet cats hunt for the thrill of stalk-and-pounce, rather than to fill their stomachs, and if living indoors hunt toys. While most cats prey on land-based mammals, some become specialized at catching birds. This may have a profound effect on the population of small birds in regions where, historically, there has been an absence of land-based predators similar to the cat.

There is a tendency among urban cat owners to deny the cat's place as nature's best-packaged small, land-based predator. Quite simply, cats hunt, and this has little to do with hunger. Biologist Robert Adamec has found that hunger and hunting are controlled by different parts of the brain. He allowed cats to see a rat while eating a favoured meal: they stopped eating to kill the rat, but returned to the meal. Because of this split, hunting is more common than many owners realize. In one German survey of the stomach contents of pet cats killed in road accidents, 40–60 per cent of home-fed cats also had the remains of prey in their stomachs.

When driven by hunger, a cat is more likely to scavenge than to hunt: it is far easier to slash through plastic bags to find a cooked chicken carcass than it is to hunt patiently for a mouse. On a full stomach, however, the most carefully bred, expensive, lovable pet cat will still be a hobby hunter, impelled to stalk and pounce.

PREFERRED PREY

Cats are opportunist hunters, taking what is on offer. Most prey upon land mammals and reptiles, but some, learning from their mothers and their experiences, become adept at catching birds.

There are local variations in cats' prey: in the German study of cats' stomach contents, the rural cats had eaten 14 different species of animals, while the urban cats had eaten only cat food and a single grasshopper. Prey also varies on a larger scale. North American cats eat mice, ground squirrels, flying squirrels, chipmunks, gophers, and robins. European cats take mice, voles, sparrows, and fledgling birds, but shrews only when they are acutely hungry. In southern Sweden, males catch more rabbits than do the smaller females. Kittens around the world, and adults in the tropics, eat spiders and insects. On sub-Antarctic islands, noddies, terns, and penguins are taken. In Australia, where cats were imported in order to control the introduced European rabbit, they will eat possums, reptiles, and ground-nesting birds.

Some cats take unusual or inedible prey in the absence of appropriate prey, or because they lack hunting experience. Cats may hunt frogs, but this is done for the thrill of the chase, and few cats will actually eat their catch. Poisonous toads are occasionally killed by cats, with painful or fatal consequences.

Tense anticipation
As it crouches ready to pounce, this Tonkinese treads its hind legs in excitement or to keep the muscles ready for action.

HUNTING METHODS

Feline senses suggest cats should hunt at dawn and dusk, and in full moonlight (*see page 60*), but this varies. In hot summers, cats hunt through the night, and in cold winters, they hunt at noon. These changes also reflect prey activity. The cat's favourite hunting strategy is to sit and wait. It locates a good hunting ground, such as a small mammal's pathway, and sits by it patiently, staring intently. When prey appears, the cat pounces. It may throw itself on its side and, pinning the prey with its forepaws, rake it with its hind claws. When hunting birds, the cat uses ground cover to stalk, slinking forwards close to the ground, freezing if it thinks it is detected, until it finally leaps, sprints, springs, or pounces.

Well-fed cats are more patient than hungry cats. Ferals are better hunters than pets, and nursing mothers are best. Studies of Swiss farm cats found that mothers took on average just over an hour and a half to capture a rodent, catching something every third or fourth pounce, while non-mothers took almost three times as long and caught every twelfth pounce.

Instinctive pounce
Pouncing is biologically built into the feline nervous system. It is an instinctive behaviour, present from the moment a kitten begins to play. This kitten is in the classic pounce pose, with ears forwards and tail high.

Practised pounce (left)
All cats, young or old, need to find a release for their hunting instinct and their energy. Rural, free-ranging cats satisfy this inclination naturally. Urban or other housebound cats need other outlets, such as small balls, feathers, and other toys, which they will "capture" and "kill" exactly like real prey.

Stealth tactics (right)
A cat crouches low and uses long grass as cover when it stalks prey. On surfaces such as manicured lawns or smooth paving, hunting is seldom productive.

Specialized hunting skills
Few cats can catch birds. This cat, playing with a bird it has killed, may not eat it, as even fewer cats can pluck feathers. Bird kills seem high because they are made in the garden during daylight, and because few birds are eaten, most being left in a visible larder. It is thought that up to six per cent of garden birds are killed by cats each year. The simple way to reduce this is by belling the cat.

The domestic cat's hunting instinct is similar to that of most other members of the cat family. In the wild, cats use similar strategies to stalk, capture, and kill, according to their abilities. Just as the domestic cat often returns to its home territory to eat its kill, many wild cats drag their prey back home, and even up a tree, where it can be consumed in safety. Wild cats consume only what they need to satisfy their hunger, saving excess food to consume later. Like the larders of corpses often built up in gardens by domestic cats, wild cats that prey on large herbivores will keep a larder of meat until it is finished, as a guaranteed food source. But while house-fed domestic cats accumulate an overflowing larder of excess meat, wild cats always consume what they kill unless there is a sudden and overwhelming supply of game.

Spoils of a successful hunt
Hunger satisfied, this leopard keeps the rest of a gazelle to eat over the following days, until it hunts again.

Whether prey is promptly eaten, or even killed, depends on the cat and the context of the hunt. If it is hungry, a cat kills its prey with a swift bite – a cat's canine teeth are perfectly formed to slip between the neck vertebrae of a small rodent, killing it instantly (*see page 68*). Many cats bring live prey back to their home territory, and may even seem to offer it to their owners. Mothers bring back prey for their young to kill, teaching them hunting skills, and these cats may regard their owners as helpless kittens.

Live prey that is carried some distance is disoriented, and easier to recapture or play with. Hunting play is a curious phenomenon: after the hunt, many pet cats literally dance with delight, taking high, curving pantomime leaps. This is most common after a dangerous prey, such as a rat, has been dispatched. Ethologist Paul Leyhausen termed this "overflow play", a cathartic release from the tension of the hunt.

It is a common behaviour in pet cats, perhaps a perpetuation of juvenile behaviour, but is rarely seen in feral cats. Tormenting of prey may also serve some as yet unknown function.

PREVENTION OF HUNTING
Cats can have devastating effects on wildlife in isolated environments. Most dramatically, the single pet cat taken in 1894 to Stephen's Island off the coast of New Zealand killed the entire world population of the Stephen's Island Wren. Domestic cats also aided the extinction of other island birds. In the Galápagos Islands, off the coast of South America, feral cats also consume vast numbers of freshly hatched baby turtles as they scrabble across the sand to the sea.

Wildlife protection agencies and legislators have been concerned about feline predation of wildlife for a long time. The American state of Illinois passed a law in the 1940s restricting cats to homes, but it was vetoed by the governor. In the 1990s many states sought to control feral cats. In Switzerland, cats must return to their homes at night. But the debate over whether cats should be prevented from hunting has been most heated in Australia, which has few natural, land-based predators.

Experimenting with prey (right)
The cat's hunting instinct provokes it into a response to anything that moves, animate or inanimate. This cat is fascinated by a frog, and is about to bat it. If the cat were hungry, it would quickly consume the frog, rather than playing with it in this way.

In 1989, a suburb of Melbourne imposed a dusk-to-dawn curfew on pet cats. In June 2000, the Federal Government published its Threat Abatement Plan, discussing nationwide ways to control the threat posed by cats.

Although we can prevent hunting, we have not removed the instinct. Increasing numbers of cats live permanently indoors. When these cats see prey through a closed window, their teeth may chatter – this is not fully understood, but does not occur in any other circumstances. Owners should play with their indoor cats and provide plenty of toys, or in the absence of live prey they may find their ankles being stalked.

COURTSHIP AND MATING

CATS ARE NOISY, PROMISCUOUS, AND ENTHUSIASTIC ABOUT SEX.
There are sound biological reasons for such torrid behaviour, all of
which derive from the cat's natural ecological niche of lone hunter.
Sociable species, such as humans, dogs, or domesticated livestock,
have ideal mating circumstances: males are constantly available.

Cats, however, evolved as lone hunters with individual territories,
and when a female comes into season it is possible that no males
will be in the vicinity. Because a female must take advantage of
any and every male available, she appears to be overwhelmingly
promiscuous when compared with other mammals.

Feline courtship is usually rather perfunctory,
with a dominant tom invoking his rights over
other males, and a receptive queen eventually
accepting his advances. The female controls the
timing of mating activity, only permitting the
male to mate when she is ready (*see page 70*).
Repeat matings occur over one or more days,
and may involve several toms.

CALLING FOR A MATE
Out of season, an unspayed female will behave
much as neutered males or females, but at the
onset of oestrus, dramatic changes occur.
The first signs of heat are increased
restlessness and a heightened
desire to rub against objects
or other animals. If allowed
out, she will urinate more
frequently and in new
places, and she may also
spray to announce her
condition to local toms
(*see page 81*). At this
stage, however, she will not
permit a tom to mate with her.

Queue jumping
*In uncontrolled or feral matings,
many males vie for the right to mate.
Toms may fight before mating, and
occasionally, as here, one may refuse
to wait patiently, but this is unusual.*

Soon after, she begins to make a plaintive and
distinctive sexual call, a vocal signal to all toms
that she is receptive. The call carries further
than scent, alerting the widest possible area.
Inexperienced owners often mistake it for pain.
   At this stage in her cycle, a queen frequently
rolls over on her back, and, if touched at the
base of her tail, she will crouch with her
hindquarters raised and her tail held to the
side, ready to mate. Often she purrs, treads
with her front paws, and stretches her body.
Even normally withdrawn females may become
lasciviously affectionate
during oestrus.

First overtures
*The female (left) sets the pace
in all sexual encounters. This
male waits cautiously. If he is
too forthright he risks a rebuff
and a vicious attack.*

All this advertising is done because the female
cat is an induced ovulator: her ovaries will not
release eggs until after mating (*see page 70*).
This is sensible given the cat's independent
nature, ensuring that eggs are not wastefully
shed when there are no males in the area.

ANSWERING THE CALL
The male cat is an opportunist. If unneutered,
he patrols his territory regularly, spraying and
responding to any scent markers and calls that
tell him that a female in the area is receptive.
When he scents a receptive female, his spraying
activity increases dramatically. Often, however,
he is not the only male who has been alerted.
   Dominant males will fight for
the right to mate. When
several toms gather, one will
dominate the others, most
often through intimidation
or strength. After this,
there is usually no further
fighting among the toms:
other males wait for a turn.
Releasing eggs after a single
mating would reward the most
dominant male, but he may not
be the best father. Queens only
release eggs after repeated matings,
so any male who mates with the
female has a chance to father a kitten.

Rejected advances
*Inexperienced males can be
impetuous. This male has
made an approach to the
female, but is rejected.
She will mate with the
male of her choice when
she is ready.*

MATING PROCEDURE

Courtship is usually perfunctory. The female allows the male to sniff her, which he does repeatedly, drawing scent into his vomeronasal organ (see page 60); any attempts to do more may be viciously rebuffed. British ethologist Roger Tabor has observed low-ranking males assiduously courting queens in early season, who then often accept the low-ranking male as the first mate, rather than the dominant male.

Mating itself is brief and tense. When she is ready, the queen assumes a crouching position with her hindquarters raised (called lordosis) and permits the male to proceed with mating. As the male mounts, he grasps the scruff of the female's neck between his jaws. This stimulates the dependent "scruff response" seen in kittens (see page 88), and subdues the queen.

The tom ejaculates almost at once. As he withdraws, his penis abrades the queen's vagina (see page 70). She cries out, and may even turn on the tom, hissing, spitting, and trying to bite. Experienced toms, who are familiar with this behaviour, will relax and groom themselves nearby until the female is receptive once more.

Swedish researcher Olaf Liberg has observed that cats will mate between 10 and 20 times each day, and that mating sessions can continue for as long as four to six days.

The first male often becomes exhausted, and may be replaced by other toms waiting in the queue. These multiple matings stimulate the cascade of hormones that induces egg release. For many cats, mating is the only time that they make physical contact with others: after mating, the male rarely has any parental role.

SURVIVAL VALUES

Feline mating rewards those males who subdue other toms, assiduously court the queen, or patiently wait for their turn. Each has excellent qualities: strength, intelligence, or persistence. Multiple matings also help to ensure a varied genetic mix. In lions, a new male claiming the breeding position in a pride often kills the cubs of the last dominant male. The frequency of infanticide in our domestic cats is unknown, but zoologist David Macdonald has observed a newly arrived male cat enter the communal nest of three queens in their absence and kill six kittens. A male believing he had kittens in the litters would have a strong disincentive for such behaviour. Maternal infanticide in cats is extremely rare, and mostly due to inexperience or stress from real or imagined threats.

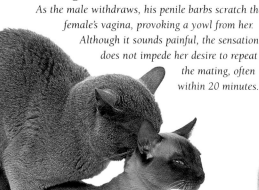

Stimulating ovulation
As the male withdraws, his penile barbs scratch the female's vagina, provoking a yowl from her. Although it sounds painful, the sensation does not impede her desire to repeat the mating, often within 20 minutes.

Post-coital grooming
After each mating, both the individuals groom themselves. Grooming has a calming effect. It begins with cleansing the genitals but then extends to the whole body before mating resumes. Serial matings will often continue all day.

Perseverence brings success
The female accepts the male and assumes a crouch. To secure his grip, subdue her, and prevent her from biting him, he grasps the skin of her neck in his jaws. He will not let go until he has completed mating – almost instantaneously – and withdrawn.

SURVIVAL IN THE WILD

Mating in wild cats is very similar to that of domestic cats. A significant difference is that wild cats seldom have the luxury of a surplus of readily available mates. The female gathers males from as wide a territory as possible, by leaving increasingly frequent scent markers and vocally proclaiming her impending heat. This is a sensible solution to the problem of being a solo hunter in a large hunting range. All of the cat family are induced ovulators. To find the best genes for her young, a female must hold on to her eggs as long as possible, then mate with the male she considers the most suitable. The onset of oestrus in wild cats varies according to location: increasing daylight hours are a major stimulus, leading to summer kittens. Once begun, oestrus cycles succeed each other until a series of successful matings produces a pregnancy.

Call of the wild
In the vastness of her range, this female bobcat seeks out a high rock: her vocal call to inform all potential mates that she is in heat can best be heard from here.

MOTHERS AND KITTENS

THE PREGNANT QUEEN BECOMES WITHDRAWN AND SEEMINGLY SERENE. As the two-month pregnancy progresses, she seeks out a safe place in which to give birth, and becomes restless as birth approaches. Mothers move their kittens frequently during their first days and, in communities, will care for each other's kittens.

A young kitten is extraordinarily precocious in its development. By three weeks its senses have matured and by seven weeks it has developed its full repertoire of behaviour. A kitten develops social bonds to other cats, to us, to dogs, or to almost any companions, if it meets them frequently between these ages.

There are few visible signs of early pregnancy. At around three weeks of gestation the nipples become firmer and "pink up". The belly grows more rotund, and progesterone (*see page 71*) induces a serene disposition.

Pregnancy takes anything from 57 to 70 days, but a few days before the birth the female grows restless and seeks a nest. She rearranges her bedding and paces until the contractions begin. Litter sizes vary, but three to six is average, and labour can take an hour or a whole day. Kittens are delivered enclosed in fluid-filled sacs. The mother licks these away, stimulating the kitten's first breath, and chews through the umbilical cord. More contractions expel the placenta, which the mother eats: she consumes all birth wastes to prevent predators from scenting the litter. At birth, production of the hormone prolactin stimulates milk production and release.

GOOD MOTHERING

Mothering ability depends in part upon genes, but also upon emotional maturity and the early experiences a cat had with her mother. Fathers as a rule play no part in kitten-care, although some breeders report "mothering" toms.

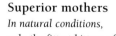

Instinctive care
This Tonkinese queen instinctively curls around her young, offering comfort and protection as well as nourishment. Her abilities depend on her maturity, her own upbringing, and her innate qualities.

On the other hand, Italian scientist Dr. Eugenia Natoli has observed that all females within a cat community may act as midwives, licking the newborns and severing umbilical cords.

There is no more fearsome aggression than that of a mother cat. Dropping the usual feline subtleties, she defends her kittens ferociously. Her other behaviour is as noble. In the first few weeks of life a kitten depends upon its mother to stimulate all body functions: her licking prompts the release of bladder and bowels.

Just as she tidied her nest at birth, she also consumes all of each kitten's body waste products, to hide her litter's presence from predators. Although she keeps her birthing den immaculately tidy in this way, she may move her kittens shortly after birth and then again when they are about four days of age. Her first instinct is to keep her litter together and safe, picking up each kitten by the scruff of its neck and carrying it to a new, safer nest. Feral cats carrying their kittens over long distances outside may move them in stages to avoid leaving any for too long.

Many behaviourists believe that mothers frequently move their newborns in this way to protect them from predators, yet some studies show that new mothers will quickly move their kittens into the just-vacated dens.

Personal care
As a lone hunter, a cat cannot depend on others for grooming. This kitten began self-grooming at three weeks.

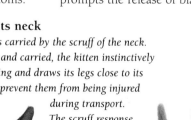

By the scruff of its neck
Kittens are always carried by the scruff of the neck. When "scruffed" and carried, the kitten instinctively stops wriggling and draws its legs close to its body to prevent them from being injured during transport. The scruff response remains throughout the cat's adult life.

Superior mothers
In natural conditions, only the fittest kittens of the best mothers survive, going on to become excellent mothers themselves later.

Displaced hunting behaviour

The hunting instinct is buried deeply in the mind of even the youngest kitten. Rather than just eat its food, this kitten prods it, as it would an animal brought back from the hunt, to check whether it is safe to approach. During their early learning period, kittens may pounce on food to "capture" or "kill" it before sniffing or eating it. Adult cats will rarely play with their food in this way.

Instinctive cleanliness

The cat is a naturally hygienic animal. A kitten's instinct to use a specific site for elimination is profound and fully developed by three weeks of age, when it starts to explore. These kittens, provided with a litter tray, investigate it and, because the texture is more natural underfoot than the hard surface of a household floor, they are inclined to use it as their toilet. Dry food and litter are sometimes confused, however.

EARLY LEARNERS

At birth, a kitten is totally helpless, unable even to regulate its own temperature. But within four days it can find its mother and crawl to her from over 0.5 m (2 ft) away. Ten days later, its co-ordination is sufficiently developed for it to use its front legs. By three weeks of age, it is tentatively standing, scenting, hearing, seeing, and responding. By seven weeks of age, it is running and leaping and somehow seems to have developed near-perfect balance.

Early actions are instinctive, but learning begins soon after birth. At one week, a kitten knows the smell of its nest and returns to it if separated. At only 18 days, it shows some signs of independence, leaving its mother to explore. It can now control urination and defecation, and it will use a litter tray if one is offered. While it still depends upon its mother for most of its grooming, it is capable of fully grooming itself by five weeks. Relationships with its mother and litter mates remain close but even now the seeds of separation are being sown.

Numerous studies have shown that kittens learn from their mothers how to negotiate obstacle courses and open door latches: without an example they cannot perform the task, or do so only by chance.

SUCKLING AND WEANING

A kitten uses heat receptors in its nose to find a teat; it will return to this teat throughout its suckling period. Milk release is stimulated by "treading" the breast with the forepaws, a behaviour some cats continue into adulthood.

Cat milk is highly concentrated in fat and protein. The kittens must compete for the most productive teats; a kitten already knows which teat is best at two days. Kittens that latch on to the most productive teats grow fastest, unless displaced by other, more dominant, litter mates. Suckling is needed for nourishment for five to six weeks, but continues beyond this age for emotional benefits. In a cat community, kittens may be suckled by half-sisters or aunts.

For the first few weeks the mother induces her kittens to suckle, but then they begin to pester her for meals. At three weeks a kitten starts to eat solid food, and by five weeks it has a full set of pin-sharp teeth (*see page 68*) and is ready to eat solid food. Those sharp teeth play a role in weaning: continued suckling becomes annoying. When kittens try to suckle, a mother may hiss and spit at them, even bat them away.

Feral mothers bring prey back to the nest and tear it up for their young. As they mature, the kittens have to do more of the tearing up for themselves. Soon live prey is brought, an invigorating learning experience. As a mother's kittens stop suckling her and move onto meat, the mother's milk dries up, and her hormonally induced urge to mother diminishes.

FULL INDEPENDENCE

In a natural situation, large litters disband by the time the kittens are six months old, while smaller litters remain together longer; their suckling seems to be less annoying, and they are allowed to continue to suckle for longer.

At the same time, play between litter mates becomes increasingly rough: between males and females it is even vicious. By six months, the mother and her daughters are seemingly so exasperated with the rough play of the male kittens that these are permanently expelled from the nest. The females form the blood-related colony that is the basic feline social group (*see page 82*). The outcast males join a "brotherhood", willing to spend time together sunning themselves, but each with its own independent territory, and each waiting its chance to father another generation.

PROLONGED DEVELOPMENT

Kittens of big cats are proportionally small at birth: although adult male Siberian tigers can weigh 350 kg (770 lb), their cubs weigh about 1 kg (2 lb) at birth. They take longer to mature: born as helpless as domestic kittens, tiger cubs remain in the den until about eight weeks of age, when they emerge and start following their mother around. Unlike domestic kittens, which become independent early, tiger cubs depend on their mothers for food for a full 18 months. Even when mature enough to hunt on its own, a young tiger continues to hunt on its mother's territory for another year before moving on to set up its own home range of 20–100 sq km (8–39 sq miles). As in the domestic cat, males play no parental role.

Staying in the nest

Although it has considerable strength and size, this tiger cub will remain dependent on its mother for safety, security, and food for at least another year.

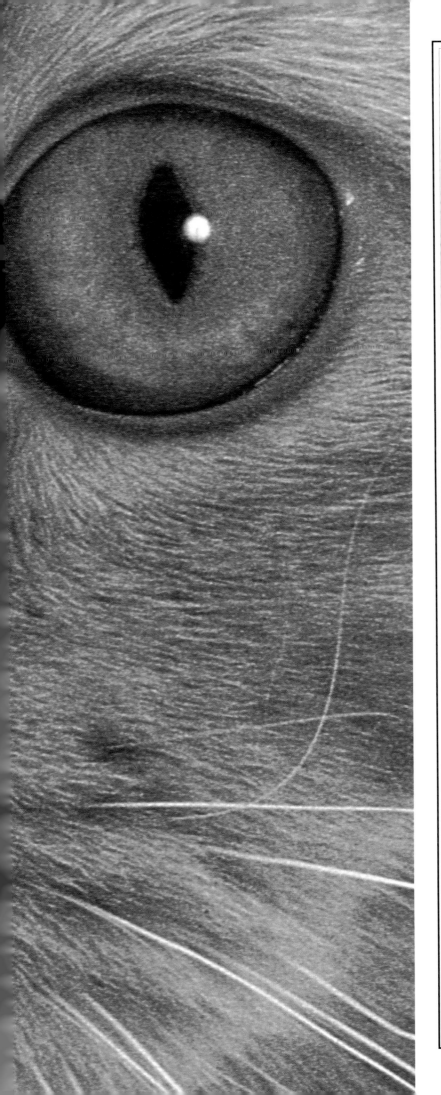

CHAPTER FIVE

DOMESTIC CAT BREEDS

UNTIL RECENTLY THERE WERE VERY FEW breeds of cat. Those that did exist were the result of environmental pressures; heavier coats in cold climates and sparse coats in hot climates, larger size where cats preyed on large game such as hares, smaller size where prey consisted of small rodents and birds. In the last hundred years, cat fanciers have created many new breeds. Some were developed by selectively breeding cats that evolved to survive in a specific environment, others by breeding for coat colour and density. Recently, breeders have perpetuated mutations such as curly coats or short legs to produce new breeds with unusual features.

ABOVE FEMALE CHOCOLATE TORTIE TONKINESE

LEFT CATS' EYES CAN BE ROUND, SLANTED, OR ALMOND-SHAPED

THE HISTORY OF SELECTIVE BREEDING

CATS HAVE BEEN KEPT IN OUR HOMES FOR MILLENNIA, BUT ONLY IN the last century or so has selective breeding been actively pursued. The first cat shows, at the end of the last century, stimulated the creation of breed clubs throughout Europe and North America. The first breeds were those that had developed naturally, but soon breeders were using their new knowledge of inheritance patterns to create a palette of coat colours and patterns in both short and long coats. This recent human intervention in feline evolution has produced some remarkable results.

Possibly the first cat show was held in the time of Shakespeare, at St. Giles' Fair in Winchester, England, in 1598. Shakespeare called the cat a "harmless, necessary creature", and this show was probably about mousing abilities as much as personality and looks. On the other side of the world, in Thailand, the *Cat Book Poems*, possibly written at the same time, recorded cats of varying colours and types. Readings of the original old Thai script imply that this is not quite the world's oldest breed register: while it identifies different types, there is no suggestion that the Thais actively bred for these types.

THE START OF BREED REGISTRIES

The first formal cat show was held in 1871 at the Crystal Palace in Hyde Park, London. It was organized by Harrison Weir, who wrote the standards for all the breeds shown and acted as one of the three judges; a Persian kitten stole the show. Most exhibitors were British gentry, so as well as breed classes, there was a special class for "Cats Belonging to Working Men".

The first cat show to attract wide attention in North America was organized by James T. Hyde in 1895 at Madison Square Garden, New York; a Maine Coon was champion. These shows, and others in mainland Europe, proved an enormous success and were the main impetus for modern selective breeding, stimulating a vogue in "pure-bred" cats.

Founding father
Harrison Weir is often described as "the father of the modern cat fancy". He raised the cat to a new status.

Lucky cats
All the cats described in the Cat Book Poems *bring different kinds of luck to their owners; they seem to have been regarded as simple varieties, not special breeds.*

Associations formed to create rules for shows: in 1887 the National Cat Club was started in Britain, with Harrison Weir as its president, and in 1896 the American Cat Club became North America's first registry to verify pedigrees. Within 11 years of Harrison Weir forming the National Cat Club, a challenger, the Cat Club, was founded. Feline politics entered the world of cat breeding and remain there today.

The role of breed registries in governing breeds is fraught with inconsistencies. It is very hard to say, for example, what exactly a breed is. Not all breeds are recognized by all registries; a cat might be recognized as one breed in one registry but another breed in a different registry, and several breeds do not even breed true to their standards. All breeders agree on the importance of their cats' health and welfare, but the finer points of breed etiquette remain a matter of constant (and lively) debate.

MODERN REGISTRIES

Cat breeding is more scientific than dog breeding in that it is often based on genetic principles. Experimentalists pursue as many new developments and traits as possible. Purists want to keep breeds as distinct, and often as limited, as possible. Each registry has a lively range of opinions between the two extremes.

Ancient or modern

The Birman has a pattern that might naturally evolve in an isolated area, but which coincides neatly with other breed patterns, and some say it is a man-made breed.

The world's largest registry of pedigree cats is the Cat Fanciers' Association (CFA), founded in 1906, which has clubs in the United States and Canada, South America, Europe, and Japan. CFA's registering philosophy is purist, allowing, for example, only four Burmese colours and four Siamese colours. Perhaps the most liberal or experimental of registries is The International Cat Association (TICA), founded in 1979 and based in North America. This registry accepts new breeds more rapidly than any other major association, even if only provisionally, and by doing so it encourages experimentation. Britain's Governing Council of the Cat Fancy (GCCF) was formed in 1910. Its policies are staid, but less so than CFA's, and its influence is worldwide: registering bodies in countries such as South Africa, New Zealand, and Australia base their decisions on GCCF rules.

Island breed
Bobbed or kinked tails are found throughout Southeast Asia, but the inevitably limited gene pool of an island was needed for the trait to gain the prevalence that gave us the Japanese Bobtail.

Character cats
There is a new tendency to breed for temperament: the famously placid Ragdoll is the most obvious example of this, but it is not alone.

Most European countries have several registries, but at least one from each country belongs to the Fédération Internationale Féline (FIFé), established in 1949. FIFé claims to be the largest feline organization in the world, and has affiliated clubs outside Europe. In policy, it is similar to Britain's GCCF: neither organization will recognize breeds like the Munchkin (*see page 213*) or the Scottish Fold (*see page 214*) and frown upon breeding for type associated with known defects, such as blue-eyed white cats, which have a high incidence of associated deafness.

OLD AND NEW BREEDS
Cat breeds can be broken down into two groups, which emerged chronologically. The first are the breeds that appeared naturally in free-breeding, although possibly isolated, feline populations. Many of these are characterized by coat colour or pattern, and genetically these are almost always "recessives" that breed true (*see page 75*). For example, the Abyssinian's ticked pattern is a recessive trait, emphasized through selective breeding. Some other breeds are characterized by distinctive mutations: the Japanese Bobtail (*see page 237*) and tailless Manx (*see page 116*) are examples. Some breeds developed naturally into types that were then formally recognized as breeds; the British, American, and European Shorthairs (*see pages 112, 120, and 126*), Norwegian and Siberian Forest Cats (*see pages 208 and 210*), and Maine Coons (*see page 202*) are in this category. A final major feature defining these early breeds is coat length. The longhairs, going under the various names of Persian, Angora, French, or Chinese cats, predate registries, but type and standards were very loose in those times.

Future directions
Africa remains an untapped source of types and genes. The type, blue eyes, and mitted bi-colour pattern of this Cameroon cat are prevalent among the street cats in its town.

More recently, breeds have been developed actively and sometimes very scientifically. Oriental Shorthairs (*see page 160*), Tonkinese (*see page 152*), Ocicats (*see page 174*), Angoras (*see page 232*), and Asians (*see page 142*) in Britain, and many others, have been created from scratch. This is the "growth" area in the cat world, with more new breeds appearing this century than in the entire history of domestic cats preceding it. Some new breeds, like the Somali (*see page 224*) or Balinese (*see page 228*), are simply longhaired versions of older shorthaired breeds. Others are coat mutations. In the wild, wavy, rexed coats die out. Through careful nurturing, this genetic "fault" has been perpetuated to create a range of breeds. Other mutations die out or are limited because they are potentially dangerous: Manx and Fold genes are dangerous when homozygous, and breeders take utmost care in their breeding programmes; free-breeding cats never exercise such caution. Extreme examples of breeding away from natural type are the hairless Sphynx (*see page 170*) and the dwarfed Munchkin. These breeds lead to the question of whether the very concept of breed standards inevitably leads to deviations from healthy norms towards potentially unhealthy extremes. Most breeds have closed registries, with no outcrossing.

Breeders argue that if they select for outward good health they can produce healthy cats even in a small gene pool. This is not always true: genetic health problems and lowered disease resistance take time to show.

WIDER EFFECTS
Selective breeding has only a peripheral effect on the species as a whole. Pedigree cats make up less than 10 per cent of the domestic cat population, even where they are most popular; in most countries, they account for less than two per cent. However, the existence of breed registries is of value for all cats, raising the profile of cats as desirable and worthwhile animals. Indirectly, these associations improve the welfare of all cats, not just pedigree breeds.

Questionable status
In most associations, the pointed kittens born in this Oriental Shorthair litter would be registered as Siamese. Some, such as the CFA, refuse to grant these kittens status as anything but "any other variety" Orientals.

COAT COLOURS

THE CAT'S ORIGINAL COAT WAS OF COLOUR-BANDED AGOUTI HAIR, designed for camouflage in the natural environment. The first mutation away from hair with bands of colour to a single colour, or non-agouti hair, was probably to black; this mutation is often seen in other cats, such as the "black leopard" or "black panther".

Other mutations occurred for red, white, and dilution of the solid colours, and all were perpetuated because the domestic cat no longer needed its wild, hunting camouflage to survive. These few, simple genetic variations created the framework for the seemingly endless variety of coat colours that exist in cats today.

All coloured hairs contain varying amounts of the two components of melanin, eumelanin and phaeomelanin. Eumelanin produces black and brown, while phaeomelanin produces red and yellow. All colours are based on the absence or presence of these pigment granules in the shafts of each hair. Pigment is made in skin cells called melanocytes, and the distribution of these cells is genetically determined. Most knowledge about coat colours is based upon observation and informed guess work rather than microscopic investigation of genes, so there are still many questions to be answered.

Cats with single-coloured "non-agouti" hair are called self or solid. Self coats are recessive: the cat must carry two copies of the non-agouti gene (see page 75) in order to conceal its "true" original tabby pattern (see page 98).

Under wraps
Black is the most dominant eumelanistic colour. It is masked by white or red, but itself masks the genetic potential for other coat colours.

DENSITY OF COLOUR
Some cats have vibrant, full-strength self coats. These come in the colours black, chocolate, cinnamon, and sex-linked red. Cats with these coats have at least one copy of the "dense" gene (D), which is dominant and ensures that each hair is packed tightly with numerous pigment globules to give the richest colour. Other cats have lighter, "dilute" coats in blue, lilac, fawn, and sex-linked cream. These cats have two copies of the dilute gene (d), which is recessive and results in fewer, more scattered globules of pigment in each hair: the effect is to create a paler shade of any of the dense colours.

Some breeders think that there is also a "dilute modifier" gene, provisionally called D^m, which is dominant over the dilute gene d, but located at a different site on the chromosome, and so able to "interact" with d. If a cat carries both the dd dilute trait and the D^m modifier gene it will have a "modified" colour: blue

Natural distinction
Dilute self colours always breed true, and so have become a defining characteristic of several naturally developed breeds, including the Russian Blue.

and lilac modify to different shades of caramel, and cream modifies to apricot. Other cat breeders dispute this, and claim that these colours are just bad blues, lilacs, or creams.

SEX-LINKED RED
There is firm evidence that the gene for red or orange colour in cats is located at a specific site on the sex-determining X chromosome. In its dominant form (O), it makes the cat red; in its recessive form (o), it lets whatever other colour the cat is carrying show through. A male cat, with an XY chromosome combination, can only ever have one copy of the gene, so if he carries one O he is red and if he has one o he will be any other colour. The female cat, because she has an XX combination, can carry two copies. She will be red if she carries two copies of O, or another colour if she carries two copies of o,

Red Burmese
Although sex-linked red existed in the East, the first Burmese in the West were brown. Red was re-created, and it is still not widely recognized.

CAT COLOURS

DENSE	DILUTE	DILUTE MODIFIER
Black $B- D-$	Blue $B- dd$	Caramel $B- d^m d^m$
Chocolate $bb\ D-$	Lilac or Lavender $bb\ dd$	Caramel $bb\ d^m d^m$
Cinnamon $b^l b^l\ D-$	Fawn $b^l b^l\ dd$	Undefined brown $b^l b^l\ d^m d^m$
Red $D-\ O/O(O)$	Cream $dd\ O-/O(O)$	Apricot $d^m d^m\ O-/O(O)$
Chocolate tortie $bb\ D-\ Oo$	Lilac tortie or Lilac-Cream $bb\ dd\ Oo$	Caramel tortie $bb\ d^m d^m\ Oo$
Cinnamon tortie $b^l b^l\ D-\ Oo$	Fawn tortie $b^l b^l\ dd\ Oo$	Undefined tortie $b^l b^l\ d^m d^m\ Oo$
Tortoiseshell $B-\ D-\ Oo$	Blue tortie or Blue-Cream $B-\ dd\ Oo$	Caramel tortie $B-\ d^m d^m\ Oo$

but unlike the male cat, a female can also be heterozygous (*Oo*). This combination makes her tortoiseshell, a mosaic pattern of both red and black, with some melanocytes obeying the *O* gene and others obeying the *o* gene. This mosaicing combination interacts with all of the other colour-controlling genes, producing torties in all the solid and dilute colours.

The distribution of mosaic colours cannot be controlled through selective breeding, so whether your tortie fulfills British standards (softly mottled) or the American preferences (distinctly patched) is purely a matter of luck. Tortoiseshell mosaicing can be so slight as to be virtually non-existent. At least one tortie was classified as black at birth, won prizes as a black, then produced red and tortie kittens and had to be re-registered as a (very poor) tortoiseshell.

EASTERN AND WESTERN COLOURS

The traditional Western cat-coat colours are black and its dilute blue, and red and its dilute cream, together with their bi-colour versions and solid white. Western breeds, such as British, American, and European Shorthairs (*see pages 112, 120, and 126*), Maine Coons (*see page 202*), and Norwegian Forest Cats (*see page 208*), began in these colours only, and some still exist in these colours only. Some breeds have even more exclusive colours, such as the Turkish Van (*see page 216*), which appears in red and cream bi-colours only. (However, other colours are now being bred, and have been accepted by some registries, including FIFé.)

The traditional Eastern colours are chocolate and its dilute lilac, and cinnamon and its dilute fawn, although many historical works of art show that Western colours also co-existed in Oriental feline populations. Cat-coat colours have now been "transposed" from one group of breeds to another. In Britain, British Shorthairs are accepted in Eastern colours and, similarly, Burmese (*see page 146*) are now often bred in "Western" reds and creams. Conservative cat associations, most notably CFA, do not accept these transposed colours in these breeds.

WHITES AND BI-COLOURS

White is dominant over all other colour genes, whether as all-over white (*W*), or as the white spotting gene (*S*) that gives us the bi-colours. White hair, unlike all other hairs, contains no colour-producing pigment at all.

Behind its snow-white exterior, the white cat is genetically coloured, and it passes on this colour potential to its offspring. White cats carry the dominant *W* gene, which masks the expression of all other colour genes. Often, a hint of a cat's underlying colour breaks through

in a "kitten cap" in the hair on the head of newborn kittens. As the kitten grows, the cap disappears to leave pure white. Deafness is sometimes associated with the *W* and *S* genes, although it is more common in white cats with blue eyes than in those with yellow or orange eyes. These white cats are different from albino white cats that have no pigment in their pink eyes: albino white is extremely rare.

Bi-coloured cats are white-coated with patches of colour – tortie-and-whites are variously classified as bi-colour or tri-colour – and come in two types. The standard bi-colour is defined as being one-third to half white, with the white principally concentrated on the underparts and legs. The Van pattern, originally associated solely with the Turkish Van, but now also seen in other longhair and shorthair cats, consists of predominant white with solid or tortoiseshell patches restricted to the head and tail. One theory is that these cats carry two copies of the white spotting gene *S*, giving them a superabundance of white.

STANDARDS FOR COLOURS

Although there are only a few genes responsible for solid colours, breed associations complicate matters by giving the same genetic colour different names, depending on the cat it appears in. This tendency is most prevalent in the patterned coats (*see page 98*) but also happens in the self colours. Lilac is called lavender in some North American associations, black Oriental Shorthairs (*see page 160*) are called Ebony, genetically chocolate Oriental Shorthairs are called Havana in Britain and Chestnut in North America, and chocolate Havana Brown (*see page 133*)

Maturing colour
Many self cats show vestigial tabby markings, known as "ghost markings" in kittenhood, especially reds, such as this young Angora. These marks often fade as the cat matures, although they may have a tendency to persist in dilute coat colours.

Van pattern
Found naturally in cat populations in the Mediterranean areas of Europe, this pattern is now widespread in carefully bred pedigree cats. First identified in cats from the Lake Van region of Turkey, it takes its name from the Turkish Van breed.

looks closer to cinnamon and is called Chestnut. Reds are often specified to be Red Selfs, perhaps because the distinction between red selfs and red tabbies is a subtle one (*see page 98*), and in the Turkish Van red-and-white is called Auburn-and-White. Tortie-and-white cats are called Calicos in some North American associations, notably CFA, because someone, at some time, perceived a resemblance to brightly printed calico cloth.

According to breed standards, the colour of the nose, lips, and paw pads (the leather) should be in harmony with the coat colour: pink in white cats, black in black cats, blue in blue cats, pink to brick-red in reds. This can vary: in some cases, the colour of the leather depends on the particular breed, or even the association.

Mosaic colour
The spotting gene S seems to have a predictable effect on tortoiseshell mosaicing. Solid tortoiseshells may have subtly mixed colours, but tortie-and-whites almost always show large, crisply distinct patches of black and red.

COAT PATTERNS

BENEATH THEIR EXTRAVAGANT VARIETY OF SHIMMERING SHADES and patterns, all cats remain tabbies in disguise. Just as the most cossetted of pet cats retains the abilities of its predatory ancestors, the hidden tabby pattern is a reminder of the cat's roots under the sophisticated exterior, an origin to which it can return at any time.

Through selective breeding, spotted, tipped, and pointed patterns are nurtured, or even created, by breeders. These are made possible by mutations in the genetics of feline coat patterns, mutations that would have reduced natural camouflage in the wild, but were no longer dangerous once the cat chose to live with us.

The domestic cat's ancestor, the African wildcat, is a striped tabby, camouflaged for hiding and hunting in the wild. The original, genetically dominant (*see page 74*) tabby pattern is inherited by all domestic cats. The hair between the tabby stripes or spots contains bands of colour, almost always light at the base and dark at the tip, which act as a camouflage. This pattern is found in other animals, including squirrels, mice, and the agouti, the rodent after which the pattern is named. The bands of colour give a "salt-and-pepper" appearance, which combines with tabby stripes to help a cat blend into its environment.

DOMINANT PATTERN

All cats inherit some form of the tabby gene, even those with solid, or "self" coats (*see page 96*). Geneticists call the dominant agouti gene *A*. Any cat that inherits *A* from at least one parent will have a patterned coat, and is notated as *A–*.

Two tabbies

The mackerel (below left) *tabby pattern is genetically dominant over the classic* (below right)*, but classic tabbies predominate in European, North American, and Australian cats.*

Cat in disguise
All selfs are tabbies in disguise: if a self cat is bred to a tabby, at least some of the kittens will be tabbies.

Solid colours exist because there is a genetically recessive alternative to agouti, called non-agouti or *a*. In cats that inherit this from both parents, which are notated *aa*, the coat appears to be a single, even colour, but careful examination may reveal disguised tabby markings. This "ghosting" is most apparent in young kittens, often disappearing with age.

There are four basic types of tabby markings in felines: mackerel or striped; classic or blotched; ticked or Abyssinian, and spotted. Although these four patterns look distinctively different they are in fact all mutated variations of the same naturally dominant tabby gene.

New spots
The patterns of new breeds, such as the Ocicat, have often been created to emulate those of wild feline species. The genetics behind these newer spotted patterns are undefined.

Pattern on pattern
Tabby markings over tortoiseshell mosaicing creates possibly the most complex of all coat patterns.

Mackerel tabby stripes are narrow, parallel, and run from the spine down the flanks to the belly. This pattern was predominant in Europe until a few centuries ago. It was superseded by the classic tabby pattern. Classic tabbies have wide stripes that form "oyster" swirls on the flanks, centred on a blotch. The distribution of this pattern in the North American and Australian cats shows that it was a popular, if accidental, export from 18th- and 19th-century Britain. Ticked tabbies are more subtle: clear markings are restricted to the head, legs, and tail, and the body is softly flecked. The African wildcat often carries overall ticking. Ticked coats appear to have spread eastwards into Asia, rather than northwards into Europe. Naturally ticked cats, often with striped legs, are found in Sri Lanka and Malaysia, down to Singapore. The Abyssinian (*see page 134*) was the first breed to sport this pattern.

Spotted tabbies have spotted bodies, often combined with striped legs and tails. Spotted patterns are formed when tabby stripes are broken up. The spots of many older European and American breeds follow mackerel tabby lines, but there are other patterns. Spots of the Ocicat (*see page 174*) fall in a blotched tabby configuration, while those of the Egyptian Mau (*see page 172*) appear random. The Spotted Mist (*see page 136*) has perhaps the most complicated combination of the newer breeds, including seemingly random spots interspersed by ticking. All of these characteristics are polygenetic: they rely upon the influence of a number of genes.

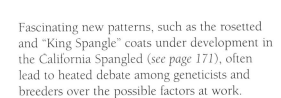

Strange smoke
*The inhibitor gene usually produces a white
undercoat, but extraordinary effects may be seen
in long coats, such as this Longhair (see page 188).*

Shaded Longhair
*The difference between a silver tabby and a silver-
shaded self is purely one of polygenetic factors, as
both these coats are genetically agouti.
The genetics behind Shaded
Goldens such as this young
Longhair (see page
192) have not yet
been defined.*

Fascinating new patterns, such as the rosetted and "King Spangle" coats under development in the California Spangled (*see page 171*), often lead to heated debate among geneticists and breeders over the possible factors at work.

Subtle shading

Hair colour is produced by skin cells that feed pigment into the hair. The inhibitor gene, or *I*, allows pigment to fill only the first part of the hair to grow. This produces a variety of subtle patterns that appear to change as a cat moves. In self cats, it creates a "smoke", with a white undercoat. In agouti coats, the colour is more restricted. Different degrees of shading give us shaded and silver colours and silver tabbies, which have appreciable colours, and tipped coats, which are merely "frosted" with colour. The names for these colours are inconsistent. Tipped coats are often, but not always, called "Chinchillas", while shaded or tipped reds are usually called shaded or tipped "Cameos". In some breeds, there is a distinction between shaded cats, with warm-toned undercoats, and silver shadeds, with clear silver undercoats.

Colourpoint patterns

The *I* gene is not the only gene to restrict colour. Restriction of colour to the extremities is called pointing. Pointed cats are light on their bodies and darker on their "points", namely their ears, feet, tail, and nose. In male cats, hair is also darker on the scrotum.

A heat-sensitive enzyme in the melanocytes, the pigmenting cells in the cat's skin, controls this pattern. Normal body temperature inhibits pigment production over most of the cat's body, but the enzyme is activated and hair pigmented at the points where skin temperature is lower. Pointing can occur in any colour or pattern. Because it is temperature-sensitive, kittens are born white, cats in cool countries have darker coats than those in warm parts of the world, and all cats' bodies darken markedly with age.

The clearest form is the Siamese (*see page 154*) pattern: an almost-white body and dark points, called "pointed" in this chapter. The Burmese (*see page 146*) pattern, usually called "sepia", shows so little difference between body and points that some do not regard it as pointed. Tonkinese (*see page 152*) or "mink" pointing is a hybrid of Siamese and Burmese pointing. Neither gene dominates, and Tonkinese can produce mink, sepia, and pointed kittens.

Pointed patterns
*Pointing genes interact with all other colour and
pattern genes. The Birman's mittens (see page 198)
have a charming myth attached to them, but the
story may simply be a breeder's inspired creation.*

True colours

The variety of feline coat colours and patterns may be more than cosmetic. Research in other animals has suggested a relationship between "tamability" and non-agouti coats. Population geneticist Dr. Neil Todd studied the distribution of feline coat colours and found ever-increasing percentages of non-agouti coats along old trade routes out of Africa and up the rivers of Europe. In Britain, the end of the routes, he found the highest percentages of non-agouti cats. Todd holds that, along these routes, human selection favoured the tamest cats. Long before the registries began, owners may already have been picking their cats by colour.

COAT PATTERNS

SELF (*aa*, non-agouti)	**SMOKE** (*aa I–*, shaded non-agouti)	**SIAMESE** (*cˢcˢ*, pointed)
TABBY (*A–*, agouti)	**SILVER TABBY, SHADED, TIPPED** (*A– I–*, shaded agouti)	**BURMESE** (*cᵇcᵇ*, sepia)
		TONKINESE (*cᵇcˢ*, mink)

All cats carry a tabby pattern, either striped (*T–*), ticked (*Tᵃ–*), blotched (*tᵇtᵇ*), or some other, as yet undefined pattern. The *aa* allele masks tabby markings: melanic pigment (*see page 96*) fills the entire hair and cats appear self or tortie. The *A–* allele shows the tabby markings. Whether a cat is *aa* or *A–* has no effect on sex-linked red colours. The difference between selfs and tabbies in these colours is one of subtle polygenetic effects that determine whether the markings are faint or strong.

The inhibitor gene *I* blocks colour production. In non-agouti cats, only the roots are white, while in agouti cats, more of the hair shaft is affected. The difference between shaded selfs and silver tabbies is polygenetic, depending on how strong the tabby markings are. Polygenetic effects probably also differentiate shaded from tipped coats, although some claim there is a "wide-band" inhibitor gene. In the sex-linked red colours, the differences between the smoke, shaded, silver tabby, and tipped cats are all polygenetic.

Technically, all of these patterns are pointed. All are heat-sensitive, with virtually all the colour (in the Siamese) or the very deepest colour (in the Burmese) concentrated in the cooler extremities, or points of the body. Colour is also slightly degraded or lightened; for example, black becomes Seal in the Siamese and Sable in the Burmese. The Tonkinese does not have its own gene: it is a softly pointed hybrid of pointed and sepia genes. It will always produce variants in these patterns, as well as its own mink pattern.

FACE SHAPE AND BODY FORM

MOST CAT BREEDS ARE NOT DEFINED BY COAT COLOUR OR PATTERN: many breeds share these attributes with each other. Breeds are more often differentiated by the characteristic shapes of their bodies and faces, and sometimes by distinctive physical characteristics such as taillessness or folded ears. Cat breeds show personality differences that are consistent to a considerable extent with these different body shapes: lean, long breeds are generally more lively and demonstrative than more compact, densely muscled breeds. These differences in breed type, from compact and cobby to sleek and sinuous, follow a general West-to-East path.

The breed differences that we see today have been developed – sometimes to extremes – from natural variations found in free-breeding cat populations. To a great extent, these original variations in type were moulded by the environments in which they evolved, and a breed's type is often a good indication of its geographical origin.

COLD-WEATHER CATS

The heaviest and most compact of domestic cats evolved through natural selection in cold climates. Many of these "natural breeds" have fairly large, rounded heads, moderately short, broad muzzles, solid bodies with broad chests, sturdy legs and round paws, and short to medium-length, thick tails. They are, in essence, built to retain as much body heat as possible.

Shorthair examples of these cobby cats are the British Shorthair (see page 112) and American Shorthair (see page 120), as well as the chunky-looking Chartreux (see page 128). Other breeds that derive from these may differ at first in just a single aspect of their body type. The Scottish Fold (see page 119) was developed through the use of British Shorthairs, and is distinguished primarily by its abnormal ears. American Shorthairs were used in the American Wirehair (see page 122), although this breed has now diverged in its looks,

Longhair looks
Over the decades, the face of the Longhair (below) has become more flattened. The degree to which this has been taken varies between registries: the popularity of extreme "peke-faces" has waned.

Old-fashioned face
The Chinchilla is the one Longhair coat colour that has generally avoided the breed's radical facial shortening: in South Africa it even has its own standard within the breed, allowing it to have a longer nose.

becoming increasingly "Oriental" or "foreign" in appearance. The development of the Manx (see page 116) was parallel to that of the British Shorthair, rather than as an offshoot; it now looks even heavier than its close relation.

The original longhaired cats, the Persians or Longhairs (see page 188), were also thickset and muscular, physical features that enabled them to withstand the harsh winters in the high mountains of Turkey, Iran, and Caucasia. The breed still retains its original powerful body; other distinguishing features, such as the flattened face, have been introduced and exaggerated through decades of selective breeding. This has often led to facial-hair staining from overflowing tears from the eyes, because the more compressed and flattened the face is, the more difficult it is for tears to drain from the eyes into the nose. Extreme flattening of the face now seems to be waning in general popularity. Other solid and robustly built longhaired cats developed naturally in northern climes.

Norwegian Forests (see page 208), Siberian Forests (see page 210), and Maine Coons (see page 202) evolved in cold climates from farm cats that lived partly outdoors. These sturdy cats have moderately long faces made for hunting. They have captured the imagination of pet owners: the Maine Coon has ousted the Longhair as the number one breed in the United States, and is Britain's second most popular breed just 20 years after it arrived.

SEMI-FOREIGN BREEDS

A second group of cats has physical characteristics that are somewhere between the muscular cats of northern Europe and the more sinuous cats that developed in the warmer climates of Africa and Asia. These lean but muscular cats are often called "semi-foreign". Breeds like the Turkish Angora (see page 220), Russian Blue (see page 132), and Abyssinian (see page 134) have slightly oval, slanted eyes set in moderately wedge-shaped heads, slender but muscular legs, oval paws, and long, gently tapering tails.

A number of new breeds have been derived from the natural semi-foreigns, and tend to have similar conformation. Some are simply new colours or coat lengths of their parent breeds, including the Nebelung (see page 219), Somali (see page 224), and the more controversial Russian Blacks and Russian Whites (see page 132).

Humble origins
The ancestors of the British Shorthair worked hard as mousers or fended for themselves in city alleyways. The result was a robust, solid build and a coat designed for warmth in cold, wet winters, characteristics of the breed to this day.

Unique departure
The Scottish Fold, like many recently developed breeds, is distinguished by a single unique, striking characteristic. Registries are strict about breed identity being unmistakable, so this trend is only likely to increase.

Some of the more recent efforts to create wild-looking cats, such as the Ocicat (*see page 174*), used semi-foreign parent breeds and reflect this in their type. Some newer breeds also fit into this type of build: American Curls (*see page 212*) have modified-wedge heads and svelte bodies, and are distinguished primarily by the ear tips. Fashion seems to favour the semi-foreign look just now: the Somali is a very popular cat in North American advertising, where it is perceived as elegant and striking without being extreme or odd-looking.

ORIENTAL BREEDS

The most dramatically slender cats are the Oriental breeds, most of which evolved in warm climates where losing excess body heat was much more important than retaining it. With their large-eared, wedge-shaped heads, fine legs, slender bodies, and long, thin tails, these cats have developed maximum body surface area for their size, in order to rid themselves of excess heat. This conformation traditionally has oval, slanted eyes, and its most classic example is the Siamese (*see page 154*). Some claim, with good evidence, that it did not always look like this, and that a stereotyped Western image of Oriental delicacy has refined a cat that never was. Others claim that the cat was always meant to look like this, and cite old breed standards, which do seem to bear this out but are open to interpretation. One breed that seems to support the theory of Western "refinement" is the Japanese Bobtail (*see page 237*), which is distinctly chunkier in Japan than it now is in North America, where it is being bred to look more typically "Oriental" and delicate. Curiously the North American breed standard for the undeniably Eastern Burmese (*see page 146*), imported from Rangoon, has increasingly emphasized solidity.

Newer breeds have been created in the West to mimic Oriental style. The Oriental Shorthair (*see page 160*), the solid-coloured version of the Siamese, was created in the West, after non-pointed shorthaired cats from Southeast Asia died out from among the original Siamese imports. In Britain this new breed was originally called Foreign, and recently became Oriental, a name change that may have been made to ease international confusion, but which also reflects the way the breed has become ever-skinnier. Other Western breeds like the Cornish Rex (*see page 166*)

Heading eastwards
The Devon Rex originated in Britain and is popular in North America, but its look is Oriental. Large ears and fine boning are not common in cats from Devon.

Warm-weather cat
Oriental breeds like the Siamese were always lightly built, but their modern type is far finer than that of free-breeding cats in Thailand.

and Devon Rex (*see page 168*) have been bred to look rather like Eastern breeds, with the types of both these breeds being more extreme in North America. The Devon Rex contributed some of this aspect to the Sphynx (*see page 170*).

Among the Oriental breeds, the Siamese remains the most popular, but less than it was in the past. It may be that, as possibly with the Longhair, some people are now repelled by the extreme look. By contrast, the Siamese-Burmese hybrids, the moderate-looking Tonkinese (*see page 152*), are soaring in popularity.

NEW DEPARTURES IN BUILD

The possibility of breeding bigger or smaller cats, either naturally or through outcrossing to other species, intrigues many. When this has been attempted, the large or small cats tend to revert back to normal domestic-cat size in the next generation.

Non-pedigree pedigree
The American Curl is most notable for its extraordinary and unique ears. Its build is semi-foreign. It differentiates itself further from many older breeds by using non-pedigree cats as outcrosses.

Unlike dogs, the domestic cat seems to have a genetically pre-determined, limited size range. Only outcrossing to another species, which is a controversial move, is likely to change this.

A few breeds are classified according to a single anatomical feature, often arguably a malformation. For example, the taillessness of Manx is associated with potentially lethal medical conditions. Other cats have extra toes, an equally odd physical characteristic. These cats exist in abnormally large numbers around Boston in Massachusetts and Halifax in Nova Scotia. While it might seem absurd to breed these individuals selectively to create a new breed, there has been at least one, albeit short-lived, attempt. In a way, this is no different than breeding selectively for any other body malformation. The single most striking change in the build of cats is the Munchkin. Taking its undeniably cute name from the diminutive race in *The Wizard of Oz*, this is a dwarfed breed: most of the bones in the body are normal, but the long bones of the legs are dramatically shortened. The domestic cat evolved to its virtual perfection through hundreds of thousands of years of natural trial and error. Interventions in its breeding that threaten such inherent and innocent flawlessness seem particularly arrogant and unwarranted.

Transatlantic fashions
In North America, Burmese have a rounded look, most noticeable in the head shape (left). The European descendants of this breed (above) have developed a typically foreign, angular look.

EYE SHAPE AND COLOUR

BEAUTY IS IN THE EYE OF THE BEHOLDER, BUT IN THE CASE OF CATS it is often in the vivid, unblinking eyes of the beheld. Cats have unusually large eyes for the size of their heads. This relationship of eye size to head size mirrors the proportions seen in most animals, including humans, during infancy: it is without doubt one of the subliminal factors that triggers our willingness to care for young animals, and cats benefit from this. Because cats' eyes are among their most appealing attributes, many breeders take great pains to breed for specific eye colour, creating a range of intense shades. Even the best photographs are unlikely to do their efforts justice.

Kittens are born with blue eyes, and their eyes change colour as they mature. Adult cats have eyes of coppery brown, orange, yellow, or green; a few remain blue due to coat-colour genes. Some cats are shown only during their early years when colour is most vivid, while others may not begin to show their best eye colour until they are over two years old.

EYE COLOURS

Wildcats have hazel or copper eyes, sometimes tending towards yellow or green. Breeding has produced a range of colours in domestic cats from sparkling blue through green to orange.

Most eye colours are not governed by coat colour, although some breed standards do link the two: silver tabbies, for example, are often required to have green eyes, but genetically they can have copper or gold eyes. The only colour that is linked to the coat colour is blue. Blue eyes are caused by forms of albinism that lead to a lack of pigmentation in both the coat and the iris, and occur in

BROWN

Green-toned eyes

Green eyes have become common in random-bred cats, and pure greens of varying shades define several breeds. The Chinchilla Longhair's sea-green eyes may be another route for breeders to blue eyes.

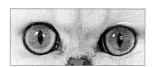

HAZEL

cats with a high degree of white in their coats. Blue-eyed white cats are often deaf, because the gene causing the lack of pigment unfortunately also causes fluid to dry up in the organ of Corti, the receptor for hearing in the cochlea, leading to deafness.

Siamese (*see page 154*) blue eyes – found by the 19th-century naturalist Peter Pallas in the Caucasus – have a different source. They are not linked to deafness, but may be associated with poor three-dimensional vision. Early Siamese often squinted to compensate; breeding has removed the squint, with no apparent loss of visual acuity.

There is at least one other rare blue-eyed gene in cats, appearing in any coat colour. It was noted in Britain in the

1960s, New Zealand in the 1970s, and in the United States, in New Mexico, in the 1980s. The genetics of these rare cats, now called Ojos Azules, are still under investigation.

EYE SHAPE

Wildcats' eyes are oval and slightly slanted. Breeds considered close to the "natural" cat, such as the Maine Coon (*see page 202*), have these "wild" eyes, although some North American standards are tending towards rounder eyes. Natural shape has been altered in two ways: eyes may be rounder, or may be more slanted. Generally speaking, old Western breeds, such as the Chartreux (*see page 128*), have round, prominent eyes. Some Eastern cats, such as the Burmese (*see page 146*), also have rounded eyes, but almond-shaped, slanted eyes are most common in foreign or Oriental breeds. This has been taken to its extreme in the Siamese and Oriental (*see page 160*) breeds.

Extremes of shape can cause problems. Prominent eyes in flat faces are prone to tear overflow and infection, while extremely slanted eyes tend to retain mucus, increasing infections.

BIRMAN BLUE

Yellow-toned eyes

These eye colours are closest to those of wildcats. Many green eyes go through an early brown or yellow stage before they mature to their adult colour. Copper eyes may "fade" with age, while yellow-gold eyes can vary greatly in appearance, depending on the ambient light.

GOLD

Blue eyes

The lack of pigment in blue eyes allows higher absorption of sunlight, used by the body to produce vitamin D. This is why blue eyes are usually found in light-starved regions; the Siamese mutation may have occurred in northern Asia and spread south with human help. Blue eyes vary considerably in depth.

BREED PROFILES EXPLAINED

ALTHOUGH THE OVERWHELMING MAJORITY OF PET CATS ARE STILL random-bred, pedigree cat breeds are popular even with those who neither breed them nor own them. Many of the people who attend cat shows each year have never owned a pedigree cat and never will. An overview of the pedigree breeds can provide an intriguing insight into the make-up of even our random-bred cats. Pedigree cats originally came from random-bred populations and sometimes have an influence on them in return through accidental breedings – as in the case of the pointed pattern introduced to the West in one breed and now found in many household cats.

A few decades ago, there were just a handful of recognized breeds. Today, there are dozens. New mutations have been adopted by some registries, new breeds have been created from existing breeds through the introduction of new colours or another coat length, and breeds from one country have been recognized in others.

BREED DESCRIPTIONS

Each breed profile includes a description of the breed's appearance and character. While the appearance of pedigree cats is highly consistent, the personality traits can vary. Much depends on an individual's experiences; cats from the most genial of breeds can be rendered nervous and even intractable by poor early socialization.

Breed histories outline the breed's ancestry and its route to acceptance by the registries. Some histories are easy to trace, but others are less distinct: older breeds may be wrapped in romantic myths, and even the exact origins of some of the newer breeds, such as the Ragdoll (*see page 200*), have been a matter of debate.

Salient facts are summarized in the Key Facts box found in each profile, which provides an overview of the history, names, breeding policy, personality, and colours of the breed.

INTERNATIONAL DIFFERENCES

Not all registries will recognize the same breeds, or the same colours and patterns in each breed. The same breed may even develop different looks in different countries, depending on the "fashions" prevalent. For example, in the Siamese (*see page 154*), colours recognized as part of the breed in Britain and Europe are recognized as a wholly separate breed by CFA, the main registry in North America.

This chapter has been made as international as possible, using cats from different countries to illustrate differences. Information on colours has been made as comprehensive as possible, listing (in roman text) the colours accepted by the main registry – GCCF for British cats, FIFé for mainland Europe, CFA for North America and Japan – and (in italic text) additional colours that occur but are either not accepted, or may be accepted in other major registries, usually TICA, which recognizes wider ranges in most breeds.

THE BREED SYMBOLS

The details on personality given in these profiles were gathered via questionnaires sent to breeders and breed clubs. They provide a guide to the tendencies across the breed, but an individual cat may not match some of these traits: one may even find some silent Siamese cats.

GREGARIOUS	SELF-CONTAINED	LITTLE GROOMING
SEDATE	ACTIVE	MODERATE GROOMING
QUIET	VOCAL	DAILY GROOMING

BREED REGISTRIES

GCCF GOVERNING COUNCIL OF THE CAT FANCY

FIFé FÉDÉRATION INTERNATIONALE FÉLINE

CFA CAT FANCIERS' ASSOCIATION

TICA THE INTERNATIONAL CAT ASSOCIATION

Understanding the profiles

Some information about a breed, such as history, names, outcross breeds, and traits can be easily summed up in a list of brief points: this information has been placed into a Key Facts box for instant reference. But, such a list of points can never give a full and fair impression of a whole breed, so greater detail is given in the breed description and history. Any additional information about build or particular colours is contained in the captions to the main and secondary photographs. Details of the exact standards that are expected of show-quality cats are contained in the annotations to the main picture.

Introductory text
Outlines physical and psychological traits of breed

Secondary pictures
Show additional colours or kittens

Breed history
Traces breed from origins to acceptance by registries

Head
Gives more detailed information on head shape

Labels
Highlight additional characteristics

Main annotations
Detail requirements of breed standards for showing

Key facts
Summarize information on history, names, breeds allowed as outcrosses, and traits

Symbols
Indicate aspects of personality and level of grooming required

Breed registries
Lists the registries that accept the breed for championship

Breed colours
Gives accepted colours, in registry names, in roman text and indicates other colours in italic text

Swatches
Where relevant, illustrate further colours or variants

Main picture
Illustrates physical characteristics, wherever possible in breed's most notable or popular colour

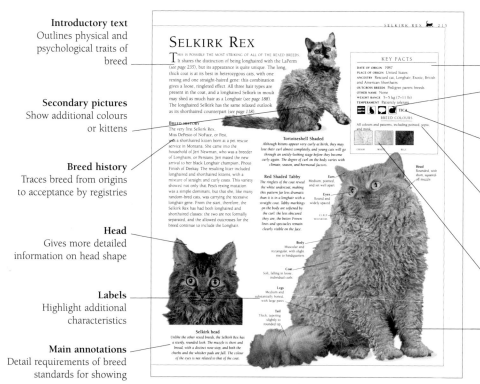

SELKIRK REX

THIS IS POSSIBLY THE MOST STRIKING OF ALL OF THE REXED BREEDS. It shares the distinction of being longhaired with the LaPerm (*see page 235*), but its appearance is quite unique. The long, thick coat is at its best in heterozygous cats, with one rexing and one straight-haired gene: this combination gives a loose, ringleted effect. All three hair types are present in the coat, and a longhaired Selkirk in moult may shed as much hair as a Longhair (*see page 188*). The longhaired Selkirk has the same relaxed outlook as its shorthaired counterpart (*see page 118*).

BREED HISTORY
The very first Selkirk Rex, Miss DePesto of NoFace, or Pesi, was a shorthaired kitten born at a pet rescue service in Montana. She came into the household of Jeri Newman, who was a breeder of Longhairs, or Persians. Jeri mated the new arrival to her black Longhair champion, Photo Finish of Deekay. The resulting litter included longhaired and shorthaired kittens, with a mixture of straight and curly coats. This variety showed not only that Pesi's rexing mutation was a simple dominant, but that she, like many random-bred cats, was carrying the recessive longhair gene. From the start, therefore, the Selkirk Rex has had both longhaired and shorthaired classes: the two are not formally separated, and the allowed outcrosses for the breed continue to include the Longhair.

KEY FACTS

DATE OF ORIGIN 1987
PLACE OF ORIGIN United States
ANCESTRY Rescued cat, Longhair, Exotic, British and American Shorthairs
OUTCROSS BREEDS Pedigree parent breeds
OTHER NAME None
WEIGHT RANGE 3–5 kg (7–11 lb)
TEMPERAMENT Patiently tolerant

BREED COLOURS
All colours and patterns, including pointed, sepia, and mink.

Tortoiseshell Shaded
Although kittens appear very curly at birth, they may lose their curl almost completely, and young cats will go through an unsly-looking stage before they become curly again. The degree of curl on the body varies with climate, season, and hormonal factors.

Red Shaded Tabby
The ringlets of the coat reveal the white undercoat, making this pattern jar less dramatic than is is in a longhair with a straight coat. Tabby markings on the body are softened by the curl; the less obscured they are, the better. Frown lines and spectacles remain clearly visible on the face.

Head
Rounded, with short, squared-off muzzle

Ears
Medium, pointed, and set well apart

Eyes
Round and widely spaced

Body
Muscular and rectangular, with slight rise to hindquarters

Coat
Soft, falling in loose, individual curls

Legs
Medium and substantially boned, with large paws

Tail
Thick, tapering slightly to rounded tip

Selkirk head
Unlike the other rexed breeds, the Selkirk Rex has a sturdy, rounded look. The muzzle is short and broad, with a distinct nose stop, and both the cheeks and the whisker pads are full. The colour of the eyes is not related to that of the coat.

SELKIRK REX 🐈 215

BREED REGISTRIES

EUROPE

Details of the main European cat registries are listed below.

GOVERNING COUNCIL OF THE CAT FANCY (GCCF)
4 – 6 Penel Orlieu
Bridgwater,
Somerset, TA6 3PG, UK
Tel: +44 (1278) 427 575
Website:
http://ourworld.compuserve.com/homepages/GCCG_CATS/

Founded in 1910, the Governing Council of the Cat Fancy was set up with the aim of keeping the registers, licensing and controlling cat shows, looking after the welfare of pedigree cats, and ensuring that the rules set up were not broken or infringed. In the earliest days of the GCCF's existence, the number of cats registered each year totalled several hundred; now it registers about 32,000 pedigree cats each year.

FÉDÉRATION INTERNATIONALE FÉLINE (FIFé)
Website: http://www.fife.org

FIFé has been described as the United Nations of Cat Federations, representing the interests of cats on a worldwide basis. Founded in 1949, its member organizations (from 40 countries) follow the same rules with regard to the breed standards, cattery names, shows, and judges. FIFé has always maintained good relations with all the major cat organizations of the world. Its official seat is in Luxembourg and its official languages are French, German, and English.

USA AND CANADA

Details of major North American cat registries are listed below. The first two listings are the main associations. The remaining registries are listed alphabetically.

CAT FANCIERS' ASSOCIATION (CFA)
P.O. Box 1005
Manasquan, NJ 08736, USA
Tel: +1 (732) 528-9797
Website: www.cfainc.org

The Cat Fanciers' Association, founded in 1906, is a non-profit making organization. In 1909, CFA published the first Stud Book and Register in the *Cat Journal*. It now holds approximately 400 shows per season world-wide. The CFA's responsibilities include litter and cat registrations, transfers of ownership, additions of cattery name suffixes, cattery name registrations, duplicated or corrected registration certificates, certified pedigrees, cattery offspring reports, reverse pedigree reports, championship, and premiership confirmations.

THE INTERNATIONAL CAT ASSOCIATION (TICA)
P.O. Box 2684
Harlingen, TX 78551, USA
Tel: +1 (956) 428-8046
Website: www.tica.org

TICA originated in 1979 when a group of ACFA members broke away to form a new organization. As it was formed after feline genetics became better understood, it has broken new ground in the way cats are registered and shown. TICA registers cats according to their genetic make-up, but shows cats according to their physical appearance.

AMERICAN ASSOCIATION OF CAT ENTHUSIASTS (AACE)
P.O. Box 213
Pine Brook, NJ 07058, USA
Tel: +1 (973) 335-6717
Website: www.aaceinc.org

Formed in 1993, the main aim of AACE is to ensure that exhibitors at shows receive a fair appraisal of their cats, while having fun in the process.

AMERICAN CAT ASSOCIATION (ACA)
8101 Katherine Ave.
Panorama City, CA 91402, USA
Tel: +1 (818) 781-5656

The ACA is the oldest existing North American cat registry, being founded to keep records in 1899. Located in California, this organization no longer holds cat shows.

AMERICAN CAT FANCIERS ASSOCIATION (ACFA)
P.O. Box 203
Point Lookout, MO 65726, USA
Tel: +1 (417) 334-5430
Website: www.acfacat.com

ACFA was the first association to accept altered cats for championship competition, and the first to require judges to pass written exams before becoming licensed. Founded in 1955, ACFA is the third largest association (after CFA and TICA) in North America.

CANADIAN CAT ASSOCIATION (CCA)
220 Advance Blvd.
Suite 101
Brampton, Ontario
L6T 4J5, Canada
Tel: +1 (905) 459-1481
Website: www.cca-afc.com

CCA is Canada's only registry. Formed in 1961, its affiliated cat clubs hold shows, mostly in Ontario and Quebec. It maintains a stud book for members.

CAT FANCIERS' FEDERATION (CFF)
P.O. Box 661
Gratis, OH 45330, USA
Tel: +1 (937) 787-9009
Website: www.cffinc.org

In 1919, a group of CFA members broke off to form the CFF. The federation currently has about 90 affiliated clubs. Most of its member clubs are located in the northeast of the United States, and therefore most of its shows are held in that region.

THE TRADITIONAL CAT ASSOCIATION, INC.© (TCA)
18509 NE 279th Street
Battle Ground, WA 98604-9717, USA
Website: http://www.traditionalcats.com

Founded in 1987, the TCA strives to preserve, protect, perpetuate, and promote traditional cats – those cats whose body styles and conformation types have given way to more extreme forms in other associations. New breeds are considered, but only healthy breeds that do not carry lethal genes. The existence of two versions of a breed – an extreme or contemporary and a traditional – is not always necessary for TCA acceptance.

UNITED FELINE ORGANIZATION (UFO)
5603 16th Street W.
Bradenton, FL 34207, USA
Tel: +1 (941) 753-8637
Email: Uforegof@tampabay.rr.com

Formed in May 1995, UFO's stated goal is to "go where no cat organization has gone before." It is dedicated to creating an association that, by a system of checks and balances, will prevent the kind of strife members have found objectionable in some other associations.

OTHER USEFUL RESOURCES

CAT FANCIERS
Website: www.fanciers.com

Cat Fanciers was founded in 1993. The Fanciers' mailing list is intended for discussions or announcements, specifically relating to showing and breeding cats. This list covers topics as diverse as feline veterinary medicine and home care, cattery management, cat show production, politics in the cat fancier world, and the history and evolution of cat breeds. The list also serves as a useful contact reference for breeders and exhibitors.

FANCIERS BREEDERS REFERRAL LIST ©
Website: www.breedlist.com

The FBRL lists more than 1,700 pedigree cat breeders in more than 20 countries. Searches can be conducted by breed, location, cattery, or breeder.

WORLD CAT FEDERATION (WCF)
Hubertstraße 280, D-45307
Essen, Germany
Tel: +49 (201) 555 724
Email: wcf@wcf-online.de

The World Cat Federation was founded in 1988 in Rio de Janeiro, Brazil. It has more than 540 member clubs worldwide. Included in its responsibilities are standards, rules, international cattery registration, training and examination of international judges, and classes, as well as international liaison. The WCF takes part in European parliament consultations in Strasbourg, including those concerning animal welfare.

INTRODUCTION TO SHORTHAIRS

Black Bombay
The Bombay was the first of the breeds to emulate a wildcat, being bred to resemble a black panther in miniature. Several more breeds of similar character have since been created.

THOUSANDS OF YEARS AGO, DOMESTIC CATS spread from Egypt to varying environments. The early African wildcat type – moderately sized, lean, with a fine coat – was not naturally suited to all the new conditions. The new varieties that evolved now form the basis of today's breeds. Survival of the fittest in the north favoured stocky individuals, with dense, insulating undercoats of weatherproof down hair to protect them in winter. In northern climates, cats developed "cobby" bodies, a type later developed into the British Shorthair (*see page 112*) and exported as the foundation stock for many of the world's cats, including the American Shorthair (*see page 120*). At the same time, the cat was also spreading east across Asia. In warm climates, natural selection favoured thinner, sparser coats and a smaller body to increase the surface-area-to-weight ratio and help to lose excess heat. These cats grew longer and leaner than their antecedents, and their smoother, thinner coats often lacked any down hair. Such cats are now called foreign or, if extremely slender, Oriental types.

Colourpoint Exotic
Some breeds are the unexpected results of breed programmes: this shorthaired Longhair was the result of an attempt to create a longhaired American Shorthair.

Ocicat
At first an accident, this breed brought a new spotting pattern into the spectrum, and is a fine example of the recent trend for breeding wild-looking cats. It requires a charitable eye to see the supposed resemblance to the ocelot.

Mutations in coat type have always occurred, but have died out without human intervention. Many shorthaired breeds have wavy or curled "rex" coats, first bred in the Cornish Rex (*see page 166*). The extreme mutation of hairlessness is often a lethal trait that naturally disappears, but with selective breeding it has formed the basis of the Sphynx's extraordinary appearance (*see page 170*).

Siamese
This breed was once defined for most people solely by its pointed pattern. Today, many breeds include this pattern, and the Siamese is recognized more by its extremely fine, elongated type, which has helped to make it one of the more controversial cats among breeders.

A new trend in breeding is that of creating a new look rather than refining what nature has already achieved. Many of these breeds emulate wildcats: the Ocicat (*see page 174*) is a typical example. The Bengal (*see page 176*) was the first breed produced by mating the domestic cat with a wildcat, the Asian leopard cat (*see page 17*). It has been followed by the Pixiebob (*see page 179*), reputedly with bobcat (*see page 14*) ancestry, and Chausies, crossed to *Felis chaus* (*see page 16*).

Korat kitten
Some of the newer breeds have achieved huge popularity in a short time, while some of the older ones, such as the Korat, still retain a certain rarity value. Known in its native Thailand for centuries, the breed was only recognized for showing in the West in 1965.

European Shorthair
Like its British and American counterparts, this type developed naturally in random-bred cats over many hundreds of years. Today it is still preserved and perpetuated by modern breeders.

EXOTIC

GENUINELY EXOTIC IN LOOKS, THIS SHORTHAIRED Longhair (*see page 188*) has the gentle personality and soft, squeaky voice of its parent breed. Exotics have the conformation of Longhairs, but a highly original coat: not quite short, but not semi-long either. Outcrossing to bring in the short coat has given the Exotic a slightly livelier and more inquisitive disposition than its antecedents; it has not, however, eliminated the anatomical problems of the face inherited from the Longhair. The dense, double coat benefits from combing twice weekly. The breed is still rare, partly because litters contain longhaired kittens (*see page 238*); the status of these kittens as Exotic or Longhair remains contentious.

BREED HISTORY

In the early 1960s, breeders of the American Shorthair (*see page 120*) attempted to introduce the Longhair's coat texture into their breed. Instead, they produced cats with the Shorthair's coat on the Longhair's cobby, compact body. Unintentionally, the "shorthaired Longhair" was born, complete with a flattened, "teddy bear" face and small, plaintive voice. To differentiate their cats from American Shorthairs, breeders called them Exotic Shorthairs, and used British Shorthairs (*see page 112*), Burmese (*see page 146*), and even Russian Blues (*see page 132*) in their breed programmes. CFA recognized the breed in 1967. In North America only outcrosses with American Shorthairs and Persians are allowed, elsewhere other breeds are still used.

Black
With a coat of lustrous black hair and eyes of brilliant gold, this colour might have been the model for many of the lucky black cats produced as promotional toys by companies over the years. Black kittens may show grey or rusty tinges in their coats, but these are not acceptable in adult show cats.

Seal Point
The pointed pattern is included within Exotic colours, rather than a separate group, as in the Longhairs. All points should be evenly matched, and the mask should extend over the entire face. Point colours develop at different rates, the dense colours, like the Seal, maturing first.

Coat
Dense, plush, standing out from body

Blue-Cream
The standard for all tortie Exotics calls for the colours to be balanced and softly intermingled, and for all four feet and the tail to contain both colours. Some distinct patches of colour are allowable, and facial blazes are permitted. Tortie patterning is inherently unpredictable.

Tail
Relatively short

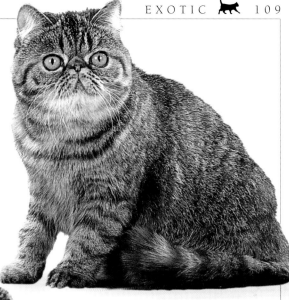

Exotic head

The Exotic has inherited some of the Longhair's flaws, such as overflowing tear ducts, constricted nostrils, and dental problems. In an effort to breed for health, the British standards for both breeds call for the upper edge of the nose leather to be below the level of the lower edge of the eye, penalizing drastically short faces. The conformation is already apparent in kittenhood, and this kitten may escape troubles.

Ears
Small, with rounded tips, set wide and low

Eyes
Large, round, and prominent

Body
Medium to large and cobby, carried low on legs

Head
Round and massive, with full cheeks

Paws
Large, round, and firm

Legs
Short, thick, and strong

MASSIVE SHOULDERS

Brown Mackerel Tabby

While the Longhair is recognized only in the classic or blotched pattern, Exotic tabbies may also be mackerel-striped or spotted. The facial markings of all three of these patterns are the same, only the body markings vary. The Mackerel Tabby should have numerous narrow vertical stripes covering the body and rings on the tail.

Blue

The colour standard for the Blue Exotic is similar to that for the blue British Shorthair, a cat involved in its creation. Comparing the two breeds shows how distinct breed conformations are, even in two relatively stockily built cats. The Exotic is the roundest of all the shorthairs, with a breed standard that calls for rounded lines in every contour from the ears to the toes. The cuddly appearance can give a deceptive illusion of softness; beneath the plush fur, the curves should be of muscle.

KEY FACTS

DATE OF ORIGIN 1960s
PLACE OF ORIGIN United States
ANCESTRY Longhairs/American Shorthairs
OTHER NAME None
OUTCROSS BREEDS Longhairs
WEIGHT RANGE 3–6.5 kg (7–14 lb)
TEMPERAMENT Gently inquisitive

    GCCF FIFé CFA TICA

BREED COLOURS

All colours and patterns, including pointed, sepia, and mink

British Shorthair (overleaf left and right) *For breed entry see page 112*

BRITISH SHORTHAIR

THIS IMPRESSIVELY BUILT CAT IS SELF-POSSESSED AND SELF RELIANT. Although undemanding and gentle, British Shorthairs are not overly tolerant of handling; judges and breeders describe them as "four-feet-on-the-ground" cats. Numerous firm guard hairs give the dense, bouncy coat a distinctive, crisp feel, and the protective undercoat insulates the cat on the frostiest day. Thick-legged and well-muscled, this breed is both compact and surprisingly heavy. Large, round eyes highlight its essentially gentle disposition, although, if prey appears, it is a highly successful hunter.

BREED HISTORY

British Shorthairs were developed in the 1800s from the farm, street, and household cats of Britain. Although it was the most-shown breed at the first British shows, and Harrison Weir, the "inventor" of the Cat Fancy bred "British Blues", the breed was in decline by the turn of the century, and had almost died out by the 1950s. It was revived by dedicated breeders who exported stock to Ireland and throughout the British Commonwealth. By the 1970s, it had arrived in the United States where, as the British Blue, its calm, loyal, enduring, and reserved manner won it immediate admirers. A curious feature separates the British Shorthair from most shorthaired cats: 40 per cent of all British Shorthairs have Type B blood, a rare trait.

British Shorthair profile
Seen here in a Red Spotted, the head is round and the neck sturdy. The standard calls for a rounded forehead, moderate nose-break, and short, straight nose.

PENCIL LINES
FROM EYES

FROWN
LINES

Blue

This is the all-time classic British Shorthair. It was one of the earliest colours, and has always remained the most popular. For a long time, the British Blue was the only British cat recognized in North America. When the breed population fell in World War II, breeders outcrossed first to Oriental cats, and then to Blue Longhairs, which were closer in type. This led to problems with fluffy coats and short noses, but the breed has long since regained form. It is still sometimes confused with the Chartreux (see page 128).

Ears
Medium in size, with round tips

Eyes
Large, round, and copper or gold in most colours

BROAD
NOSE

Head
Round face with full cheeks

SHORT, STURDY
NECK

Chest
Good width in shoulders and broad chest

SHORT, STRONG
LEGS

Paws
Compact, firm, and round

Silver Spotted Tabby

This striking pattern was also one of the earliest, appearing in the 1880s. There are silver versions of all the tabby colours and patterns, but the black remains one of the most popular. As in other breeds, the Silvers have hazel, rather than copper, eyes.

Red Classic Tabby

The original tabby was the Brown, less often seen today, but Reds also made an early appearance. Ginger-reds are common in the non-pedigree cats of Britain, but a century of breeding has modified this colour to deep, tawny shades.

UNBROKEN
NECKLACES

KEY FACTS

DATE OF ORIGIN 1880s

PLACE OF ORIGIN Great Britain

ANCESTRY Household, street, and farm cats

OTHER NAME Tipped colours once called Chinchilla Shorthairs

WEIGHT RANGE 4–8 kg (9–18 lb)

TEMPERAMENT Genial and relaxed

 GCCF FIFé CFA TICA

BREED COLOURS

SELF AND TORTIE
Black, Chocolate, Red, Blue, Lilac, Cream, Tortoiseshell, Chocolate Tortie, Blue Tortie, Lilac Tortie, White, (Blue-, Odd-, Orange-Eyed)

SMOKE AND TIPPED
Colours are as for self and tortie colours, with the addition of Golden Tipped

TABBY (CLASSIC, MACKEREL, SPOTTED)
Brown, Chocolate, Red, Blue, Lilac, Cream, Tortie, Chocolate Tortie, Blue Tortie, Lilac Tortie

SILVER TABBY (CLASSIC, MACKEREL, SPOTTED)
Colours are as for standard tabbies

BI-COLOURS
All self and tortie colours with white

POINTED COLOURS
All self, tortie, and tabby colours

RED SPOTTED TABBY | BLUE SPOTTED TABBY | BLACK SMOKE | BROWN CLASSIC TABBY

NOSE LEATHER PINK

PAW PADS PINK

Orange-Eyed White

This colour was developed from the Blue-Eyed White, which appeared at the turn of the century. A perfect white with no yellow tinge is rare. Blue- and Odd-Eyed Whites can show a congenital deafness (see page 96), and the standard requires deep blue eyes in an attempt to avoid the problem; orange-eyed cats do not tend to suffer from this form of deafness.

Tortoiseshell

A difficult colour to breed, this was nevertheless one of the very first to be recognized. The standard for the Tortoiseshell pattern in this breed calls for an even mingling of colours without obvious patches, in contrast to the Tortoiseshell standard in the American Shorthair (see page 120). Brindling or tabby markings are faults.

FACIAL BLAZE ACCEPTABLE

PAW PADS MAY BE MOTTLED

Body
Cobby and strong, carried low on legs

Coat
Dense, with crisp feel

Tail
Short and thick, with blunt tip

THICK COAT "BREAKS" EASILY

COAT DEVELOPS RUSTY TINGE IN SUNLIGHT

Black

Superstitions about black cats have cut both ways (see page 30), but in Britain a black cat has been fully rehabilitated and is treated as a lucky omen. Even so, these cats fell into obscurity after World War I. The clear golden eye colour is only rarely seen among non-pedigree cats, which are usually green- or hazel-eyed.

NEWER BRITISH SHORTHAIR COLOURS

SINCE THE BREED WAS FIRST DEVELOPED, OVER A CENTURY AGO, THE RANGE of British Shorthairs has increased enormously. Some of these appeared before World War II, but since then there has been an even greater development of colours and patterns. In the 1950s, numbers were low, and breeders outcrossed surviving British Shorthairs with Longhairs (*see page 188*) to ensure the survival of the blue coat. Although now rare, the Longhair influence means British Shorthair matings still occasionally produce kittens with fluffy coats. The Longhairs also brought in the tipped coat, resulting in a new category of colours. More recently, adventurous matings have created delightful new coat colours, which indicate outcrossing to Oriental cats. These are not universally accepted outside Britain. The Siamese (*see page 154*) gene, however, brings not only its pattern but the possibility of a Blue-Eyed White without deafness.

Cream
Although Creams, which are dilute Red selfs, appeared from the turn of the century onwards, they were not recognized until the 1920s. Breeders did not know how to produce them, and early Creams were "hot", resembling Reds. Good, "cool" Creams, as far as possible from red, took a great deal of work.

Chocolate kitten
The gene for this rich brown colour came originally from Oriental cats. Crossbreeding brought the colour into the Longhairs, from which it was introduced into the British Shorthair. Because of its outcross origins, this colour is not universally recognized outside Britain.

Lilac kittens
The dilute version of the Chocolate, the Lilac was introduced into the British Shorthair by outcrossing with Longhairs. Both blue and lilac are shades of grey, but they are quite distinct in good examples.

Black-tipped
This was known until 1978 as the Chinchilla Shorthair because it was developed through matings with Chinchilla Longhairs. Longhair heritage is still apparent in kittens before they acquire their adult coats.

Black and White Bi-colour
Although there have been bi-colour British Shorthairs from the earliest days of the breed, the show standard originally specified that the markings should be symmetrical, a standard almost impossible to meet. A revision to allow less rigid distribution of colour in the one third to half white coat brought more popularity. At first, only Black, Blue, Red, and Cream were allowed.

NOSE LEATHER SHOWS TWO COLOURS

Blue-Cream
Although both the dilute Blue and the Tortoiseshell were among the earliest colours recognized, this dilute tortoiseshell did not achieve the same status until the 1950s; in part this was because the genetics behind it were not understood.

BALANCE OF COLOURS MUST BE EVEN

Seal Colourpointed kitten
These colours have only been fully accepted in Britain during the 1990s, and they are still not allowed elsewhere. The result of outcrossing to Siamese, they still resemble the traditional British Shorthair in all respects other than their exotic pattern.

PALE EYES DEEPEN WITH MATURITY

Tortoiseshell and White
Unlike the solid Tortoiseshell, this pattern has distinct patches of red and black. For some as yet unknown reason, the bi-colour gene affects the sex-linked red gene, making mingled colours almost impossible to achieve in tortie-and-whites. The colour balance of the coat should be one third to half white, as in other bi-colours.

THICK, CRISP COAT

Blue-Cream Colourpointed
As with the solid colours, pointed torties should show evenly mingled colours on their points. Ideally, each point shows a mix of colours. Despite the Oriental influence, these cats tend to have the placid nature of their British forebears.

MANX

THE LACK OF A TAIL IS THIS BREED'S MOST OBVIOUS VISIBLE CHARACTERISTIC, but its "bunny-hop" gait is just as unique. If there is any single word that encapsulates the Manx it is "round" – round-bodied, round-eyed, round-rumped, round-headed, even round-pawed. Slow-maturing Manx come in a huge range of colours and patterns, although pointed patterns are not accepted by all associations.

Black and White
A common combination in random-bred British cats, this coat must show no blurring or speckling at all in a show cat. This "stumpy" has a vestigial tail, making it unsuitable for showing.

Local emblem
Manx appear on stamps in their homeland: this is one of the first.

Cats may be "rumpies" (with no tail, just a dimple at the base of the spine), "stumpies" (with short tails), and "tailies" (with almost natural, although usually kinked, tails). Stumpies and tailies make excellent pets, with retiring but friendly personalities, but show cats are all rumpies. This classification has health implications, because the Manx gene carries potentially lethal consequences. Rumpy-to-rumpy breedings can result in Manx Syndrome, in which kittens die at or soon after birth, or develop fatal bowel or bladder problems by four months of age. Responsible breeders select cats carefully for matings, and will not part with kittens until they have reached the four-month milestone in good health.

BREED HISTORY

The Manx originated on the Isle of Man in the Irish Sea. Taillessness occurs occasionally in feline populations as a spontaneous mutation: in large populations it will usually disappear, but in isolated groups, such as those on islands, it has a good chance of surviving. This is how the Manx, Japanese Bobtail (*see page 164*), and Kurile Island Bobtail (*see page 236*) developed. The dominant Manx mutation is not that of the Asian cats, belying myths that they are related. The traditional Manx was rangier than today; now it is bred for roundness, "like a hairy basketball with legs" as one breeder put it. Manx have been shown in Britain since the late 19th century, and reached North American shows in 1899. It was recognized by CFA in the 1920s.

Rump
No rise of bone or cartilage discernible when stroked

Coat
Thick and double, with quality more important than pattern

PADDED LOOK
TO COAT

Manx head
The full-cheeked face of the Manx shows its roots in the typical random-bred British cat. The head should be broad and rounded, with a straight nose and a firm chin.

KEY FACTS

DATE OF ORIGIN Pre-1700s

PLACE OF ORIGIN Isle of Man

ANCESTRY Household cats

OTHER NAME None

OUTCROSS BREEDS None

WEIGHT RANGE 3.5–5.5 kg (8–12 lb)

TEMPERAMENT Mellow and even-tempered

   GCCF FIFé CFA TICA

BREED COLOURS

Colours are as for British Shorthair
All colours and patterns

| BLUE TABBY | WHITE | RED | SILVER TABBY |

EARS ANGLED OUTWARDS

Brown Classic Tabby and White

The Manx's hindlegs are markedly longer than its forelegs, although they are often carried bent. This contributes to its rounded look and gives it its characteristic gait. In the US, this "bunny-hop" gait is considered a defect. The deep flanks of the breed provide a canvas to display tabby markings, especially the bullseye motif.

Ears
Fairly tall and high set, with rounded tips

Brown Tortie Tabby

The Manx is one of the stoutest of all cat breeds. The GCCF showing standard calls for the cat to have "good breadth of chest" and "flanks of great depth", while the CFA standard states that the proportion of the body to the legs should be such that, taken together, they describe a square. The show standards of all breed registries call for a completely tailless individual, with a definitely rounded rump that is slightly higher than the shoulder when the cat stands squarely.

Eyes
Colour in keeping with coat colour

Head
Large and round, with medium length nose

Tortoiseshell

As in all breeds, the British prefer a more softly mingled tortoiseshell pattern than the North American registries. This individual is a stumpy, with a short, stubby tail, and is therefore unsuitable for showing in any registry.

Body
Solid and compact, with short back

Legs
Relatively short and powerful

SELKIRK REX

REXING IS APPARENT IN THE SOFT, THICK, PLUSH COAT OF THIS breed from the minute a kitten is born, but it then disappears, to reappear at eight to ten months. While the coat, in which all hairs are curled, needs routine grooming, excessive combing and brushing, especially after bathing, straightens the hair. In body conformation, this patient and relaxed breed most closely resembles the British Shorthair, especially in its leg length. It also comes in two versions, the plush shorthair and the more dramatic longhair (*see page 215*). There are known debilitating medical conditions associated with some other rexed breeds: it is not known yet whether there are any such problems with the Selkirk Rex. The Selkirk is unlike most other Rexes in that the trait is dominant, so outcrossing to broaden the gene pool still produces 50 per cent rexed litters.

BREED HISTORY

Every year, many kittens are born at shelters. In 1987, a female calico kitten was born at For Pet's Sake, a pet rescue service run by Kitty Brown in Montana. This kitten, one of a litter of seven, was the only one to have curly hair and curly whiskers. Miss DePesto of NoFace, as she was called by the breeder who adopted her, Jeri Newman, was bred and three in her litter of six had curly coats, indicating that the rexed coat was genetically dominant. It is assumed that Pest herself was the source of this genetic mutation. Further breedings, including one back to her son, revealed that she carried the recessive genes for long hair and pointing. Jeri named the breed after the nearby Selkirk Mountains, and it is recognized by TICA.

Head
Rounded, with distinct stop to nose

Ears
Medium and pointed, set well apart

Eyes
Round and widely spaced

CURLY WHISKERS

CURL ON NECK PROMINENT

Body
Medium in build, with good musculature

Coat
Thick, medium length, in loose curls

CURL ON BODY VARIABLE

Tail
Thick, tapering to rounded tip

Legs
Medium length and substantially boned

KEY FACTS

DATE OF ORIGIN   1987
PLACE OF ORIGIN   United States
ANCESTRY   Rescued domestic cat, Longhair, Exotic, British and American Shorthair
OUTCROSS BREEDS   Longhair, Exotic, British and American Shorthairs
OTHER NAMES   None
WEIGHT RANGE   3–5 kg (7–11 lb)
TEMPERAMENT   Patiently tolerant

CFA TICA

BREED COLOURS

All colours and patterns, including pointed, sepia, and mink

RED TABBY AND WHITE	BLACK AND WHITE	SILVER SHADED

Selkirk head
The Selkirk has a Western look, rather than the Oriental style favoured for other rexes. The full-cheeked face has round eyes and a short muzzle.

Black Smoke
Like other rexed breeds, the Selkirk shows smoke and shaded colours to great advantage. Outcrossing to pedigree breeds allows a new breed like the Selkirk to take advantage of the decades of breeding that have gone into producing traits such as deep copper eyes, achieving them in a shorter time itself.

SCOTTISH FOLD

OLDED EARS ENSURE THAT THIS BREED IS IMMEDIATELY ARRESTING,
but it also has a distinctive, rounded look, with a short neck,
round head, and compact body. Those unique ears are due
to a dominant gene that causes varying degrees of fold.
The first Fold had what is now called a "single" fold,
where the ears bend forwards; today's show-quality
cats have tight "triple" folds. Straight-eared cats are
still essential for breeding healthy Folds. The breed
has a placid personality, and its undemonstrative
behaviour suits its reserved appearance.

BREED HISTORY

Susie, the Fold's founding mother,
was a farm cat born in Tayside, in Scotland.
Local shepherd William Ross and his wife Mary
were given one of Susie's kittens, named Snooks.
Bred to a British Shorthair (see page 112), she
produced Snowball, a white male who was
shown locally. In 1971, Mary Ross sent some
Folds to Neil Todd, a geneticist in Newtonville,
Massachusetts. Development continued in the
United States, using British and American
Shorthairs (see page 120), and Folds were fully
recognized by 1994. In Britain, the problems of
crippled homozygous Folds prevent recognition.

Brown Classic Tabby
*The distinctive ears of the Fold are present from birth,
although how folded they are may not be apparent
until later. The skeletal problems associated with
homozygosity for this gene are not apparent in kittens;
the abnormal bone growth shows up with maturity.*

Blue Tortie Tabby and White
*The scoring system for Folds places
most of the emphasis on body type.
After the ears, the most important
part of the body is the tail: any
shortening or stiffness indicates
skeletal problems. All aspects of the
Fold standard call for a healthy,
well-proportioned, supple cat.*

KEY FACTS

DATE OF ORIGIN 1961
PLACE OF ORIGIN Scotland
ANCESTRY Farm cat, British and American Shorthairs
OUTCROSS BREEDS British and American Shorthairs
OTHER NAME None
WEIGHT RANGE 2.5–6 kg (6–13 lb)
TEMPERAMENT Quietly confident

   CFA TICA

BREED COLOURS

All colours and patterns, including pointed, sepia, and mink

Ears
Folded, with rounded tips

Eyes
Large and rounded

Head
Rounded, with broad, short nose

Body
Medium, sturdily built, and flexible along spine

Coat
Short and dense

Legs
Must not be short

Tail
Long, tapering tail preferred

Fold head
*The ears of a Scottish Fold should be "set in a cap-like
fashion", flat against the head. Small, tightly folded ears
are the ideal. The face should have a sweet expression.*

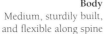

AMERICAN SHORTHAIR

THIS EASY-GOING, SELF-SUFFICIENT, AND NO-NONSENSE CAT has won friends in both the home and the showring in the United States and Canada; elsewhere, it is uncommon. American Shorthairs can be quite large, and the full-cheeked face and robust, muscular body exude strength. The CFA standard specifies, as should all standards, that "Any feature so exaggerated as to foster weakness" brings disqualification in a show. American Shorthairs still share features with household cats, and the aim of breeding is to produce kittens that have the best of these qualities. Until relatively recently, any non-pedigree cat that met the standard could be registered, and these cats widened the gene pool: this practice is no longer allowed.

Household origins
The ancestors of this breed were often recorded in naive scenes, such as this one by an unknown artist of the late 19th century.

BREED HISTORY

Domestic cats arrived in North America with the first settlers, whether in British settlements from Newfoundland to Virginia, the French possessions from Nova Scotia to Louisiana, or the Dutch colony in what is now New York. Many of these cats were brought on voyages to keep down rats on ships, but some remained in the New World as pioneers for the world's largest domestic cat population. This new land brought new environmental pressures to bear: the descendants of surviving cats developed thick, hard coats, dense enough to protect them from moisture, cold, and superficial injuries. With more natural predators, the cats evolved to be bigger than European felines. In the early 1900s, some American breeders realized that their cat's characteristics should be preserved in a breed. The first litter, born in 1904, was from a mating of American and British Shorthairs. The breed was called the Domestic Shorthair until 1965, when it acquired its present name. In 2000, the TCA recognized the breed.

Silver Tabby
The Silver Tabby is a popular colour, with dense, inky black markings on a background of sterling silver. In 1965, a Silver Tabby won the United States Cat of the Year award, prompting the breed's name change from the original "Domestic Shorthair" to its present "American Shorthair".

Ears
Medium-sized and well spaced, with rounded tips

GENTLY CONCAVE NOSE

Head
Large, slightly longer than it is wide

Eyes
Large, rounded, and very slightly tilted

Chin
Vertically in line with nose

Neck
Medium length and muscular

Body
Solid, powerful, and muscular

Coat
Short, thick, and hard in texture

Legs
Medium length and heavily muscled

Tail
Medium length, thick at base

BROWN EYES
LATER TURN GOLD

Brown Classic Tabby kittens
In CFA, only the blotched and striped tabby patterns are accepted, while TICA accepts the spotted and ticked patterns as well. The blotched tabbies were prevalent among the breed's emigrant ancestors; blotched tabby populations show a high correlation with old trading routes.

Blue-Cream
Females, as tortoiseshells virtually always are, may be much lighter than males, and have a less "heavy" face. The standard for the American Shorthair, like most other North American standards, calls for clearly defined patches of unbrindled colour.

BRILLIANT GOLD
EYES PREFERRED

TAIL TAPERS TO
BLUNT END

KEY FACTS

DATE OF ORIGIN 1900s
PLACE OF ORIGIN United States
ANCESTRY Domestic cats
OUTCROSS BREEDS None
OTHER NAME Once called Domestic Shorthair
WEIGHT RANGE 3.5–7 kg (8–15 lb)
TEMPERAMENT Easy-going

   CFA TICA

BREED COLOURS

SELF AND TORTIE COLOURS
Black, Red, Blue, Cream, White, Tortoiseshell, Blue-Cream *All other self and tortie colours*

SMOKE COLOURS
Black, Cameo, Blue, Tortoiseshell, Blue-Cream *All other self and tortie colours, except White*

SHADED AND TIPPED COLOURS
Colours are as for self and tortie colours except White

TABBIES (CLASSIC, MACKEREL)
Brown, Red, Blue, Cream, Brown Patched, Blue Patched *Spotted and ticked patterns, all other self and tortie colours*

SHADED TABBIES
Colours and patterns are as for standard tabbies

BI-COLOURS (STANDARD AND VAN)
Colours are as for self and tortie colours except White

SMOKE, SHADED, AND TIPPED BI-COLOURS
Black Smoke, Cameo Smoke, Blue Smoke, Tortoiseshell Smoke, Shaded Cameo, Shell Cameo with White *All other smoke, shaded, and tipped colours with white*

TABBY BI-COLOURS
All tabby colours with white

SILVER TABBY BI-COLOURS
Silver Tabby, Cameo Tabby, Silver Patched Tabby with White
*All other silver tabby colours with white
Any other colour or pattern except those that indicate hybridization, such as chocolate, lilac, the Himalayan pattern, or agouti ticking*

Black Smoke
Buster Brown, the first all-American cat to be registered in the breed, in 1904, was a Black Smoke of street origins. In repose, the smoke coat colours look almost indistinguishable from those of the selfs, but in motion, the undercoat shimmers softly.

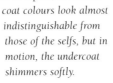

TABBY LINES ON
CREAM AREAS

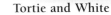

Tortie and White
This coat pattern is generally known as Calico – or Dilute Calico if it is a blue tortie and white – in North America. This odd name has been in use since the earliest cat shows, and comes from the pattern's resemblance to the bold patterns once commonly printed on calico fabric.

AMERICAN WIREHAIR

THIS BREED'S MOST NOTABLE CHARACTERISTIC IS, OF COURSE, its coat. Distinctive to the touch, it feels like stroking an astrakhan hat. Every hair is thinner than usual and crimped, hooked, or bent, giving the overall appearance of "wiring". The most prized coat is dense and coarse, but a kitten born with a coat that appears to have ringlets may have only a wavy coat at maturity, while coats that are only lightly wired early in life may continue to develop throughout the cat's first year. A cat with curly whiskers is highly valued. The American Wirehair is a relaxed and friendly breed. Its advocates say it is rarely destructive and enjoys being handled.

BREED HISTORY

This breed descends from a kitten born in 1966 on a farm in Verona, in upstate New York. Local breeder Joan O'Shea obtained both this male kitten and his normal-coated sister. Through a careful breeding programme, she determined that this was a dominant mutation. Successive crossings with American Shorthairs developed the breed, and for a while its standard, apart from the wired coat, was the same as that of the Shorthair. Distinctive qualities in the shape of the head, body, and tail consistently occurred in litters, however, and a breed standard was written for the Wirehair in 1967. Although there are now American Wirehairs throughout the United States and Canada, this rare breed remains unrecognized in most other regions of the world.

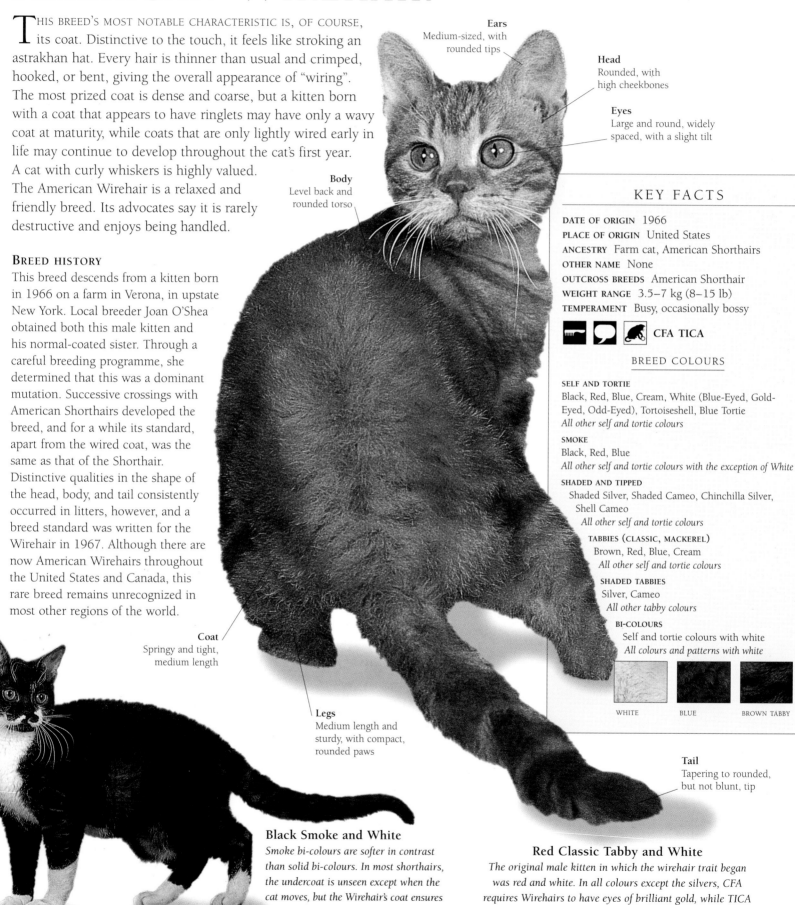

Ears
Medium-sized, with rounded tips

Head
Rounded, with high cheekbones

Eyes
Large and round, widely spaced, with a slight tilt

Body
Level back and rounded torso

Coat
Springy and tight, medium length

Legs
Medium length and sturdy, with compact, rounded paws

Tail
Tapering to rounded, but not blunt, tip

KEY FACTS

DATE OF ORIGIN 1966
PLACE OF ORIGIN United States
ANCESTRY Farm cat, American Shorthairs
OTHER NAME None
OUTCROSS BREEDS American Shorthair
WEIGHT RANGE 3.5–7 kg (8–15 lb)
TEMPERAMENT Busy, occasionally bossy

CFA TICA

BREED COLOURS

SELF AND TORTIE
Black, Red, Blue, Cream, White (Blue-Eyed, Gold-Eyed, Odd-Eyed), Tortoiseshell, Blue Tortie
All other self and tortie colours

SMOKE
Black, Red, Blue
All other self and tortie colours with the exception of White

SHADED AND TIPPED
Shaded Silver, Shaded Cameo, Chinchilla Silver, Shell Cameo
All other self and tortie colours

TABBIES (CLASSIC, MACKEREL)
Brown, Red, Blue, Cream
All other self and tortie colours

SHADED TABBIES
Silver, Cameo
All other tabby colours

BI-COLOURS
Self and tortie colours with white
All colours and patterns with white

WHITE — BLUE — BROWN TABBY

Black Smoke and White
Smoke bi-colours are softer in contrast than solid bi-colours. In most shorthairs, the undercoat is unseen except when the cat moves, but the Wirehair's coat ensures that a little white always shines through.

Red Classic Tabby and White
The original male kitten in which the wirehair trait began was red and white. In all colours except the silvers, CFA requires Wirehairs to have eyes of brilliant gold, while TICA recognizes no relationship between eye and coat colour.

AMERICAN CURL

THIS ELEGANT AND SWEET-TEMPERED BREED COMES IN TWO COATS, long and short. The shorthaired American Curl has taken longer to develop, because the original Curls (*see page 212*) were all longhaired: the shorthair gene is dominant, so a great many shorthaired Curls carry a hidden longhair gene and produce longhaired kittens. The extent of curl to the ears is of profound importance to breeders. The curl is graded in three stages: cats with ears just turned back (first degree) become pets, those with more curl (second degree) are used for breeding, and those with full crescents (third degree) are shown. The curled ears should be handled with care: uncurling them risks damaging the cartilage.

EAR FURNISHINGS DESIRABLE

Curl development
All American Curls are born with straight ears. When they are two to ten days old the tips begin to curve; they then curl and uncurl until they "set" permanently at about four months.

BREED HISTORY
For over a decade, the cat has been America's most popular pet, and California is where the most active breeding programmes exist. This new breed is the result of a genetic mutation that occurred in a black, longhaired, stray kitten who was named Shulamith. Half of her kittens also showed this unusual feature, a genetically dominant characteristic, and were distributed to form a breeding programme. All members of this shorthaired breed descend, like their longhaired cousins, from this mother and her kittens, and are accepted on a showing standard that differs from the Shorthair only in coat.

Silver Ticked Tabby
The Curl standard calls for distinct striping on the face, legs, and tail; in other breeds this pattern should be free of such bars. TICA recognizes this pattern in the silvered colours, while CFA does not.

Head
Modified wedge with gentle curves

Ears
Curving at least 90° in smooth arc

Eyes
Upper rim oval, lower rim rounded

Body
Semi-foreign shape, moderately muscled

Coat
Soft and close-lying, with minimal undercoat

Tail
Equal to length of body, wide at base, tapering to tip

Legs
Medium length and build

KEY FACTS

DATE OF ORIGIN 1981
PLACE OF ORIGIN United States
ANCESTRY American household cat
OTHER NAME None
OUTCROSS BREEDS None
WEIGHT RANGE 3–5 kg (7–11 lb)
TEMPERAMENT Quietly affable

 CFA TICA

BREED COLOURS

SELF AND TORTIE
Black, Chocolate, Red, Blue, Lilac, Cream, White, Tortoiseshell, Blue-Cream
All other self and tortie colours

SMOKE
Colours as for self and tortie, except White and with the addition of Chocolate Tortie
All other self and tortie colours

SHADED AND TIPPED
Shaded Silver, Shaded Golden, Shaded Cameo, Shaded Tortoiseshell, Chinchilla Silver, Chinchilla Golden, Shell Cameo, Shell Tortoiseshell
All other self and tortie colours

TABBIES (CLASSIC, MACKEREL, SPOTTED, TICKED)
Brown, Red, Blue, Cream, Brown Patched, Blue Patched
All other colours

SILVER TABBIES (CLASSIC, MACKEREL)
Silver, Chocolate Silver, Cameo, Blue Silver, Lavender Silver, Cream Silver, Silver Patched
All other standard tabby colours, spotted and ticked patterns

BI-COLOURS (CLASSIC AND VAN)
Black, Red, Blue, Cream, with White, Calico, Dilute Calico
All other colours with white

TABBY BI-COLOURS
Colours are as for standard tabbies

SELF AND TORTIE POINTS
Seal, Chocolate, Flame, Blue, Lilac, Cream, Tortie, Chocolate Tortie, Blue-Cream, Lilac-Cream
All other colours, sepia and mink patterns

LYNX (TABBY) POINTS
As for self and tortie points, excepting Red
All other colours, sepia and mink patterns

MUNCHKIN

ALTHOUGH BREEDERS CLAIM THAT THE DWARFISM OF THESE CATS causes no problems, the Munchkin, a short-legged cat in longhair (*see page 213*) and shorthair versions, has stirred up unprecedented controversy. The breed has had to undergo rigorous health investigations to gain the level of acceptance it has, but many breeders still feel that Munchkins are "uncatlike". While their playful personalities are unmistakably feline, it cannot be denied that the breed represents a radical departure from normal feline anatomy. Those who see the cat as a lovable pet may well come round to this breed: those attracted to the cat's self-sufficient image are unlikely to approve of it.

KEY FACTS

DATE OF ORIGIN 1980s
PLACE OF ORIGIN United States
ANCESTRY Household cats
OUTCROSS BREEDS Household cats
OTHER NAME None
WEIGHT RANGE 2.25–4 kg (5–9 lb)
TEMPERAMENT Appealing and inquisitive

BREED COLOURS

All colours and patterns, including pointed, sepia, and mink

RED CLASSIC TABBY   TORTIE AND WHITE   BLACK

Munchkin head
The breed standard is for a cat of moderation. The head should be neither rounded and jowly nor angular and foreign. All colours and patterns are allowed.

Red Spotted Tabby
In some ways, the Munchkin resembles a squirrel, both in its gait and in its tendency to sit up on its haunches with its forepaws held in front of it. This characteristic pose contributes to the image of the breed as curious and comical.

BREED HISTORY

The Munchkin has been bred for a little over a decade in North America, and from the start attracted a high level of attention. After health investigations, TICA granted it "new breed" status in 1995. Some breeders are working with pedigree breeds, producing rexed and curled-ear Munchkins. A short-legged, curly coated type, called a Skookum, was produced by crossing a LaPerm with a Munchkin in 1996.

Brown Spotted Tabby
Colour and pattern are relatively unimportant in the breed standard. Build alone accounts for half the points. Part of the Munchkin's appeal may be that an adult cat's proportions resemble those of a kitten, whose legs are usually fairly short in relation to the body.

Ears
Moderately large, wide at base, and upright

Head
Almost triangular, with nose of medium length

Eyes
Walnut-shaped, large, and slightly tilted

Neck
Thick and firmly muscled

Body
Medium-sized, substantially muscled, neither svelte nor cobby

Tail
Set high, medium-thick, tapering to rounded tip

Legs
Short but not misshapen, well muscled

Paws
Round, compact, and pointing outwards

SNOWSHOE

Named for its characteristic white mittens, this breed combines the pointing of the Siamese (*see page 154*) with white spotting, giving it crisp white paws. Two patterns exist: the mitted, with limited white, and the bi-colour, with more white on the face and body. The white mittens may be inherited from the breed's American Shorthair (*see page 120*) ancestry, or they may come from its Siamese side; white toes were a known fault in early Siamese. Snowshoes are gregarious and affectionate, and although they are talkative, they have soft voices.

BREED HISTORY

In the 1960s, Philadelphia breeder Dorothy Hinds-Daugherty began to cross her Siamese with American Shorthairs. The resulting hybrid faced some opposition from Siamese breeders at first, partly due to fears that the spotting might find its way into Siamese bloodlines after decades of breeding to eradicate this early fault. The dramatic pointing pattern was also almost a trademark of the Siamese breed at that time, although it is recognized in many breeds today. The Snowshoe remained little known until the 1980s, when it was recognized by TICA. It has since gained wider popularity, but remains rare. It is not recognized by other major registries.

Snowshoe face

The facial pattern decides if a Snowshoe is mitted or bi-colour. An inverted "V" of white makes the cat a bi-colour, anything less makes it a mitted. The amount of white on the body determines whether it is a good or bad example of either pattern.

Ears
Medium to medium-large, continuing lines of face

Head
Broad modified wedge, with slight stop in profile

Eyes
Medium-sized, oval, and blue

Coat
Short, smooth, and close-lying, with no noticeable undercoat

Legs
Length in proportion to body, with medium musculature

Paws
Medium-sized, oval, and white in both patterns

Seal Bi-colour kitten
The white areas must not exceed two-thirds of the total body area. The body should be coloured, showing subtle shading to lighter underparts; there should be no isolated white spots.

Mitted Blue
The mitted pattern is the "classic" Snowshoe. White paws ideally stop at the ankles on the forelegs and below the hocks on the hindlegs, and the amount of white must not exceed one-third of the body. There may be some white on the face. Sex-linked red colours and cinnamon colours do not occur in the Snowshoe.

Body
Moderate size and musculature, semi-foreign build

Tail
Medium thickness, tapering slightly

EUROPEAN SHORTHAIR

A CAT WITH A WHOLE CONTINENT TO CALL HOME MIGHT EASILY BE assumed to have wide popularity, but the European Shorthair is less known than either its British Shorthair (*see page 112*) or American Shorthair (*see page 120*) counterparts. Over the years since the breed was established, it has become less cobby than the British type, with a slightly longer and less heavily jowled face, perhaps reflecting the typical feline type of warmer mainland European countries more than it did in the past. It has many of the same basic traits as the British cats, however, being a strong, hardy cat with an all-weather coat. Its personality tends to be calm and affectionate, and it is a relatively quiet breed.

BREED HISTORY

Until 1982, European Shorthairs were classified with British Shorthairs. FIFé then gave the breed its own category, and it began effectively as a "ready-made" breed, with a full range of colours, established type, and breeding stock with known histories. Despite this advantage, or maybe because it is so similar to the British and American Shorthairs, the European Shorthair does not seem to have caught the imagination of breeders, and remains rare. The breed is now being selectively bred, with no British Shorthair crosses permitted in the pedigree. It is not recognized by GCCF or major breed registries outside Europe.

Ears
Medium-sized and upright, with rounded tips

Head
Triangular to rounded, with well-defined muzzle

Eyes
Large, round, and well spaced, with colours to match coat

MUSCULAR NECK

Body
Medium to large, well muscled, but not cobby

Coat
Short and dense, standing away from body

Legs
Medium length and well muscled

Black Silver Mackerel Tabby
The European Shorthair is bred in the three "traditional" tabby patterns of classic, mackerel, and spotted. The ticked pattern, not historically present in Europe, is not recognized. The Black Silver Tabby is popular because of the vibrantly contrasting colours in its coat. The markings should be symmetrical on both sides of the body.

Tail
Medium length, thick at base, tapering to rounded tip

Tortoiseshell Smoke

In the European Shorthair, the patches of black, red, and cream in the tortoiseshell pattern should be broken into clearly defined patches, rather than subtly intermingled. Because the coat does not lie flat, the white undercoat of a Smoke can be seen and has the effect of "watering down" the colours slightly.

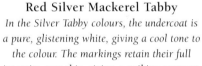
GOLDEN OR COPPER IN COLOUR

KEY FACTS

DATE OF ORIGIN 1982
PLACE OF ORIGIN European mainland
ANCESTRY Household cats, British Shorthairs
OUTCROSS BREEDS None
OTHER NAME None
WEIGHT RANGE 3.5–7 kg (8–15 lb)
TEMPERAMENT Intelligent and reserved

   FIFé

BREED COLOURS

SELF AND TORTIE COLOURS
Black, Blue, Red, Cream, Tortie, Blue Tortie, White (Blue-, Odd-, Orange-Eyed)

SMOKE COLOURS
Colours are as for self and tortie, except White

TABBY COLOURS (CLASSIC, MACKEREL, SPOTTED)
Brown, Blue, Red, Cream, Tortie, Blue Tortie

SILVER TABBY COLOURS (CLASSIC, MACKEREL, SPOTTED)
Colours are as for standard tabbies

BI-COLOURS (STANDARD AND VAN)
All self and tortie colours with white
Smoke, tipped, and tabby colours with white

BROWN TABBY   BLUE-CREAM   SEAL POINT
AND WHITE

Red Silver Mackerel Tabby

In the Silver Tabby colours, the undercoat is a pure, glistening white, giving a cool tone to the colour. The markings retain their full intensity over this, giving a striking contrast. This cat is heavily silvered, with colour restricted to the ends of the hairs.

Blue Point

The European Shorthair, just like the British Shorthair, has "imported" the pointing pattern of the Siamese breed (see page 154). The body tends to show more shading than that of the flat-coated Siamese, and darkens markedly with age.

Cream Shaded Cameo Tabby

The inhibitor gene does not always produce a silvery-white colour. Sometimes the undercoat retains a creamy tone, and these tabbies are called "shaded" rather than "silver". Although creams are bred to look "cooler" than reds, a Cream Shaded may look warmer than a Red Silver.

CHARTREUX

A KEEN OBSERVER OF LIFE, RATHER THAN AN IMPULSIVE PARTICIPANT, the Chartreux is a tolerant breed, less talkative than most, with a rather high-pitched miaow and an infrequently used chirp. Its short legs, stockiness, and dense, close coat mask its true size. This is a big, powerful, late-maturing cat. Although a good hunter, it is not a fighter: individuals tend to withdraw from conflicts rather than become aggressive. There is an intriguing naming system for Chartreux: each year is designated by a letter (omitting K, Q, W, X, Y, and Z), and cats' names begin with the letter determined by the year of birth. For example, cats born in 1997 have names beginning with N.

BREED HISTORY

Possibly originating in Syria, the Chartreux's ancestors would have arrived in France by ship. By the 1700s, the breed was described by the naturalist Buffon as the "cat of France" and given a Latin name, *Felis catus coeruleus*. After World War II it effectively became extinct, and was re-established by outcrossing survivors with blue Longhairs (*see page 188*) and British Blues (*see page 112*). The Chartreux reached North America in the 1970s, but is not bred in many European countries. FIFé assimilated the Chartreux and British Blue under the name Chartreux during the 1970s, and all British and European Blues were described as Chartreux for a time: these breeds are now distinct again.

Blue male
Male Chartreux are much larger than females, with a heavier build, although they should never become cobby. They also develop pronounced jowls with age, which adds to the broadness of the head.

COAT "BREAKS"
OVER CURVES

Coat
Short to medium length,
with dense undercoat

Colour
Even, bright blue-grey,
from ash to slate in tone

SLIGHT
IRIDESCENT
SHEEN

Tail
Thick at base,
tapering to
rounded tip

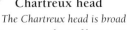

WARM-TONED PADS

Chartreux head
The Chartreux head is broad but stops just short of becoming a sphere. The muzzle is relatively narrow, but the rounded whisker pads and heavy jowls prevent it from looking too pointed. The expression should be sweet, with a slight smile.

The patriotic pet

The Chartreux was the chosen companion of both French president Charles de Gaulle and author Colette, seen here with two of her pets. Colette's choice of the Chartreux was as much a part of national pride as her continued presence in Paris throughout the German occupation.

Chartreux kitten

Chartreux can take up to two years to mature fully: below this age, allowance is made for a coat that is finer and silkier than the ideal. In very young cats the eye colour may also be less brilliant: it brightens as the cat matures, and then fades again with age.

Ears
Medium-sized, set high on head

Head
Broad but not round, with high forehead

Eyes
Large and round, gold or copper in colour

SLATE-GREY NOSE

Neck
Short and heavy

Chartreux build

The Chartreux build is neither cobby nor slender: it has been termed "primitive". Other descriptions include a "potato on matchsticks" due to the relatively fine-boned legs. In fact, the Chartreux today looks much the same as its forebears, because the breed standard is kept close to historical descriptions of the breed.

BROAD CHEST

Body
Robust and densely muscled

Legs
Short and sturdy, but not thick

Paws
Small relative to body, round in shape

KEY FACTS

DATE OF ORIGIN Pre-18th century
PLACE OF ORIGIN France
ANCESTRY Household cats
OUTCROSS BREEDS None
OTHER NAME None
WEIGHT RANGE 3–7.5 kg (7–17 lb)
TEMPERAMENT Calmly attentive

    FIFé CFA TICA

BREED COLOURS

SELF COLOURS
Blue

Chartreux (overleaf left) *For breed entry see above*

RUSSIAN SHORTHAIR

THE ORIGINAL OF THIS GROUP IS THE SLIGHTLY RESERVED AND immensely dignified Russian Blue. A cautious cat, it is sensitive to changes in its environment and controlled in its activities with strangers. Its most vivid features are its thick, lustrous coat and its emerald-green eyes. The soft, dense, insulating, double coat is unique in feel, and described in the British breed standard as "the truest criterion of the Russian". The Russian's trademark eye colour is of more recent origins; the first Russian Blues exhibited in the West, at the Crystal Palace show in England in 1871, had yellow eyes. It was not until 1933 that breed standards called for eye colour to be "as vividly green as possible". This is a gentle breed, among the least destructive of all cats, and an ideal indoor companion.

Lucky Blue
The Blue is considered lucky in Russia, and images, such as this 19th-century engraving, are given to brides.

BREED HISTORY

Legend says that the Russian Blue descends from ships' cats brought from the Russian port of Archangel to Britain in the 19th century. Russian Blues are mentioned by name in Harrison Weir's 1893 book *Our Cats*, but from the Russian Revolution in 1917 until 1948, they became known as Foreign Blues. The modern Russian Blue contains bloodlines from British Blues (*see page 112*), and even from Blue Point Siamese (*see page 154*), consequences of Swedish and British efforts to revive the breed in the 1950s. Black and white versions have been developed in New Zealand and Europe; these are accepted in Britain, but not by FIFé or North American fancies.

KEY FACTS

DATE OF ORIGIN Pre-1800s
PLACE OF ORIGIN Possibly Russian port of Archangel
ANCESTRY Domestic cats
OTHER NAME Archangel Cat, Foreign Blue, Maltese Cat, Spanish Blue, Russian Blue
OUTCROSS BREEDS None
WEIGHT RANGE 3–5.5 kg (7–12 lb)
TEMPERAMENT Reserved and shy

 GCCF FIFé CFA TICA

BREED COLOURS

SOLID COLOURS
Black, Blue, White

Ears
Large, pointed, and set vertically

Eyes
Large, almond-shaped, and widely spaced

LONG, GRACEFUL OUTLINE

Head
Longer from ears to eyes than from eyes to nose

Body
Well muscled, but never cobby or heavy

Coat
Double, with very dense undercoat

Russian Blue
The original, and to some the only, Russian, this colour is an even blue with a silvery sheen. This sheen gives the coat a luminous appearance, and many breeders say that the less the cat is brushed, the more radiant the coat becomes.

Legs
Long, but not delicate

BLACK NOSE LEATHER

Tail
Moderate length and thickness, tapering to rounded tip

Russian Black
The Russian Blue bred true for centuries because the dilute colour is recessive, never masking other colours. Black and White Russians are a recent development, and regarded by many as controversial. Least welcome of all are the blue-pointed "Russians", a result of past Siamese outcrosses, and being worked on by a very few breeders.

Russian Shorthair (previous page) *For breed entry see above*

HAVANA BROWN

ELEGANT AND GRACEFUL AS ITS OUTWARD APPEARANCE IS, the Havana Brown is a very physical breed. It is, for example, an excellent climber. Although this rare breed's origins are the same as those of the Oriental Shorthair (*see page 160*), it has developed to resemble the Russian Blue (*see opposite*). The Havana Brown stands high on its legs and is heavy for its size. Kittens and young adults have ghost tabby marks that disappear with age, leaving an even, rich shade of brown. Over the past decade, the breed has struggled to exist. One reason for this is thought to be the end to outcrossing in 1974. To help overcome the problem, the CFA voted in 1998 to re-allow outcrossing.

BREED HISTORY

During the 1950s, British cat breeders developed a solid chocolate of Siamese type. The colour was called Havana, but the breed was registered in Britain as Chestnut Brown Foreign. Havana Browns were exported to the United States, where breed development produced Quinn's Brown Satin of Sidlo, which is now found in the background of all North American Havana Browns. Chestnut Brown Foreigns continued to be imported into America and registered as Havana Browns until 1973, when CFA accepted the Oriental Shorthair breed: from then on these imports were registered as Chestnut Oriental Shorthairs. Ironically, the Oriental Shorthair colour called Chestnut in North America is now called Havana in Britain, leading to some confusion.

Havana head
The long head narrows to a slim muzzle with a pinch just behind the whisker pads. In profile, the chin is strong, giving the muzzle an almost squared appearance.

Lilac
Russian Blues were used in the creation of the Havana Brown, introducing the recessive dilute trait. This means that Lilacs appear occasionally. They are recognized by TICA, but not by CFA. The colour is warmer than that in other breeds.

KEY FACTS

DATE OF ORIGIN 1950s
PLACE OF ORIGIN Great Britain and United States
ANCESTRY Chocolate Point Siamese, black household cat, Russian Blue
OTHER NAME None
OUTCROSS BREEDS Black or blue domestic shorthairs, Oriental Shorthairs in any colour except fawn, cinnamon, or pointed
WEIGHT RANGE 2.5–4.5 kg (6–10 lb)
TEMPERAMENT Sweet and sociable

 CFA TICA

BREED COLOURS

SOLID COLOURS
Chocolate *Lilac*

Ears
Large, wide-set but upright

Eyes
Oval in shape, green in colour

Head
Long and slender, with narrow muzzle

Whiskers
Must be brown or lilac to complement coat colour

Body
Medium length, carried level

Colour
Even shade of warm mahogany

TAPERED TIP

Tail
Medium length and thickness

Paws
Oval and compact

Chocolate
The standard for this shade of brown varies considerably across the different breeds in which it is found. In the Havana Brown, it is defined as a warm brown, tending towards red-brown; to some eyes, it comes close to cinnamon. A dark, sable-type coat is a serious fault.

ABYSSINIAN

THE ALMOST TRANSLUCENT COAT PATTERN OF THIS BREED IS due to a single gene, first noted in the Abyssinian. This gene gives each hair several dark bands, evenly dispersed on a lighter background, resulting in a striking "ticked" coat pattern. Abyssinians' ears sometimes have caracal-like tufts, which add to their striking appearance. Although often almost silent, Abys' personalities are far from quiet; they become attached to their owners, and demand attention and play. They are natural athletes, climbing and investigating almost any available object or person. They can suffer inherited forms of retinal atrophy, a blindness more common in dogs.

BREED HISTORY

The Abyssinian's ticking is a perfect camouflage in the dry, sunburnt habitat of North Africa. The founding cats, including one called Zula, were brought to Britain from Abyssinia (now Ethiopia) after the Abyssinian War in 1868. There is a strong similarity between these first Abys and some ancient Egyptian images, and it is likely that the ticked mutation occurred thousands of years ago. Accepted in 1882, the breed was almost extinct in Britain in the early 1900s, but by the 1930s it was established in the US and France. Today it is the fifth most popular breed in North America. Standards differ internationally: Europeans have a more foreign shape and a wider range of colours.

Blue male
This dilute form of the Usual has underparts of pale oatmeal, which stands out well against the warm, blue-grey ticking of the coat on the rest of the body. Like the Usual, the Blue has dark paw pads.

CLEAR FROWN LINES

Abyssinian head
The face (below left) is a wedge. A slight nose break is seen in profile (below right). Rounded, almond-shaped eyes are green, hazel, or amber.

DARK TIPS

"MASCARA" LINES

Lilac kitten
One of the newer additions to the range of Abyssinian colours, the Lilac is the dilute shade of the Chocolate. Both colours were introduced through outcrosses to Oriental cats in the 1970s, and they are still not accepted by the more traditional registries.

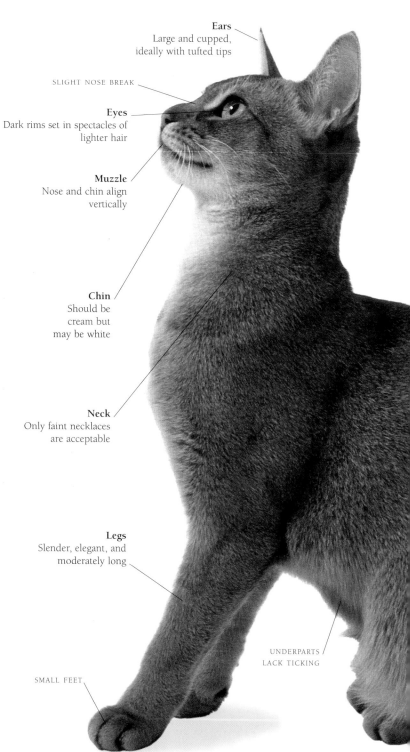

Ears
Large and cupped, ideally with tufted tips

SLIGHT NOSE BREAK

Eyes
Dark rims set in spectacles of lighter hair

Muzzle
Nose and chin align vertically

Chin
Should be cream but may be white

Neck
Only faint necklaces are acceptable

Legs
Slender, elegant, and moderately long

UNDERPARTS LACK TICKING

SMALL FEET

CHOCOLATE
TICKING

APRICOT BASE
COAT

Red or Sorrel

Once called Red in all registries, this colour is now more usually called Sorrel in Britain. It is not, however, the true sex-linked Red, but the recessive light brown that is more commonly known as cinnamon in other breeds.

Body
Medium-sized and muscular

Coat
Close-lying, fine but not soft

Ticking
At least four bands of darker colour required

PINK PADS

KEY FACTS

DATE OF ORIGIN 1860s
PLACE OF ORIGIN Ethiopia
ANCESTRY Ethiopian household and street cats
OUTCROSS BREEDS None
OTHER NAME Colour names vary internationally
WEIGHT RANGE 4–7.5 kg (9 16 lb)
TEMPERAMENT Attention-demanding

   GCCF FIFé CFA TICA

BREED COLOURS

TABBIES (TICKED)
Ruddy, Red (Sorrel), Blue, Fawn
Chocolate, sex-linked Red, Lilac, Cream, Chocolate Tortie, Cinnamon Tortie, Blue Tortie, Lilac Tortie, Fawn Tortie

SILVER TABBIES (TICKED)
All standard ticked tabby colours

CREAM   CHOCOLATE

EARS ARE
TICKING COLOUR

TICKING LINE ON TAIL

Fawn kitten

This colour is the dilute version of the Sorrel and was once called Cream, but it is generally less bright in appearance than a true sex-linked Cream. In some associations only these colours are accepted, but in Britain true reds and creams have also been bred and are accepted for showing.

Ruddy or Usual

The historical Abyssinian colour, the Ruddy is genetically black agouti, the colour most commonly called "brown" in other tabby patterns. In the early days, it was similar to the coat of the wild rabbit, and earned the cat the name of "hare cat" or "rabbit cat" for a while. The colour is still called lièvre, or hare, in French, but selection for rufism has led to a warmer, reddish base colour today. In Britain, Ruddy was known as Normal or Usual until the 1970s; now it is simply called Usual.

Tail
Gently tapering, and the same length as the body

TIP IS TICKING
COLOUR

Usual Silver

With the warm base coat lightened to white, the Usual becomes a sparkling Silver. The inhibitor gene, which produces this effect, seems to have been present in the early Abyssinians, naturally or through early crossing. These colours have now been re-established in Britain, but are not universally accepted in North America.

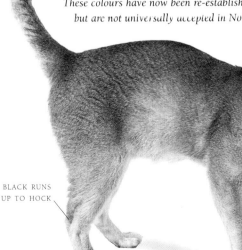

BLACK RUNS
UP TO HOCK

AUSTRALIAN MIST

THIS IS THE FIRST BREED THAT HAS BEEN DEVELOPED ENTIRELY IN Australia. Originally named Spotted Mist, this breed has a playful and home-loving nature. Gentleness of spirit is prized, and the standard penalizes any aggression on the show bench. Similar in some ways to the Asian Spotted Tabby (*see page 145*), its appearance is moderate in all ways: medium in size, foreign in build but not extremely so, with a short, but not close-lying coat. The delicate markings create a misty appearance and account for the breed's name; the background ticking is essential to this effect. The breed is recognized in six colours, some of them unique in appearance. Full colour takes a year to develop, and in some colours the delicate spots can be hard to see.

BREED HISTORY

Dr. Truda Staede in New South Wales developed a programme to produce an indoor-loving, people-oriented breed, with the Burmese shape and a spotted tabby coat. The Burmese contributed conformation, companionability, and four of the six Australian Mist colours. The Abyssinian added two more colours, the essential ticking, and a lively disposition. Domestic tabbies gave the spots and offset the tendency towards early sexual maturity. In January 1980, the first half-Burmese, quarter-Abby, quarter-domestic tabby kittens were born. Fully recognized in Australia, this breed is still very rare and remains unknown elsewhere.

Ears
Medium-sized, upright or slightly flared in set, with rounded tips

Head
Rounded wedge, with smooth, flowing lines

Body
Medium-sized, with moderate muscling

Eyes
Deep gold in colour, almond-shaped, and slightly tilted

Coat
Short, stands out softly from body

WHITISH UNDERPARTS

Legs
Medium length, neither stocky nor slender

Paws
Small and round

Peach
Salmon-toned markings on a warm, pinkish-cream ground make this the closest thing yet to a pink cat. Underneath, it is a dilute cinnamon, but this coat shows the effect that different combinations and polygenic influences can have.

Tail
Medium length, tapering slightly to rounded tip

WHITE LIPS AND CHIN

LEGS ARE STRIPED

Gold
One of the colours brought in from the Abyssinian ancestry, this is genetically a cinnamon. The mixture of genetic influences in the Australian Mist has transmuted the coat to gold and bronze markings on a cream ground.

KEY FACTS

DATE OF ORIGIN 1975
PLACE OF ORIGIN Australia
ANCESTRY Abyssinian, Burmese, Non-pedigree shorthairs
OUTCROSS BREEDS None
OTHER NAME Spotted Mist
WEIGHT RANGE 3.5–6 kg (8–13 lb)
TEMPERAMENT Lively and harmonious

BREED COLOURS

TABBY (SPOTTED AND TICKED TOGETHER)
Blue, Brown, Chocolate, Gold, Lilac, Peach

SINGAPURA

PEACEFUL, EVEN RETIRING, BY TEMPERAMENT, THESE CATS ARE AT the smaller end of the feline scale. Singapuras are distinctive in their one recognized coat, a ticked tabby in a colour called Sepia. The Singapura's temperament and physical attributes are widely reputed to be a natural result of selective pressures. In Singapore, most cats are feral and nocturnal. Cats that attract the least attention are more likely to breed successfully, leading to small size, a quiet voice, and a retiring disposition. Western cats are larger than Singapore's "drain cats", either because of genetic differences or better diet. On the basis of this and other factors, some people believe that Singapore's feral stock was the inspiration for the breed, rather than its sole genetic founders.

Singapura kittens

The coats of Singapura kittens seldom meet the breed standard for adults: they are longer in relation to the small bodies, and also show less developed ticking. Even quite young kittens, however, show the "cheetah lines" that run from the inner corner of the eye to the whisker pads.

BREED HISTORY

This new breed's name is Malaysian for Singapore, from where Hal and Tommy Meadows brought cats to the United States in 1975; all registered Singapuras today originate from their breeding programme. Within seven years of embarking on planned breeding, the Singapura was recognized, and it received its first formal championship recognition in 1988. The breed has reached Europe, but neither FIFé nor the GCCF recognize it, and there is controversy about its origins. Tommy Meadows also bred Burmese (*see page 146*) and Abyssinians (*see page 134*), and there are claims that these were used to create the Singapura. Others maintain that this is untrue, and allows unscrupulous breeders to sell fake Singapuras. Because there are still less than 2,000 Singapuras, some believe that the closed register could work to its detriment, while allowing outcrosses to these other breeds could help the Singapura.

Sepia Agouti

This is the only colour in which the Singapura is bred. Genetically, it is a sable ticked tabby, a combination uniting the Burmese allele with the tabby pattern first known in the Abyssinian. The ticking is brown, on an ivory ground, with tabby markings showing on the face and inner legs.

Ears
Wide and deeply cupped, set at a slightly outward angle

Head
Rounded, with straight nose and broad, blunt muzzle

Coat
Short and close-lying, with at least two bands of ticking

Eyes
Hazel, green, or yellow outlined in black, almond-shaped and slightly tilted

Body
Medium length and build

Tail
Slightly shorter than body, slender, with blunt tip

UNDERPARTS PALE

Legs
Strong but not stocky, slight barring on insides allowed

Paws
Small and oval, with brown pads, dark hair between toes

KEY FACTS

DATE OF ORIGIN 1975
PLACE OF ORIGIN Singapore and United States
ANCESTRY Disputed
OUTCROSS BREEDS None
OTHER NAME None
WEIGHT RANGE 2–4 kg (4–9 lb)
TEMPERAMENT Affectionately introspective

 GCCF CFA TICA

BREED COLOURS

TABBY (TICKED)
Sepia Agouti

KORAT

MEDIUM-SIZED, WITH A SEMI-COBBY BODY AND A SILVER-BLUE COAT, the Korat is similar to the Russian Blue (*see page 132*) in size and colour. Korats, however, are a densely muscled and rounded breed, with a single rather than double coat and peridot-green rather than emerald-green eyes. The large, prominent eyes give it an innocent expression, but this is a strong-willed, even pushy, cat. While Korats are playful and trainable, they do thrive on having their own way. They have an excellent capacity to grumble. Very rarely, individuals carry neuromuscular disorders, called GM1 and GM2. In 2000, an international programme began to eliminate these inherited diseases by molecular testing.

Korat kitten
Young Korats lack both the silver tipping and the brilliant eye colour of adults, both of which may take two years to reach their full intensity.

BREED HISTORY

The *Cat Book Poems* of the Ayutthaya Kingdom (AD 1350–1767) describe the silver-blue Si-Sawat from Korat, a remote, high plateau in northeast Thailand. The first Korat in the West may have been entered in English cat shows in the 1880s as a solid blue Siamese (*see page 154*). Modern Korats were introduced into the United States by Mrs Jean Johnson in 1959; they were recognized in the United States in 1965. The first pair was imported into Britain in 1972, and the breed was recognized in 1975. It remains rare everywhere.

Ancient breed
The Thai Cat Book Poems distinguish the Si-Sawat and its luck-bringing properties from those of other blue cats recorded in it.

Head
Large, flat forehead and firm, rounded muzzle

Ears
Large, wide at base, with rounded tips

Eyes
Luminous, prominent, and rounded

Body
Strong and semi-cobby build

Coat
Glossy, fine, and close-lying

Korat head
Oversized eyes and a heart-shaped face described by gentle curves give the Korat a much softer appearance than its compatriot breeds, but it has every bit as much personality.

Blues only allowed
As is to be expected from a breed that developed in a warm climate, the Korat has dispensed with the unnecessary insulation of an undercoat, and the single, silver-blue coat is short. Some lilacs have been bred in mainland Europe, but they are unlikely ever to be accepted.

COAT "BREAKS" OVER SPINE

KEY FACTS

DATE OF ORIGIN Pre 1700s
PLACE OF ORIGIN Thailand
ANCESTRY Household cats
OTHER NAME Si–Sawat
OUTCROSS BREEDS None
WEIGHT RANGE 2.5–4.5 kg (6–10 lb)
TEMPERAMENT Demanding and opinionated

   GCCF FIFé CFA TICA

BREED COLOURS

SELF COLOURS
Blue

Tail
Heavy at base, tapering to rounded tip

MEDIUM LENGTH

Paws
Oval and compact

BOMBAY

THIS MAJESTIC BREED HAS A COAT LIKE JET-BLACK PATENT LEATHER, a pleasant and distinctive voice, and a gregarious personality. Like its Burmese forebear (*see page 146*), the Bombay thrives on human company. It is a real heat-seeker, most often found in the lap of a heat-emitting human. The coat is almost maintenance-free: an occasional rubdown with a chamois, or even your hand, is all that is needed to keep its sheen and texture. The breed standards can be difficult to meet: the brilliant copper eye colour is thought to be the result of two genes, one for colour and the other for intensity, and can fade or turn slightly green with age. Litters are large, but the Bombay remains rare, especially outside North America: it is sometimes confused with the British Bombay (*see page 144*).

KEY FACTS

DATE OF ORIGIN 1960s
PLACE OF ORIGIN United States
ANCESTRY Black American Shorthairs and Sable Burmese
OTHER NAME None
OUTCROSS BREEDS None
WEIGHT RANGE 2.5–5 kg (6–11 lb)
TEMPERAMENT Actively inquisitive

CFA TICA

BREED COLOURS

SELF COLOURS
Black, *Sable*

Bombay kitten
The full depth of colour and texture of the Bombay coat can take up to two years to develop, and allowance is made for this in the breed standards.

BREED HISTORY

In the 1950s, Kentucky breeder Nikki Horner embarked on an attempt to create a "mini black panther" from black American Shorthairs and Sable Burmese. By the 1960s, she had produced cats with shiny black coats, muscular bodies, rounded heads, and copper eyes. The Bombay was first recognized in 1976. Over the years the look has diverged from that of the Burmese, and the Bombay is no longer a "black Burmese". Litters still produce sable-brown kittens with sepia pointing, although these are now rare.

Head
Rounded in all aspects, with a short to medium muzzle

Ears
Wide at the base with rounded tips

Eyes
Large, rounded, and widely spaced

Body
Medium, semi-cobby, and surprisingly heavy

Tail
Medium length and thickness

PATENT-LEATHER GLOSS

Bombay head

Development of the breed has led some breeders to feel that the head is now closer to the American Shorthair than it was originally. Because of the Burmese heritage, some Bombay lines suffered the Burmese head malformation: breeding to avoid this may account for the present look.

Coat
Close lying, with satin-like texture

Legs
Medium length and sturdy

ROUNDED PAWS

Bombay Black
The glossy, Black Bombay has always had a unique look. The type came from its Burmese heritage, and the solid, even colour from its American Shorthair antecedents. Because the sepia pointing gene of the Burmese is recessive, Sable kittens still occasionally appear.

Abyssinian (overleaf left and right below) *For breed entry see page 134* **Burmilla kitten** (overleaf right above) *For breed entry see page 142*

THE ASIAN GROUP

Although they share the same origins for the most part, shaded, smoke, self, tabby, and Tiffanie (*see page 227*) Asians have sufficiently distinct breeding for GCCF to designate them a group rather than a breed. Asians had the first breed standard that allowed points to be awarded at shows for temperament.

ASIAN SHADED

These striking cats, which were the founding members of the group, are also commonly known as Burmillas. Their type is inherited from their Burmese (*see page 148*) parentage and the shaded coat comes from their Chinchilla Longhair (*see page 188*) side. Like all Asians, Burmillas are less boisterous than the typical Burmese, but more sociable than Longhairs. While Burmillas are undoubtedly attention-seekers, they remain polite, moderate in their vocal demands, and are remarkably even-tempered.

BREED HISTORY

In 1981, an unauthorized alliance took place in London between a Burmese and a Chinchilla Longhair, resulting in attractive shaded silver kittens. After consultation with relevant bodies, owner Miranda von Kirchberg began a planned breeding programme. The original litter was of Burmese type, and initial policy was to breed back to Burmese every other generation. Both in Britain and elsewhere in Europe, other Burmese-Chinchilla crosses were made: backcrossing still helps to enlarge the genetic base. The Burmilla was recognized by GCCF in 1989 and FIFé in 1994: bloodlines in the two associations differ.

Brown Silver Shaded
A striking characteristic of the Burmilla is its natural dark "eyeliner" or "mascara". Noting also the line of colour around the nose and the almost red, rather than pink, nose leather, these cats tend to look like they have just returned from a full make-up.

Lilac Silver Shaded
The Burmilla, or Asian Shaded, includes both shaded and tipped cats, although the coat should not be so lightly tipped that it appears white. The undercoat should be pale, almost white in Silvers. Tabby markings are restricted to the face, legs, and tail, and broken necklaces.

Body
Medium-sized and firmly muscled

Shading
Varies from medium-heavy to tipping

Tail
Medium to long, tapering to rounded tip

Ears
Medium to large, widely spaced, angled slightly outwards

Head
Short wedge, gently rounded on top

Eyes
Neither almond-shaped nor round, any colour from gold to green

Coat
Short, fine, and close-lying

Legs
Medium length with oval paws

ASIAN SMOKES

Smokes are the halfway-house between the Shaded Asians (*see opposite*) and the solid Asian Selfs (*see page 144*). They combine the white undercoat of the former, caused by the inhibitor gene, with the solid coat caused by the non-agouti allele in the latter. They come in all the colours of the selfs, and Burmese colour restriction is also permitted: the Smoke already differs from the Burmese enough by virtue of its white undercoat for there to be no question of confusing the two breeds. With a coat slightly longer than that of the Burmese, the smoke undercoat shows in a faint gleam over curves when the cat is still, and a gentle shimmer when it moves. This coat type appeared in the second generation of crosses from Shaded Asians.

Black Smoke

In many cases it can be hard to tell a sepia Smoke from a solid Smoke, because Smokes naturally tend to be darker on the short hair of their "point" areas. Black Smokes, however, are not sepia pointed; if they were, the colour would also be affected, becoming a deep sable-brown.

ANY COLOUR FROM GOLD TO GREEN

UP TO HALF OF HAIR-LENGTH IS WHITE

KEY FACTS

DATE OF ORIGIN 1981

PLACE OF ORIGIN Great Britain

ANCESTRY Burmese, Chinchillas, non-pedigrees

OUTCROSS BREEDS Burmese, Chinchillas with restrictions

OTHER NAME Smokes were once called Burmoires

WEIGHT RANGE 4–7 kg (9–15 lb)

TEMPERAMENT Relaxed and engaging

    GCCF

BREED COLOURS

BURMILLA OR SHADED (SOLID, SEPIA)
Black, Chocolate, Red, Blue, Lilac, Cream, Caramel, Apricot, Tortoiseshell, Chocolate Tortie, Blue Tortie, Lilac Tortie, Caramel Tortie

SILVER SHADED
Colours and patterns are as for shaded

SMOKE (SOLID, SEPIA)
Black, Chocolate, Red, Blue, Lilac, Cream, Caramel, Apricot, Tortoiseshell, Chocolate Tortie, Blue Tortie, Lilac Tortie, Caramel Tortie

SELF (SOLID)
Bombay, Chocolate, Red, Blue, Lilac, Cream, Caramel, Apricot, Tortoiseshell, Chocolate Tortie, Blue Tortie, Lilac Tortie, Caramel Tortie
Sepia colours

TABBY (ALL PATTERNS IN SOLID, SEPIA)
Brown, Chocolate, Red, Blue, Lilac, Cream, Caramel, Apricot, Tortoiseshell, Chocolate Tortie, Blue Tortie, Lilac Tortie, Caramel Tortie

SILVER TABBIES
Colours and patterns are as for standard tabbies

Chocolate Smoke

Although Asian Smokes are genetically non-agouti cats, they should show ghost tabby markings, which give the coat the look of watered silk. Silvery frown marks are apparent on the forehead, and there may be lighter "spectacles" or "clown marks" around the eyes.

ASIAN SELFS

INDEPENDENT OF THE BREEDING PROGRAMMES THAT WERE INITIALLY underway in the Asians, black Burmese-type cats, called Bombays (*see page 139*), were being bred in the 1980s. Although the Bombay development programme had used non-pedigree cats, it was then integrated into the continuing development of the Asian group as a whole. Other self colours were also developed, and there is now a wide range of colours. The sepia pattern is not allowed for showing in any Asian self colours: this helps to preserve the identity of these cats as distinct from the Burmese. Self breeders, in particular, tend to see selfs as distinct from shaded cats, and some do not use cats from Chinchilla lines in their breeding programmes. There are some breeders, however, who hold to a philosophy that all colours are one breed.

Black Tortie

As in the Burmese, the standard for tortoiseshell Asians allows for colours to be either subtly mingled or dramatically blotched. Distinct facial blazes and solid legs and tails are all accepted. The colour should be even to the roots of the hair: in the Black Tortie the standard calls for jet black mingled with varying shades of red. The eye colour may range from gold to green.

PATCHES ARE DARK
AND LIGHT RED

DARKER
COLOUR
ON FACE

Burmese variant Cream

The recessive gene for Burmese pointing is still present in the gene pool of these cats. A sepia-pointed self such as this one cannot appear in cat shows, but it may still be used in breeding programmes because this pattern is allowed in other cats within the Asian group.

Bombay

Characterized by a sleek black coat, the Bombay is one of the original Asian selfs. This colour should not be confused with the North American breed with which it shares its name (see page 139). The type of the British Bombay resembles that of the European Burmese (see page 148); the American Bombay stands alone as both breed and type.

ASIAN TABBIES

AS WELL AS THE RECESSIVE SELFS AND SMOKES, TABBIES ALSO BEGAN to appear in the second generation of Asian crosses. They appeared because the Chinchilla Longhair (*see page 188*), one of the group's founding cats, has an agouti coat, although its tabby markings are rendered invisible by the inhibitor gene. It was then found that some breeders had also been bringing Abyssinians (*see page 134*) into their programmes to introduce the ticked gene, and this work was also incorporated into the Asian breed. The result is that the Asian tabby is one of the few cats recognized in Britain in which all four tabby patterns are bred; the ticked pattern is the most popular of them. The slightly disparate origins of these various coat patterns were another factor in the unusual decision to grant the Asians status as a group, rather than as a single breed.

EDGES OF EARS
ARE DARK

STRONG TABBY
MARKINGS

PENCIL MARKINGS
ON CHEEKS

Black Ticked Tabby

Ticked tabbies are allowed in both standard and silvered coats, in all the Asian colours. The Asian Ticked Tabby standard allows no unbroken necklaces, but it does call for tabby markings on the legs and tail. The intensity of the colour may be reduced in silver tabbies, but there should always be at least two bands of darker ticking on every hair. Underparts are paler than the rest of the coat, but should harmonize in colour.

NECKLACES MUST
BE BROKEN

TABBY MARKINGS
ON LEGS

Blue Ticked Tabby kitten

The blue in Asian standards is medium to dark, rather than the light colour preferred in many other breeds. This helps to provide good contrast in the tabby colours. The ticked pattern is the most commonly found tabby pattern in Asians, and tabbies are less common than the other coat types.

DARK TIP

BLACK RUNS UP
TO HOCK

AMERICAN BURMESE

RICHLY COLOURED AND WIDE-EYED, THIS BREED HAS BEEN DESCRIBED AS "bricks wrapped in silk". Although it is fond of human company, the Burmese is less vocal or demonstrative than other Oriental breeds. The North American Burmese differs from its European counterparts (*see page 148*), in that its standard emphasizes roundness, notably in the shape of the head. The extremely round "contemporary" look was born out of Good Fortune Fortunatus in the 1970s; unfortunately, so was the Burmese head fault, an inherited deformity of the skull that is often lethal or requires euthanasia. As late as the 1980s, Sable was the only colour universally accepted. Other colours had occurred since early breeding programmes, but were known within CFA as Mandalays. TICA recognizes a wider range of colours.

Contemporary head
The "Contemporary" is one of the three types of Burmese recognized in the US today. This look ruled the CFA until 1995, when their top three cats were of the more rounded "Traditional" type. The Classic is the third type.

BREED HISTORY

The Burmese begins with Wong Mau, a brown cat from Rangoon in Myanmar (formerly Burma), brought to the United States with US Navy psychiatrist Joseph Thompson in 1930. She founded a breed of which, ironically, she was not a member. Thompson bred Wong Mau to a Siamese (*see page 154*), the most similar breed, and bred the kittens back to Wong Mau. Three types emerged: Siamese-pointed; dark brown with minimal pointing (the first true Burmese); and a dark body with darker points, like Wong Mau. She herself was a Burmese-Siamese hybrid, and a natural Tonkinese (*see page 152*).

Champagne
In the slightly idiosyncratic terminology of the Burmese colours, CFA puts this colour in its "dilute" category; it is in fact the brown more generally referred to as chocolate. Some degree of darkening in the mask area is almost unavoidable in this colour; the main coat is a warm, even, honey brown.

Body
Medium-sized, muscular, and compact

Coat
Short and fine, with satin-like texture and glossy shine

CLOSE-LYING COAT

TAIL IS DARK

LEGS SHOW LEAST POINTING

Sable
This was the first, and for a long time the only, Burmese colour. It was recognized in the 1930s, and ruled in splendid isolation for decades. Kittens are often born lighter than the standard requires, their coats darkening with maturity. The standard calls for a solid-coloured cat, with the genetic masking or pointing so minimal as to be imperceptible.

EYES BROWN IN YOUNG CATS

Blue
The genetic dilute of the Sable, the American Blue Burmese is not the cool, frosted colour of many other blue cats. The standard calls for warm, fawn undertones, in marked difference to the standard for the European Burmese.

Tail
Medium in length

EARS TILT FORWARDS

ROUNDED TIPS

Eyes
Rounded in shape
and golden in
colour

Ears
Medium-sized
and widely spaced

Head
Pleasingly rounded,
with full cheeks

Muzzle
Short and
broad, with a
rounded chin

WELL-DEVELOPED NECK

SLIGHT MASK APPARENT

COAT LOOKS FROSTED

Legs
Moderate in boning
and length

ROUND PAWS

KEY FACTS

DATE OF ORIGIN 1930s
PLACE OF ORIGIN Myanmar (formerly Burma)
ANCESTRY Temple cats, Siamese crosses
OUTCROSS BREEDS None
OTHER NAME Some colours previously Mandalay
WEIGHT RANGE 3.5–6.5 kg (8–14 lb)
TEMPERAMENT Friendly and relaxed

CFA TICA

BREED COLOURS

SEPIA
Sable, Champagne, Blue, Platinum
All other self and tortie colours

SABLE TORTIE (NOT CFA)	RED (NOT CFA)	CARAMEL (NOT CFA)	CINNAMON (NOT CFA)

Frost or Platinum

*This is the colour referred to in other
cats as lilac or lavender: the Burmese
names for it are unique to this breed.
As with the Blue, the tone of the colour
is warmer in the Burmese than other
breeds, with pale fawn undertones. The
shade is also lighter than in other
breeds, accounting for the metallic
name, which is less used today.*

Burmese ancestor

*The Thai Cat Book Poems include
copper-coloured "Thong Daeng" cats,
which are possibly ancient Burmese.
Mutations are rarely completely isolated:
Siamese pointing occurred elsewhere in
Southeast Asia – Wong Mau carried it –
and the Burmese pattern may similarly
have been known in Thailand.*

EUROPEAN BURMESE

THIS BREED HAS DIVERGED INTO TWO TYPES, ALMOST TWO BREEDS, on either side of the Atlantic. While the American side of the family (*see page 146*) has developed into a rounded cat, European breeders, followed by those in South Africa, New Zealand, and Australia, have opted for a well-muscled but more angular shape. This more Oriental look has a moderately wedged-shaped head, oval eyes, and long legs. Both planned and accidental matings expanded the colour range to ten: more than CFA recognize, but lacking the Cinnamon and Fawn accepted by TICA. Regardless of colour or type, the Burmese is ideally suited to living in active households.

BREED HISTORY

The European Burmese is descended from the American breed: American cats were imported into Europe after World War II, and the brown was recognized in 1952 by GCCF. Here, the story takes a new turn. Firstly, the Europeans preferred a more Oriental look, and secondly, they were interested in a wider range of colours. GCCF, for example, took only eight years to recognize the Blue, while in North America CFA took decades to do so. This seems to have started a trend: the wide range of colours seen in the European breed was developed by the introduction of the red gene (from a red-point Siamese, a tortie-and-white farm cat, and a ginger tabby), and in the 1970s this range was broadened to create tortie versions of all the recognized colours. A further change came in 1996, when FIFé amended its breed standard to allow green eyes.

Burmese head
This Lilac shows the typical European Burmese head and eyes. Burmese eye colour can appear very different in different lights. A golden yellow is much preferred: South African and Australian cats are particularly noted for their excellent eye colour.

Brown Tortie
Deep brown with shades of red and cream, this is the tortie version of Sable. In the Burmese, tortie patterns may be quite dramatically blotched, as in this cat. Facial blazes and solid areas of dark or light colour are also allowed.

HARD BODY

SOLID-COLOURED
LEGS PERMITTED

ROUNDED TIP

HINDLEGS LONGER
THAN FORELEGS

Legs
Slender legs in proportion to body

Chocolate
This colour is more accurately detailed as a shade of "milk chocolate" in the breed standard. An even overall colour is preferable, but elusive. Although the nose leather matches the coat, the paw pads are pink, shading to brown.

Blue
The British breed standard recognizes the Burmese as a pointed cat, but requires that the shading be slight, gradual, and confined to the face. The underparts may also be lighter than the rest of the coat. The standard for the Blue calls for a soft colour, with a distinct silver sheen showing on any rounded areas of the body.

Lilac Tortie

This subtly mingled lilac-and-cream coat is not rated above the more blotched mixtures in the Burmese, although it would be in other breeds in Britain. Because tortie results from the presence of the red gene, it is unknown in the more traditional American cats.

NOSE LEATHER MOTTLED

Red

A tangerine shade is preferred in Burmese, with less rufousing than is seen in other breeds. Exposed skin may show some freckling, and tabby markings are tolerated on the face, although they are a fault elsewhere. The short Burmese coat will not camouflage vestigial tabby markings, so a good Red is to be prized.

EARS TILT FORWARDS

Head
Short, blunt wedge

Ears
Medium-sized, rounded at tip

Body
Strong, muscular, surprisingly heavy

Colour
All colours must be solid to roots

Eyes
Large, widely spaced, and slightly slanted

WIDELY SET

Muzzle
Blunt, with deep chin

Underparts
May be slightly lighter than rest of coat

Coat
Short and fine, lying close to body

OVAL PAWS

KEY FACTS

DATE OF ORIGIN 1930s
PLACE OF ORIGIN Myanmar (formerly Burma)
ANCESTRY Temple cats, Siamese crosses, household cat crosses
OUTCROSS BREEDS None
OTHER NAME None
WEIGHT RANGE 3.5–6.5 kg (8–14 lb)
TEMPERAMENT Friendly and relaxed

GCCF FIFé

BREED COLOURS

SELF AND TORTIE COLOURS
Brown, Chocolate, Red, Blue, Lilac, Cream, Brown Tortie, Chocolate Tortie, Blue Tortie, Lilac Tortie.

CHOCOLATE TORTIE   CREAM   BROWN   LILAC

Burmese (overleaf left and right) *For breed entry see pages 146 – 9*

TONKINESE

SOME BREEDERS DISPUTE WHETHER THESE CATS CAN IN FACT BE called a breed: as a hybrid of the Burmese (*see page 146*) and the Siamese (*see page 154*), Tonkinese inevitably produce variants in the pointing patterns of both their parent breeds. But this is hardly the first breed to produce variants, and the softly pointed "mink" pattern is not the only distinguishing feature of the breed. Their type is a blend of their parent breeds, less angular than one but lighter than the other, and their temperament has all the lively curiosity and affection of an Oriental breed, without the more vociferous traits. They also possess an attractive and distinctive eye colour that is often referred to as "aqua". With its pleasing looks and friendly temperament, it is small wonder that the "Tonk", although it has yet to achieve full championship status in some registries, is already one of the more popular breeds.

BREED HISTORY

Some breeders believe that the "Chocolate Siamese" of the 1880s were in fact Tonkinese-type Siamese-Burmese hybrids, but this is unprovable. The first documented Tonkinese seen in the West was Wong Mau, the mother of the Burmese breed imported from Rangoon. Her natural hybrid characteristics were bred out of her offspring, however, and not until the 1950s did work begin to recreate this blend through controlled breeding programmes. Early work was carried out in Canada, and the breed was first recognized by the Canadian Cat Association. Today it is accepted by all major registries, although there is still considerable variation in the acceptance of colours. British authorities, for example, accept all the colours that already exist in the Burmese. There have been two attempts to produce a longhair of this pattern, the Silkanese and the Himbur, but neither gained the popularity of the Tonk.

Brown or Natural

Called Sable in the Burmese and Seal in the Siamese, this colour has two names in the Tonkinese, being called Natural in North America and Brown elsewhere. It should be a light brown with darker seal points and matching seal nose leather and paw pads. The type of the Tonk is well muscled and medium-foreign. There is a slight difference between North American and European cats, perhaps reflecting the difference in Burmese type: European Tonks tend to be slightly more angular than their transatlantic counterparts.

Ears
Slightly taller than wide, with broad base and oval tips

Head
Moderate wedge with slight nose break and whisker pinch

Eyes
Upper edge oval, lower edge rounded, blue-green in colour

Pattern
Darker points merge gently into lighter body

Body
Medium to long, muscular, surprisingly heavy

Coat
Short, silky, and close-lying

Tail
Equal to body in length, neither heavy nor thin

Legs
Slim, well muscled, and in proportion, with oval paws

Chocolate Tortie
The pointing is less apparent when overlaid with tortie or tabby coat patterns, but the mask and legs should still be darker than the body. Tonkinese eyes may be a light aqua blue, as in this cat, but Siamese-blue eyes are not allowed.

Cream
The standard for this colour in the Tonkinese calls for a "rich, warm" tone, shading to paler cream. The points may be less even than in other colours, with the legs being lighter than the mask and tail. As in all breeds, it is difficult to eliminate tabby markings completely from Creams and Reds, and slight markings are allowed in cats that are excellent in other respects.

KEY FACTS

DATE OF ORIGIN 1960s
PLACE OF ORIGIN United States and Canada
ANCESTRY Burmese and Siamese
OUTCROSS BREEDS Burmese and Siamese
OTHER NAME Once called Golden Siamese
WEIGHT RANGE 2.5–5.5 kg (6–12 lb)
TEMPERAMENT Sociable and intelligent

 GCCF CFA TICA

BREED COLOURS

SELF AND TORTIE MINK
Brown, Chocolate, Red, Blue, Lilac, Cream, Brown Tortie, Chocolate Tortie, Blue Tortie, Lilac Tortie Cinnamon, Fawn, pointed and sepia patterns

TABBIES (ALL PATTERNS)
Colours are as for self and tortie

BLUE   BROWN TABBY   CHOCOLATE   LILAC TORTIE

Lilac
The body of the Lilac is a pale, dove grey with a pinkish cast, and the points are a darker shade of the same colour. Kittens are often lighter than adults, although they should develop distinct pointing fairly early. Eye colour can range from light blue to the green seen in this individual, but any tinge of yellow is unacceptable. The exact shade is not related to the coat colour.

Red
The light red body shades to darker points, although the legs may not show much shading. Slight freckling may appear on the nose leather, lips, eyelids, ears, or pads of Reds and Creams: this is not penalized in mature cats. Reds and Creams are not universally accepted in North America.

SIAMESE

POSSIBLY THE WORLD'S MOST INSTANTLY RECOGNIZABLE CAT BREED, the Siamese is also one of the more controversial. The first cats often had crossed eyes and kinked tails; the early breed standards even required these traits, as well as legs that were "a little short". Since then, selective breeding has altered the cat considerably – crossed eyes and kinked tails are now rare, and regarded as a serious fault – but the build of the cat is a matter of dispute. GCCF Siamese have a svelte body, long, slim legs, and a long head with slanting eyes and a fine muzzle; North American cats take this look to greater extremes. All Siamese still have the gregarious, chatty nature for which they are famous: this is the most vocal breed, even strident in its talking. It is also one of the most sexually precocious, often mating by five months of age.

BREED HISTORY

Siamese originated in a mutation in Asia over 500 years ago. The mutation may have been widespread: in the late 1700s, the naturalist Pallas described a white-bodied cat with dark ears, feet, and tail in central Asia. In Thailand, however, these cats were revered by Buddhist monks and royalty. They made their way to the West in the late 19th century, appearing at a British show in 1871. Originally the Siamese included solid colours: these are now known as Oriental Shorthairs (*see page 160*). New colours have been bred (*see page 156*), and the longhaired Balinese (*see page 228*) has won recognition. The popularity of the breed peaked in the 1950s, although it still remains the most popular CFA shorthair. In 1987, a group of breeders formed TCA to breed "Old-Style" or "Applehead" Siamese (*see page 180*).

Prestigious pet
The Siamese has long been among the preferred pets of stars from Jean Cocteau to Anna Pavlova, seen here with her kitten. Even today, the Siamese counts celebrities such as actress Anjelica Huston among its fans.

Lilac Point
In 1896 a cat was disqualified from a show for being "not quite blue" – some shows did accept Blue Points. It is possible, but by no means certain, that the cat was an unrecognized Lilac Point. The American standards call for a white body, while the English allows slight shading.

Chocolate Point
With points the colour of milk chocolate on a ground of ivory, this colour was not accepted in Britain until the 1950s. It may have been carried recessively from the start: dark Chocolate Points could have passed for poor Seal Points.

GLOSSY SHINE

Body
Medium-sized, long, and svelte

Coat
Very short and fine, with no undercoat

FINE TIP

Tail
Long, tapering, and free of kinks

Legs
Hindlegs longer than forelegs

Paws
Small and oval

SIAMESE 155

Changing looks
This 1930s lithograph by Jacques Nam, although slightly stylized, shows a Siamese far heavier than the show cats of today. Artists' works are open to interpretation, but old photographs and films also suggest a less etiolated build.

Siamese head
The long head should have good width between the ears, tapering down to a fine muzzle. In profile, the nose is straight, with only the slightest change in direction at eye level. The head is carried on a slender, elegant neck.

PRICKED TIPS

WIDE BASE

Ears
Follow lines of face outwards

Eyes
Oriental in shape, slanted in set

DEEP, BRILLIANT BLUE

Head
Long, narrowing to muzzle in straight lines

Mask
Clearly defined, covering whole face

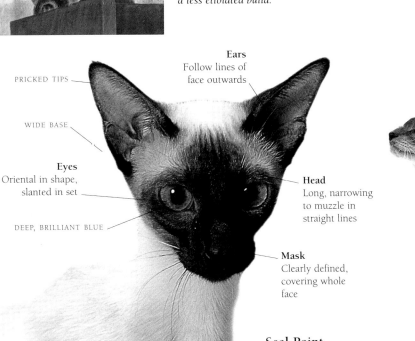

Blue Point
Records of Blue Points go back at least as far as 1903: given that Thailand's other famous cat, the Korat (see page 138) is blue, it is likely that the dilute gene came to the West with the Thai cats, rather than being introduced after arrival. The blue of the Siamese points is lighter than that found in solid-coloured coats.

Seal Point
The classic colour, this is the Siamese that has appeared in films, advertisements, and cartoons. Genetically, this colour is black, translated into a dark seal-brown by the Siamese pointing gene. For a time, this was the only Siamese accepted: other colours existed, but were usually classified as "any other variety", and records are unclear as to what they may have been. For some, this colour remains the only "true" Siamese.

Legs
Slim, in proportion to body

Points
All dense and matching in shade

Original standard
The Thai Cat Book Poems (see page 94) show a short-legged cat with a pure white body and dark ears, tail, and paws only: this is because the Siamese is paler in warm climates. The first Siamese to reach the West were Seal Points that resembled the book's renditions.

KEY FACTS

DATE OF ORIGIN Pre-1700s
PLACE OF ORIGIN Thailand
ANCESTRY Household and temple cats
OTHER NAME Royal Cat of Siam
OUTCROSS BREEDS None
WEIGHT RANGE 2.5–5.5 kg (6–12 lb)
TEMPERAMENT Energetically enterprising

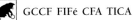 GCCF FIFé CFA TICA

BREED COLOURS

SIAMESE POINT COLOURS
Seal, Chocolate, Blue, Lilac,
COLOURPOINT SHORTHAIR POINT COLOURS (IN CFA)
Red, Cream, tortie and tabby versions of all colours
Cinnamon, Fawn, Smoke, Silver and parti-colour versions

CHOCOLATE TORTIE POINT | CREAM TABBY POINT | LILAC TORTIE TABBY POINT | CHOCOLATE TABBY POINT

NEWER SIAMESE COLOURS

NEW COLOURS AND TABBY STRIPES HAVE AUGMENTED THE original four Siamese colours (*see page 154*) since the 1960s. "Shadow-points" and silver-points also exist, but have never won wide recognition. These pointed cats live under two names. In Britain and mainland Europe, all colours are classified as Siamese, and pointed Oriental Shorthairs are also classified and registered as Siamese. In North America, TICA also follows this method, but CFA classifies only the eumelanistic colours as Siamese and calls all others Colorpoint Shorthairs, relegating pointed Orientals to the "any other variety" category. One body of opinion suggests that, as with the Asian Group, there could be an Oriental Group, with pointed, solid, shorthair, and longhair divisions, but this strategy is unlikely to be widely accepted. Regardless of name, these cats are Siamese both in spirit and in conformation. Actively inquisitive, often resoundingly vocal in their demands, they are absurdly gregarious in their relationships with people. Like the Siamese, these cats live long lives, but they may be more prone than other breeds to inherited heart disease (*see page 66*), and amyloidosis, which causes lethal deposits in body organs.

Chocolate Tabby Point

Tabby points have been known since the turn of the century, but were generally ignored. A litter of Seal Tabby Points was shown in 1961 in Britain, bringing renewed interest, more active breeding, and acceptance a few years later. In North America, these cats are called Lynx Points.

SOLID-COLOURED TIP

Red Tabby Point

Tabby versions of all the point colours are now generally accepted. In all, especially the Red and Cream, the legs may be paler than in solid points. Shading on the body should be no more than in solid points, but it should follow tabby patterning.

Cinnamon Point

The Cinnamon Point is one of the newest colours. The body is ivory and the points a warm cinnamon-brown; as in the Chocolate Point and the Caramel Point, the legs may be slightly paler than the other points.

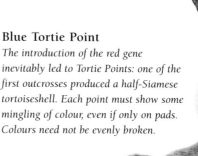

Blue Tortie Point

The introduction of the red gene inevitably led to Tortie Points: one of the first outcrosses produced a half-Siamese tortoiseshell. Each point must show some mingling of colour, even if only on pads. Colours need not be evenly broken.

Lilac Tabby Point

The points should all show tabby markings, but these should not extend over the body; in particular, the frown lines on the forehead should not extend over the rest of the head. The Lilac Tabby Point has pinkish-grey markings on a magnolia ground, with faded lilac or pink nose leather and paw pads.

Fawn Point

Another very recent colour, this is the dilute of the Cinnamon. The standard calls for "rosy mushroom" points against a magnolia body colour. As in many lighter colours, the legs may be paler than the mask and tail.

Cream Point

Crosses to Red Tabby Longhairs brought this gene into the Siamese in the 1930s, but the controversial colour faded into obscurity. It was recreated in the late 1940s, but the loss of body type from the outcrossing was a problem for a time. They gained general recognition in the mid-1960s.

FRECKLING ON FACIAL SKIN ALLOWED

Seal Tortie Tabby Point

In these cats, the colouring may be mottled, mingled, or patched. The degree of patching is not a prime consideration, as long as it is present; the tone of the colours is more important.

COAT SHADING DEEPENS WITH AGE

Siamese (overleaf left and right) *For breed entry see pages 154 – 7*

ORIENTAL SHORTHAIR

EVERY OWNER CAN TELL YOU THAT THIS BREED'S FAVOURITE PLACE is between you and your book, or newspaper, or keyboard. These active and athletic cats are also outrageously gregarious – breeders call them shameless flirts. In physique and temperament, the Oriental is a Siamese (*see page 154*), but in solid coat colours. There is some dispute over the status of the pointed Orientals that do crop up: most fancies classify them as Siamese, but the CFA in North America does not allow this. Like Siamese, Orientals can suffer from inherited heart problems, but they have impressively long lives, belying their reputation as a delicate breed.

BREED HISTORY

The historical *Cat Book Poems* (*see page 94*) show Siamese or Thai cats in varied colours. Today, over half of their descendants are self or bi-colour, and fewer than a quarter are pointed. There were self cats among the first Siamese to be brought to the West, but in the 1920s the Siamese Club of Britain vetoed "any but blue-eyed Siamese", and numbers declined, although blacks and blues may have been bred in Europe until World War II. Work on a solid chocolate in Britain in the 1950s led to the Chestnut Brown Foreign, recognized in 1957 and the origin of the Havana Brown. In 1995, CFA combined Oriental Shorthairs and Longhairs (*see page 234*) into one breed group called Orientals; a move that caused controversy.

Brown Spotted Tabby

Oriental Spotted Tabbies were once called Maus, but they were confused with the Egyptian Mau breed in the United States (see page 172). Their spots should be round, clear, and evenly distributed.

KITTENS MAY HAVE SOLID SPINAL LINE

NECKLACES ARE ACCEPTABLE

Oriental Blue

Blue cats were imported from Thailand in the 1800s, but these could also have been Korats (see page 138). There could be no confusing the breeds now: the Oriental Blue has the characteristic elongated build and slanted eyes of the breed.

TRIANGULAR FACE

Ears
Large, pricked, continuing lines of head

EARS WIDE AT BASE

Eyes
All Orientals have green eyes

Head
Long, triangular wedge, described by straight lines

SLANTED, WIDELY SET EYES

LONG, SLENDER NECK

Legs
Long and slim, but muscular

SMALL, OVAL PAWS

Tail
Long, tapering to point

Havana or Chestnut Brown

This rich, warm-toned brown, genetically a chocolate, was called Havana by early breeders, but then recognized as the Chestnut Brown Foreign, reverting to its present name in 1970. It is still known as Chestnut Brown in the United States, where the Havana Brown is a separate breed.

NOSE AND CHIN ALIGN

Body
Medium-sized, long, and svelte

Coat
Very short, fine, and glossy

Lilac or Lavender

The dilute of the Havana, this colour was one of the first to be developed in the 1960s. It was at first called Lavender, a term still used in North America. As with all dilute colours, any faint tabby markings are more easily seen, so good colour is quite an achievement.

COLOUR OF COAT MUST BE EVEN

HINDLEGS LONGER THAN FORELEGS

COAT IS FROSTY, PINKISH GREY

Foreign White or Oriental White

This colour is the last one where different names still exist. It is called Oriental White internationally, and in most countries it may have green or blue eyes. In Britain, only blue eyes are allowed, and the designation "Foreign" has been retained to acknowledge this.

Oriental Black

In sleek, solid black, the Oriental build resembles nothing so much as an Art Nouveau poster come to life. The entire cat must be solid, jet black, from the tips to the roots of the coat and from the eye rims to the paw pads. Eyes of brilliant, unflecked green stand out dramatically against this background, making a very striking cat.

KEY FACTS

DATE OF ORIGIN 1950s
PLACE OF ORIGIN Great Britain
ANCESTRY Siamese, Korat, Persian, Shorthairs
OUTCROSS BREEDS Siamese and Colourpoint
OTHER NAME Previously called "Foreigns" in Britain
WEIGHT RANGE 4–6.5 kg (9–14 lb)
TEMPERAMENT Devoted and demanding

GCCF FIFé CFA TICA

BREED COLOURS

SELF AND TORTIE
Black, Havana, Cinnamon, Red, Blue, Lilac, Fawn, Cream, Caramel, Apricot, Foreign White, Black Tortie, Chocolate Tortie, Cinnamon Tortie, Blue Tortie, Lilac Tortie, Fawn Tortie, Caramel Tortie

SMOKE, SHADED, AND TIPPED
Colours are as for self and tortie, excepting White

TABBIES (ALL PATTERNS)
Colours are as for self and tortie

SILVER TABBIES (ALL PATTERNS)
Colours are as for standard tabbies

SILVER SPOTTED TABBY   BLUE AND WHITE   CHOCOLATE TORTIE   RED MACKEREL TABBY

NEWER ORIENTAL COLOURS

NEW COLOURS AND PATTERNS CONTINUE TO BE DEVELOPED within the Oriental Shorthair breed. Many of the genes for these cats have been present since Our Miss Smith, the breed's Siamese founding mother, gave birth to green-eyed, brown-coated kittens in the 1950s. Since then, other genes have been intentionally added. The smoke coat, which has a shimmering presence only apparent in the short, close-lying fur when a cat is moving, and other silvered types resulted from the accidental crossing of a Siamese to a Chinchilla Longhair (*see page 188*), an outcross that also introduced the longhair gene responsible for the Oriental Longhair (*see page 234*). Today there are over 50 recognized coat colours, and this number will probably continue to increase. Bi-colours are accepted by North American associations, but not in Britain; matings were made in Britain in an attempt to create a "Seychellois" breed, with Van patterning on an Oriental body, but the programme never reached fruition.

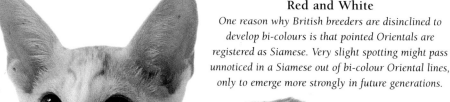

Red and White
One reason why British breeders are disinclined to develop bi-colours is that pointed Orientals are registered as Siamese. Very slight spotting might pass unnoticed in a Siamese out of bi-colour Oriental lines, only to emerge more strongly in future generations.

Red
As in other Oriental breeds, sex-linked red colours made a late appearance, even though this gene does exist in the cat populations of Asia. The coloration should be as even as possible, but tabby markings are tolerated in otherwise good cats.

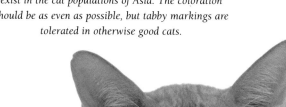

FRECKLING
ALLOWED

Apricot
A recent addition to the Oriental colour range, Apricot should be a reddish cream. It is darker and warmer in tone than a true cream. The genetics behind this colour are still disputed among breeders.

TABBY RINGS

LONGER HAIR
ON BELLY THAN
ELSEWHERE

Chocolate Silver Tabby
*In Silver Tabbies the colour is restricted to
the tips of the hairs and the undercoat is
white. Although the contrast between
the markings and the base is quite
clear, the colour may be
so restricted that the
pattern cannot be
distinguished on
the flanks.*

WHITE
UNDERPARTS

Cinnamon
*Although the cinnamon colour was not present
in the earliest Orientals, it was one of the
earlier colours added to the range. The gene
was introduced in the 1960s, by crossing
Havanas to Sorrel Abyssinians (see page 134),
which are genetically cinnamon.*

Chocolate Classic Tabby
*The earliest Oriental Tabbies were of the
spotted pattern, but since then classic,
mackerel, and ticked patterns have all been
added to the range. Oriental Tabbies often
have white lips and chins, but this should
not extend over the muzzle or throat.*

WHITE
LIPS

JAPANESE BOBTAIL

PLAYFUL AND AFFECTIONATE, THIS BREED IS A DISTINCTLY ATTRACTIVE companion. The Japanese Bobtail's ancestors were portrayed in ancient art (*see page 38*), and superstition may have played a role in the perpetuation of its most notable characteristic, its short, 8–10 cm (3–4 in) tail. In Japan, a cat with a bifurcated tail – one with two tips – was considered a demon in disguise (*see page 31*). Cats with normal tails may have been persecuted, but those with short tails left alone, which would have led to more successful breeding among Japan's short-tailed cats. The Bobtail is seen everywhere in Japan as the famous *Maneki-neko*, or beckoning cat, popular as a good-luck symbol (*see page 31*).

BREED HISTORY

In Japan, legend holds that cats arrived in the Japanese Archipelago from China in AD 999, and that their ownership was restricted to aristocrats for the next five centuries. In reality, cats had arrived hundreds of years earlier, and ownership was more general (*see page 24*). Among the original feline immigrants from mainland Asia there were individuals with stumpy or "bobbed" tails. Within the restricted gene pool that existed in Japan, the recessive gene for the bobbed tail flourished in the cat population. In 1968, American breeder Elizabeth Freret established the first breeding programme outside Japan. Recognized in North America, the breed is not yet accepted in Britain.

Red Tabby and White

The Bobtail has undergone some changes since its arrival in the West. Like this male, the original cats were less delicate in appearance, with broader faces and shorter legs. Breeders have worked to refine these features, and the breed is now quite distinct from simple bobtailed cats from Japan.

KEY FACTS

DATE OF ORIGIN Pre-19th century
PLACE OF ORIGIN Japan
ANCESTRY Household cat
OUTCROSS BREEDS None
OTHER NAME None
WEIGHT RANGE 2.5–4 kg (6–9 lb)
TEMPERAMENT Vibrantly alert

    FIFé CFA TICA

BREED COLOURS

SELF AND TORTIE COLOURS
Black, Red, Tortoiseshell, White
All other self and tortie colours, pinted, mink, and sepia
TABBY COLOURS (ALL PATTERNS)
All colours and patterns
BI-COLOURS
Black, Red, Tortoiseshell, Red Tabby white
All colours and patterns, except those that show evidence of hybridization

BROWN MACKEREL TABBY BLACK AND WHITE

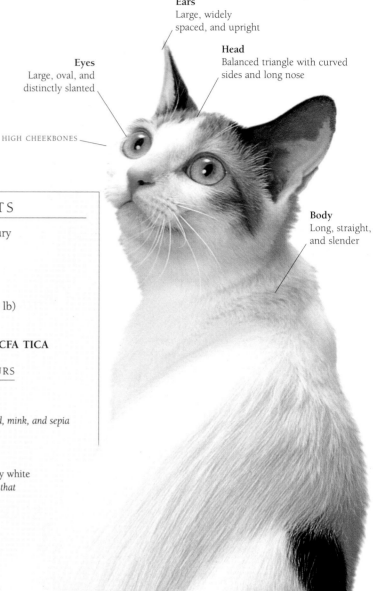

Ears
Large, widely spaced, and upright

Head
Balanced triangle with curved sides and long nose

Eyes
Large, oval, and distinctly slanted

HIGH CHEEKBONES

Body
Long, straight, and slender

Coat
Medium-short, with negligible undercoat

Legs
Long and slender, but not dainty

Tail
Short, with fanned hair resembling pom-pom

Mi-ke

Tortoiseshell and White Bobtails, known as Mi-ke in Japan, are the most prized coat colour and pattern. Odd-eyed Mi-ke cats are even more valued than their blue- or gold-eyed counterparts. The cat should resemble a porcelain figurine, with a pure white coat and minimal splashes of rich colour.

LAPERM

For centuries, rex mutations have occurred and then vanished into the random-breeding feline population. The advent of breed registries changed the situation. Since the first major rexed breeds, the Cornish (*see page 166*) and the Devon (*see page 168*), became established, many more have appeared. The LaPerm is undoubtedly the most quirkily named of these, and in some ways the oddest in expression. Cats are born with fur, either straight or curly, but at some time in their lives, usually during infancy, they lose it, becoming completely bald. The new coat that grows after this is thick and silky, and often curlier. Also, unusually for a pedigree breed, the standard describes them as working cats and "excellent hunters". There are both longhaired (*see page 235*) and shorthaired versions: the soft hair of the shorthairs may be more wavy than curly.

BREED HISTORY

In 1982, a working farm cat in The Dalles, Oregon, produced a litter of six kittens that included a single bald kitten. In spite of this disadvantage, the kitten survived, and at the age of eight weeks she finally grew a coat. But this coat, unlike that of her littermates, was curly and soft to the touch. Linda Koehl, the owner and founder of the breed, named this kitten Curly. Over the next five years, Koehl produced more curly coated kittens, which became the basis of the breed. The gene is dominant, so wide outcrossing to increase the gene pool can be done while still producing reasonable numbers of rexed kittens. The breed has been granted "new breed" status by TICA.

KEY FACTS

DATE OF ORIGIN 1982
PLACE OF ORIGIN United States
ANCESTRY Farm cats
OUTCROSS BREEDS Non pedigree cats
OTHER NAME Also called Dalles LaPerm
WEIGHT RANGE 3.5–5.5 kg (8–12 lb)
TEMPERAMENT Affectionate and inquisitive

BREED COLOURS

SOLID COLOURS
All colours/patterns, including sepia, pointed, and mink

SILVER TORTIE TABBY | BLUE | BLUE-CREAM | CINNAMON SILVER

Ears Medium-sized, with wide base and rounded tip

Eyes Large, almond-shaped, and slightly slanted

Head Broad, modified wedge, with prominent muzzle

FULL WHISKER PADS

Neck Long, and carried vertically

Body Medium-boned, muscular, and heavy for size

Coat Short, thick, and silky, with moderate undercoat

Legs Medium length and well muscled

Paws Medium-sized and round

Tail Long and tapering, with wavy hair

Brown Tabby kitten
With its softly wedge-shaped head, the LaPerm has a foreign look, especially apparent in kittens. Most individuals will go through their bald stage as kittens; straight-coated kittens may become curly after this drastic moult.

Red Tabby
Rexed coats tend to obscure the clarity of tabby markings, and the LaPerm is no exception. Frown lines on the forehead and "mascara" lines on the temples and cheeks remain clear, as do the rings on the tail and bars on the legs, where the hair is shorter or less curly. There is no relationship required between the eye and coat colour.

CORNISH REX

EXTROVERT AND CURVACEOUS, WITH WASHBOARD WAVES OF HAIR, the Cornish Rex is a show-stopper. The coat lacks guard hairs and is gloriously soft to touch, much like cut velvet. But the breed also has a distinctive physiognomy: dramatic ears are set high on a relatively small head, and the arched body is set on fine, lean legs. Although the coat is the same on both sides of the Atlantic, the conformation is slightly different. British cats are less delicate in appearance than their American relations, which have a "tucked-up" torso. This makes them look much like the feline equivalent of a racing hound, an impression in keeping with the breed's lively behaviour. Contrary to popular opinion, it should be stressed that no breed of cat, including the Cornish Rex, is hypoallergenic.

BREED HISTORY
In 1950, Serena, a farm cat from near Bodmin, Cornwall, had a litter with one curly-haired male kitten, Kallibunker. Her owner, Nina Ennismore, recognized this as similar to the "rex" mutation in rabbits. Breeding Kallibunker back to his mother confirmed that the trait was recessive. Descendants were crossed to British Shorthairs and Burmese. In 1957, the Cornish Rex arrived in the United States, where Oriental Shorthair and Siamese lines were introduced. A similar rexed breed was known in Germany, developed from a stray adopted by breeders in 1951. Crossing with the Cornish produced rexed kittens, revealing that both breeds had the same mutation; the German breed has since dwindled. The Cornish Rex is not related to any other rexed breeds.

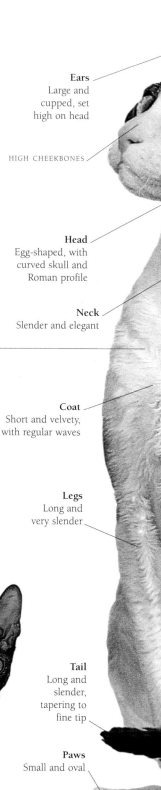

Ears
Large and cupped, set high on head

HIGH CHEEKBONES

Eyes
Oval, set at slight slant

Head
Egg-shaped, with curved skull and Roman profile

Neck
Slender and elegant

Body
Small to medium and muscular, with fine boning

Coat
Short and velvety, with regular waves

Legs
Long and very slender

Tail
Long and slender, tapering to fine tip

Paws
Small and oval

Tortoiseshell
In finely built breeds, such as the Cornish Rex, the breed standards favour female cats. This Tortoiseshell shows the typical Cornish arch of the spine and the upward tuck of the belly required in the North American registries' breed profiles.

HINDLEGS LONGER THAN FORELEGS

SLENDER, TUCKED TORSO

Black Smoke and White
The absence of guard hairs makes the white undercoat of Smoke and Silver colours particularly apparent in the Cornish Rex. This cat shows the typical North American conformation, with dramatically large ears, finely chiselled facial features, and a racy build.

Ears
Large, wide at base, with rounded tips

EARS SET HIGH ON HEAD

Head
Medium length, with rounded muzzle and strong chin

Eyes
Medium-sized and oval

British profile
In contrast to the curves of North American cats, the British standard for the Cornish Rex head requires a wedge shape with a flat skull, a curved forehead, and a straight nose. British cats also have smaller ears.

White
This male typifies the moderately foreign look of the British type. While the build should be slender, the body is strong and distinctly muscular, and the legs should not be overly long and fine. The original Cornish Rex, Kallibunker, had a distinctly foreign appearance, but the look of the breed was defined by the cats used in the early development of the breed, as well as later breeding choices on each side of the Atlantic.

Body
Medium-sized, hard and muscular

Coat
Short, plushy, and silky

Tail
Long, fine, and tapering

HAIR CURLS ON TAIL

Paws
Small and oval

Red Smoke
Cornish Rexes carry their bodies very high on their legs, with the spine in a gentle arch. The waves of the coat are most apparent over the back and rump. The red gene was present in the breed from Kallibunker's own offspring.

KEY FACTS

DATE OF ORIGIN 1950s
PLACE OF ORIGIN Great Britain
ANCESTRY Farm cat
OTHER NAME None
OUTCROSS BREEDS None
WEIGHT RANGE 2.5–4.5 kg (6–10 lb)
TEMPERAMENT Enterprising acrobat

GCCF FIFé CFA TICA

BREED COLOURS

All colours and patterns, including pointed, sepia, and mink

| CINNAMON SILVER | TORTIE AND WHITE | CHOCOLATE POINT |

DEVON REX

STARTLING, DRAMATIC EYES AND STRIKINGLY OVERSIZED, LOW-SET ears give the Devon Rex the look of an elfin clown. Its coat ripples, unlike the waves of the Cornish Rex (*see page 166*). Good breeding has greatly improved the coat: it now matures in four months rather than a year, and is very rarely patchy. Due to outcrossing with a variety of breeds, including the Persian, in the 1960s, longhairs sometimes occur. The coat has led to claims that the Devon is non-allergenic, but this can never be guaranteed. Breeders are unanimous that Devons never sit around looking elegant and bored; they are seldom elegant and always find life amusing: this, together with their coat, earned them the nickname "poodle cats". Typical breeder descriptions are "lovable slobs" and "love babies". Perhaps they should add "extraterrestrials": a Devon was the chosen pet in the otherworldly epic *Dune*. Superman also rescued one from a tree in the 1984 film, so Hollywood certainly likes this breed.

Silver Tortie Tabby

Wide outcrossing in the development of the Devon Rex gave it almost limitless patterns and colours. The curl shows up shaded colours well, and softens the markings.

BREED HISTORY

In 1960, Beryl Cox found a curly coated cat near an old mine in Devon, southwest England. A mating to a local female gave a normal litter with one curly coated kitten, named Kirlee, showing that the gene was recessive; Kirlee's parents were almost certainly closely related, and inbreeding was needed to perpetuate the Devon Rex. The Coxes bred Kirlee to several Cornish Rex females, but all the offspring had straight hair; the Devon Rex gene is a different mutation, and the breeds have been developed as distinct types. The Devon was recognized in Britain within the decade; in North America it was not separated from the Cornish until 1979. Unfortunately, early inbreeding brought to light a genetic spasticity syndrome, which is being studied as part of the Feline Genome Project.

Brown Tabby

The tabby markings on a Devon Rex will be most evident on the legs, where the hair is both shorter and less curly. All kinds of tabby patterns are allowed, but the ticked type is less often seen.

Ears
Large and very wide at base, tapering to rounded tips

EARS SET WIDE APART

BASE OF EARS "JUGGED"

GUARD, AWN, AND DOWN HAIRS ALL PRESENT

SLENDER NECK

Head
Wedge-shaped, with full cheeks and well-defined chin

CRINKLED WHISKERS MAY BREAK EASILY

BROAD CHEST

BODY CARRIED HIGH ON LEGS

Legs
Long and slender

Paws
Small and oval, with pads of any colour

KEY FACTS

DATE OF ORIGIN 1960
PLACE OF ORIGIN Great Britain
ANCESTRY Feral and household cats
OUTCROSS BREEDS British and American Shorthairs until 1998
OTHER NAME Nicknamed "poodle cats"
WEIGHT RANGE 2.5–4 kg (6–9 lb)
TEMPERAMENT Appealing clowns

 GCCF FIFé CFA TICA

BREED COLOURS

All colours and patterns, including pointed patterns

WHITE   BLACK SMOKE AND WHITE

Red Tabby

This Red Tabby was a top show cat. He was kept unneutered for breeding, and has developed the "stud jowls" that are often seen in older toms. Neutered Devons keep the classic elfin face, and never have this very full-cheeked appearance.

STUD JOWLS

Devon Rex profile

The short, wedge-shaped profile, seen here in a Blue, has a curving forehead, and a well-defined stop to the nose. The whisker pads are prominent; the whiskers themselves are coarse and brittle, and are easily broken.

Cream Point

The pointed colours are the result of crossbreeding with Siamese cats in the early development of the breed, now no longer allowed. The broad-chested body conformation of the Devon and its very distinctive "bulldog" stance are quite different from the Siamese, however. These colours were called "Si-Rex", a term less often used today.

Body
Slender, but hard and muscular

Coat
Very short, soft, and rippled or swirling

Black Smoke

The first Devon Rex to be bred, Kirlee, was a Black Smoke. Because of its curly coat, the Devon shows the smoke shading better than any straight-coated shorthair will, and the darker the colour is, the more striking the contrast with the silvery white undercoat will be. An Oriental look has been a characteristic of these cats since the early days of the breed, but the elfin appearance of the face can belie the solidity of the body.

Tail
Long and tapering, the tail must be well covered but not bushy

UNDERPARTS DOWNY BUT NOT BALD

HINDLEGS LONGER THAN FORELEGS

Blue-Cream and White

The broad chest of the Devon Rex can be clearly seen here. Combined with its quite slender legs, this can give an appearance of bow-leggedness, but the legs do not really curve. There are no skeletal deformities associated with the breed.

SPHYNX

HAIRLESS CATS HAVE APPEARED WORLDWIDE AT DIFFERENT TIMES. The most successful is the Sphynx. Not truly hairless, the Sphynx is covered in short, silky, "peach-fuzz" down. Without the insulating protection of a coat, Sphynx cats are vulnerable to both heat and cold and must be housed indoors. Each empty hair follicle has an oil-producing gland. With no hair to absorb the oil, Sphynx cats need daily rubbing down with a chamois.

BREED HISTORY

The first Sphynx, Prune, was born in 1966, but his line died out. In 1978, a longhaired cat with a hairless kitten were rescued in Toronto. The kitten was neutered, but his mother subsequently had other hairless kittens. Two were exported to Europe, where one was bred to a Devon Rex. Hairless offspring resulted (implying that this recessive gene may have some dominance over the Devon gene). One, nicknamed E.T., was acquired by Vicki and Peter Markstein in New York and bred again to a Devon Rex. Today, the breed is recognized by TICA and is accepted in the miscellaneous class by the CFA: many other registries fear potential health problems. In Britain, GCCF registers the Sphynx to ensure the gene is not carried into Devon Rex lines.

Blue-Cream and White

Sphynx show colours on their skin just as other breeds do on their hair; with no tweezing out of stray white hairs possible, these cats show the naked truth. The natural absence of hair apart from "peach-fuzz" down is vital: any evidence of hair removal is strongly penalized in shows.

Peterbald

Developed in St. Petersburg, Russia, the Peterbald is a new hairless breed. The first kittens, born in 1994, were the result of outcrossing an Oriental female with a Don Hairless male. Unrelated to the Sphynx, the gene that determines hairlessness is different in both breeds.

Body
Medium-sized and muscular

Tail
Long, strong, and whip-like

Ears
Very large, wide at base, and open

Head
Slightly longer than it is wide, with high cheekbones

FINE HAIR ALLOWED ON EARS

Eyes
Large and slightly slanted

Coat
Apparently hairless, but has fine down

POWERFUL NECK

Legs
Firm and muscular

Paws
Oval, with long toes and thick pads

Body
Rounded, hard, and muscular

Tail
Tapering and whip-like

Sphynx head
The jugged ears, large eyes, and elfin face of the Sphynx show the influence of the Devon Rex. Its whiskers, if there are any, are often brittle and broken.

KEY FACTS

DATE OF ORIGIN 1966
PLACE OF ORIGIN North America and Europe
ANCESTRY Non-pedigree longhair
OUTCROSS BREEDS Devon Rex
OTHER NAME Once also called Canadian Hairless
WEIGHT RANGE 3.5–7 kg (8–15 lb)
TEMPERAMENT Mischievous and playful

TICA

BREED COLOURS

All colours and patterns, including pointed, sepia, and mink

WHITE   BLACK   TORTOISESHELL

CALIFORNIA SPANGLED

THIS SOCIABLE AND ACTIVE CAT IS A SOLIDLY MUSCULAR BREED. The body is long and lean, but surprisingly heavy for its size. The round head is reminiscent of many small wild cats, and the dense double coat was developed to mimic leopard markings, with spotting over the back and sides, and striping between the ears and down the neck to the shoulders. The wild appearance of the California Spangled indicates that the diversity of coats in the cat family has not been lost in the domestic cat. The breeding programme has produced a kitten black at birth, except for its face, legs, and underbelly, maturing to show a coat that is similar to that of the rare African king cheetah.

BREED HISTORY

This breed is the creation of Californian Paul Casey, who set out to create a wild-looking spotted cat without wild cat bloodlines. Using non-pedigree cats from Asia and Cairo and a range of pedigree breeds, he produced the desired cat. It was launched in 1986 through the catalogue of a department store, a move unpopular with other breeders. "Spangled" is an ornithological term for spotted. Rosetted and ringed coats mimic those of the ocelot, margay, and jaguar. The breed has not as yet gained wide recognition at home or abroad.

Gold kitten
All spotted tabby California Spangled kittens are born with their spots. Kittens with the "snow leopard" pattern are born white, and those of the "King Spangled" pattern are born black. Blue eyes will become green or golden as the kitten matures.

Royal mutation
The spots of the African king cheetah are joined together in rosettes. The same mutation has come to light in the Spangled genepool. These King Spangled cats are not shown, but are being used in breeding studies.

Ears
Upright, with rounded tips, set far back on head

DARK RIMS TO EYES

Eyes
Oval and slightly slanted

Head
Wild in appearance, with broad, high cheeks and full whisker pads

Blue
The spots on a California Spangled vary in shape from round to oval or triangular: the wilder the appearance of the coat, the better. Investigation of patterns and development is still ongoing. The Blue is not as cool as it is in many other breeds: in the paler areas some rufousing may be seen.

KEY FACTS

DATE OF ORIGIN 1971
PLACE OF ORIGIN United States
ANCESTRY Abyssinian, Siamese, British and American Shorthairs, Manx, Longhairs, African and Asian street cats
OUTCROSS BREEDS None
OTHER NAME None
WEIGHT RANGE 4–8 kg (9–18 lb)
TEMPERAMENT Gently sociable

BREED COLOURS

TABBY (SPOTTED)
Black, Charcoal, Brown, Bronze, Red, Blue, Gold, Silver

SNOW LEOPARD
Colours and pattern are as for standard tabbies

BROWN   SILVER   GOLD

Body
Long and muscular

Coat
Short and sleek, but not close-lying

Legs
Medium length and muscular, with tabby bars

Tail
Tapering from heavy base

EGYPTIAN MAU

MAUS LOOK SIMILAR IN MANY WAYS TO THE CATS FEATURED on ancient wall paintings and scrolls in Egypt, and in fact Mau is the Egyptian word for cat. Body and face are both moderate in form, and the coat has a spotted pattern in shades of the original brown colour. Only the eyes do not tally: early portraits have wild-looking eyes, while the modern Mau has wide, round eyes with a decidedly worried look. According to fanciers, the Egyptian Mau can run at speeds of more than 48 kilometres per hour (30 miles per hour), making it the fastest breed of domestic cat.

BREED HISTORY

While all domestic cats trace back to ancient Egyptian ancestors, the Mau is possibly the breed that resembles them the most, and its ancestors can still be seen on Egyptian streets. Nathalie Troubetskoy, an exiled Russian, was taken with the spotted markings of street cats in Cairo, and imported a female to Italy to mate with a local tom. In 1956, she travelled to the United States, where the kittens were registered and shown the following year. The breed received full recognition from CFA by 1977, and is also shown in TICA, but remains almost unknown in Europe. It has often been confused with Oriental Spotted Tabbies (see page 160) in Britain, as these were called Maus for a time.

Smoke
Mau Smokes differ from smokes of other breeds. Instead of being selfs with no tabby markings, they are distinctly tabbies. Cats of this genotype are usually called shaded, but they are also usually much lighter than Mau Smokes. The white undercoat shows just enough to provide contrast, and no more.

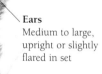

HINDLEGS LONGER THAN FORELEGS

Unique spots
Breed standards for the Egyptian Mau claim that it is the only natural spotted breed. The various definitions of what is a "natural breed" make this a difficult claim to assess, but it is certainly true that the Mau's spotting is unique. Ideally, the spots, which may be of any shape or size as long as they are clear, should be distributed randomly on the torso, not following the configuration of either marbled or mackerel stripes.

Head
Medium, rounded wedge without flat planes

Ears
Medium to large, upright or slightly flared in set

YELLOW CAST IN YOUNG CATS

Mau face
Moderate in shape, the face is neither rounded nor wedge-shaped. The nose is the same width from brow to tip, and the muzzle flows smoothly from the lines of the head. Strong "mascara" lines accentuate eyes.

Eyes
Large, rounded, and gooseberry-green in colour

Nose
Smooth curve from tip to forehead, with no distinct break

Paws
Small, very slightly oval, with long toes

LEGS ARE STRIPED

DARK TIP

Tail
Medium length,
tapering from
base to tip

Body
Medium-sized,
well muscled, neither
cobby nor foreign

Coat
Fine and silky, close-
lying, but not too short

Legs
Medium length, with
good musculature

Ancient model
*In ancient Egyptian art, such as
this depiction of Ra as a cat in
the Book of the Dead, cats are
striped or spotted. The blotched
mutation became prevalent in
the rest of the world, but
spotting remained common
in the cat's original home.*

Bronze
*There are only three colour forms of
the Egyptian Mau, and the Bronze is
the most ancient-looking. Markings
on the flanks are random, but spots
along the spine run in symmetrical
lines, often joining into a dorsal
stripe at the start of the tail.*

KEY FACTS

DATE OF ORIGIN 1950s
PLACE OF ORIGIN Egypt and Italy
ANCESTRY Egyptian street cats, Italian
domestic cats
OUTCROSS BREEDS None
OTHER NAME None
WEIGHT RANGE 2.25–5 kg (5–11 lb)
TEMPERAMENT Friendly and intelligent

FIFé CFA TICA

BREED COLOURS

SELF
Black, Blue

SMOKE
Black, *Blue*

TABBY (SPOTTED)
Bronze, *Blue*

SILVER TABBY (SPOTTED)
Silver, *Blue*

OCICAT

MORE THAN JUST ANOTHER NEW BREED WITH UNUSUAL SPOTTING, the Ocicat is an excellent blend of the attributes of its Siamese (*see page 154*) and Abyssinian (*see page 134*) blood. Playful and curious, with a love of laps, Ocicats enjoy company, respond well to early training, and are not suited to prolonged solitude. They are muscular and surprisingly solid; males are much larger than females. The most distinctive feature of the breed is its spotting. The distribution should follow the classic tabby pattern, swirling around the centre of the flank. Show quality Ocicats must have perfect spots: cats with imperfect spotting are sold as pets, but numbers are still limited.

BREED HISTORY

The Ocicat is a happy accident. Virginia Daly of Berkeley, Michigan, crossed a Siamese with an Abyssinian, aiming to develop an Aby-pointed Siamese. The kittens looked like Abys, but when one was bred to a Siamese, the litter included not only Aby-pointed Siamese but also an odd, spotted kitten. Daly's daughter, noting its resemblance to the ocelot (*see page 14*), called the kitten an "ocicat". This first Ocicat was neutered and sold as a pet, but the mating was repeated, producing Dalai Talua, the foundation female of this still-rare breed. Another breeder, Tom Brown, helped to continue the Ocicat's development, introducing American Shorthairs (*see page 120*), and in 1986 the breed received its first recognition, from TICA.

The Ocicat head

The facial markings of Ocicats should be clear and detailed. The frown lines on the forehead are balanced by "mascara lines" on the temples and cheeks, and the dark-rimmed eyes are surrounded by "spectacles".

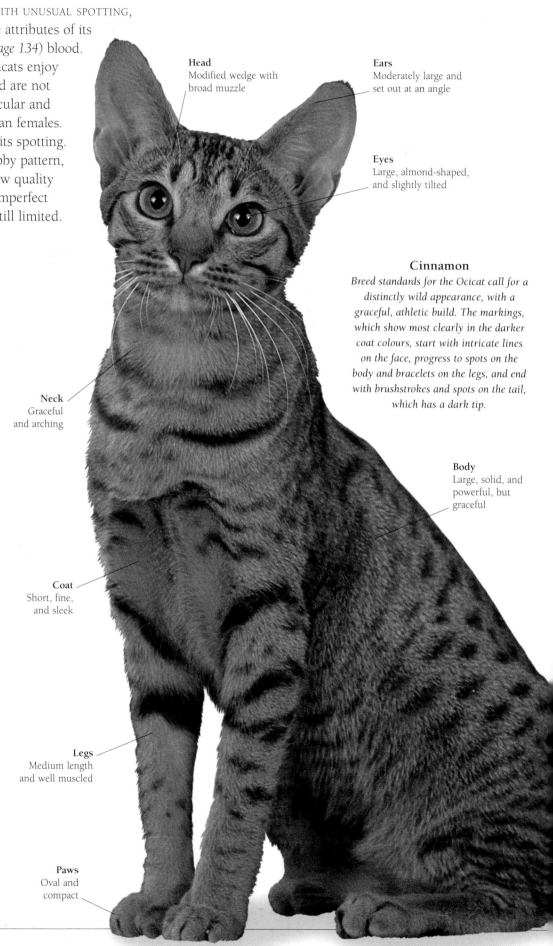

Head
Modified wedge with broad muzzle

Ears
Moderately large and set out at an angle

Eyes
Large, almond-shaped, and slightly tilted

Cinnamon
Breed standards for the Ocicat call for a distinctly wild appearance, with a graceful, athletic build. The markings, which show most clearly in the darker coat colours, start with intricate lines on the face, progress to spots on the body and bracelets on the legs, and end with brushstrokes and spots on the tail, which has a dark tip.

Neck
Graceful and arching

Body
Large, solid, and powerful, but graceful

Coat
Short, fine, and sleek

Legs
Medium length and well muscled

Paws
Oval and compact

Lilac or Lavender
This dilute version of the Chocolate has lilac spotting on a background of pale buff or ivory. Allowance is made in the breed standards for the slightly softer appearance of the tabby markings that is almost inevitable in the dilute colours.

BROKEN NECKLACES DESIRABLE

Tawny or Brown
Genetically a brown tabby, this colour is called Tawny by CFA. It has black or dark-brown spotting on a warm, ruddy background, showing the rich rufousing of its Abyssinian ancestors.

KITTENS HAVE DEEP EYE COLOUR

Cinnamon Silver kitten
Ocicat markings take time to develop. Kittens and young cats may show solid lines along the spine, which resolve into separate spots as the cats mature.

Tail
Fairly long and slender

Tip
Colour of coat is shown most truly here

Blue Silver
The silver gene came into the Ocicat from American Shorthair outcrosses during the early development. All silver colours have a white ground, which shows off the markings dramatically.

KEY FACTS

DATE OF ORIGIN 1964
PLACE OF ORIGIN United States
ANCESTRY Siamese, Abyssinian, American Shorthairs
OTHER NAME Oci
OUTCROSS BREEDS Abyssinian until 2005
WEIGHT RANGE 2.5–6.5 kg (6–14 lb)
TEMPERAMENT Social and responsive

FIFé CFA TICA

BREED COLOURS

TABBIES (SPOTTED)
Tawny or Brown, Chocolate, Cinnamon, Blue, Lavender, Fawn
Classic and mackerel patterns

SILVER TABBIES (SPOTTED)
Colours are as for standard tabbies
Classic and mackerel patterns

SELFS
Colours as for tabbies

SMOKES
Colours as for tabbies

SILVER   FAWN

BENGAL

STILL RARE WORLDWIDE, THIS BREED HAS A DISTINCTIVELY THICK and luxurious coat. Because of the wildcat origins (*see page 182*) of the breed, a dependable temperament is a vital feature of breeding programmes. Curiously, although Bengal numbers are still relatively small, breed clubs abound. In Britain, for example, there are three separate breed clubs, although there are only a few hundred cats. Early breeding introduced some undesirable genes, for dilution, long hair, and spotting, but also the Siamese coat pattern, which has resulted in the extraordinary "Snow" shades. With its robust good looks, the Bengal's future is promising.

BREED HISTORY

The first mating of Asian leopard cat (*see page 16*) with a domestic cat at Jean Sugden's Californian property in 1963 was accidental. Ten years later, Dr. Willard Centerwall at the University of California continued this hybridization in order to examine the Asian leopard cat's resistance to feline leukemia virus (FeLV). The research was disappointing, but out of these beginnings appeared the Bengal. Dr. Centerwall passed eight of his hybrids to Jean Sugden, now remarried as Mill. The first Bengal, Millwood Finally Found, was registered by Jean Mill in 1983. Initially this was a nervous feline family, but continued development has led to a more outgoing breed. Early crosses were to non-pedigrees, but when the leopard-like coat appeared, individuals were crossed with an Indian street cat and Egyptian Maus (*see page 172*).

"SCARAB" LINES

GOLD, GREEN, OR HAZEL EYES

Bengal head
Slightly longer than it is wide, the Bengal's face has high cheekbones and a full, broad muzzle. The chin is strong, and widely set canine teeth help to produce pronounced whisker pads. Frown lines and broken streaks of colour cover the head, and the "puffed" nose leather is pink outlined in black. In profile, there is a gentle curve from forehead to nose, rather than a break.

Blue-Eyed Snow Marbled

The Snows come from pointed lines in the non-pedigree cats used in the Bengal's development. Breed registries are controlled to avoid such unexpected occurrences, but Bengal breeders have used this happy accident to create truly stunning cats. The colour restriction should give an impression of "pearl dusting" to the coat; the pattern for these cats is as for their full-coloured counterparts.

Brown Marbled

This pattern should resemble the coat of a wild cat, not the blotched or classic tabby pattern, and it should be distinct, but not symmetrical. The markings should be clear, but, uniquely to the Bengal standard, cats should show three shades of colour: the base, the dark markings, and darker outlines.

KEY FACTS

DATE OF ORIGIN 1983
PLACE OF ORIGIN United States
ANCESTRY Asian leopard cat, household cats, Egyptian Mau, Indian street cat
OUTCROSS BREEDS None
OTHER NAME Once called Leopardettes
WEIGHT RANGE 5.5–10 kg (12–22 lb)
TEMPERAMENT Elegantly conservative

  GCCF FIFé TICA

BREED COLOURS

SELF
Black

TABBIES (SPOTTED, MARBLED)
Brown, Snow

Head
Relatively small,
rounded wedge

Ears
Short, with wide
base and rounded
tips, no tufts

Tail
Thick and even,
with rounded tip

Eyes
Large and oval, with
slightly slanted set

Brown Spotted
*The first coat to be stabilized, the Brown Spotted
resembles that of the Asian leopard cat, right down to
the light-coloured "ocelli" on the back of each ear. The
base colour is buff and the markings are deep brown or
black. Facial features are outlined in black, and the
spots on the coat should be large, forming rings or
rosettes, and randomly distributed; any resemblance to
the vertical stripes of the mackerel pattern, the pattern
behind most spotted tabbies, is avoided.*

LARGE, BROAD NOSE

THICK, MUSCULAR NECK

BROAD
CHEST

Body
Large, very
muscular,
and sleek

FUR IS SHORT
TO MEDIUM
LENGTH

Coat
Dense, soft
to touch

HINDLEGS
LONGER THAN
FORELEGS

Legs
Strong and muscular,
medium length

LARGE, ROUNDED PAWS

AMERICAN BOBTAIL

THERE WERE, UNTIL RECENTLY, ONLY A COUPLE OF BOBTAILED or tailless breeds. But in the last few years, breeds from the former Soviet Union such as the Kurile Island Bobtail (*see page 236*) have become better known, and two new bobtail breeds have been registered in North America. The American Bobtail was the first of these. The genetic background of the breed is uncertain: bobcat parentage is unconfirmed, but both Manx (*see page 116*) and Japanese Bobtail (*see page 164*) genes may be present, as both completely tailless "rumpies" and cats with kinked tails appear. In contrast to the Manx breed standard, these rumpy Bobtails are not showable: the Bobtail should have a short tail that stops just above the level of the hocks.

BREED HISTORY

This breed can be traced back to a random-bred bobtailed tabby kitten that was adopted from an American Indian reservation in Arizona by John and Brenda Sanders from Iowa. Early work with the breed aimed to produce bobtailed cats with a pattern similar to that of the Snowshoe but the cats became inbred and unhealthy. Later work, led by Reaha Evans, reintroduced more colours and patterns, and improved the health of the breed. CFA and TICA have both granted American Bobtails "new breed" status.

Classic Tabby longhair kitten
There are both longhaired and shorthaired divisions for the American Bobtail: because the longhair gene is recessive, there are fewer of these. The coat is semi-long, and longer "mutton-chops" on the cheeks are desirable. The Bobtail coat does not mat easily, although it appears shaggy.

KEY FACTS

DATE OF ORIGIN 1960s
PLACE OF ORIGIN United States
ANCESTRY Uncertain
OUTCROSS BREEDS Non-pedigree cats
OTHER NAME None
WEIGHT RANGE 3–7 kg (7–15 lb)
TEMPERAMENT Friendly and inquisitive

BREED COLOURS

All colours / patterns, including sepia, pointed, and mink

FAWN AND WHITE   BLUE TABBY   RED TABBY   WHITE

Eyes
Oval and angled, with heavy brow

Ears
Medium-sized, wide at base, and set high on head

Spotted Tabby shorthair

The American Bobtail breed standard calls for a brawny, hearty, and wild-looking cat, with a strong head and a hunting look to the eyes. American Bobtails are a slow-maturing breed, and may take as long as three years to reach their full potential. The shorthair coat is long enough to stand away from the body and appear slightly shaggy.

Head
Broad, modified wedge with curved contours

Coat
Resilient and double

Body
Semi-cobby, with substantial muscling

Tail
Must be present, but stops above hock

Legs
Heavy, with large, round paws

PIXIEBOB

DOMESTIC-SIZED CATS WITH A WILD APPEARANCE HAVE BECOME increasingly popular over the last two decades: it was perhaps inevitable that North American breeders would eventually develop a cat that resembled the native bobcat (*see page 14*). Despite their wild looks, Pixiebobs are claimed to have the temperament of faithful dogs; breeders recommend that owners consider carefully before acquiring one, as they do not readily change homes, and are often happiest as a single-cat "ruler of the roost". The wild look is described as "essential to the uniqueness of this breed".

BREED HISTORY

The origins of the Pixiebob are thought to be wild bobcat to domestic "barn cat" breedings that took place in rural areas, although DNA profiles have not provided supporting evidence. The assumed hybrid kittens are called "legend cats" because of the claims of wildcat ancestry. Two such cats were acquired in 1985 by Carol Ann Brewer in Washington State, who bred them to produce Pixie, the founding cat who gave her name to the breed. The breed was first recognized by TICA almost 10 years later, and is not recognized by CFA or known outside North America. "Legend cats" are still found in rural areas, and may be refined to produce Pixiebobs: no other pedigree breeds or wildcats can be used in this process.

KEY FACTS

DATE OF ORIGIN 1990s
PLACE OF ORIGIN North America
ANCESTRY Domestic cat, possibly bobcat
OUTCROSS BREEDS Brown tabby non-pedigrees
OTHER NAME None
WEIGHT RANGE 4–8 kg (9–18 lb)
TEMPERAMENT Quiet but affectionate

  TICA

BREED COLOURS

TABBY (SPOTTED, ROSETTE)
Brown
Any other tabby pattern, other colours

Brown Spotted Tabby kitten
Most kittens acquired for breeding have been given, rather than sold, to breeders by farmers who would usually have culled them. The hybrids seem to inherit the tractable natures of their mothers, rather than the wild characteristics of their putative fathers.

Brown Spotted Tabby
Any brown spotted tabby of suspected "legend cat" heritage can be used in breeding Pixiebobs. The look of these cats can vary, due to the parentage of domestic cat and possible regional varieties of bobcat. Coat markings are regarded as secondary to the physical conformation.

Pixiebob head
The facial markings must be strong in Pixiebobs, with "mascara" lines on the cheeks and light "spectacles" around the eyes. Lynx tips on the ears are desirable, but not essential. The lips and chin should be creamy white.

Ears
Wide at base, slightly rounded, and set far back on skull

Eyes
Prominent brows above medium-sized, deep-set eyes

WIDE NOSE

Paws
Polydactyls are accepted

Legs
Long and substantial

Head
Wide, medium to large pear shape, with slightly curved contours

Body
Medium to large in size, substantial in build

Coat
Luminous quality, with much ticking

Tail
Generally carried low

TRADITIONAL BREEDS

BEAUTY IS IN THE EYE OF THE BEHOLDER. AND IN THE CAT SHOW world the beholder is often the judge. While breed standards appear to be very specific, they are open to interpretation. By the 1970s, the US show Siamese had evolved from its original shape, exemplified by Kim Novak's cat in the 1958 film *Bell, Book and Candle*, into the long, lithe, fashion-model shape that it is today. Some Siamese breeders, feeling that the breed was being bred to an extreme, began breeding back to the cat's older style. This was the foundation of the "traditional" cat movement, which led to the 1987 formation of the Traditional Cat Association (TCA) by Diana Fineran. The TCA, with its head office in aptly named Battle Ground, Washington, fiercely guards the copyright on its breed standards for over 30 breeds. Success in restoring the Siamese to its less radical shape has led other breeders to return to standards that they feel are less extreme and healthier. Today, other breeds, including Burmese, Longhairs, and Abyssinians, are bred to "traditional", as well as contemporary standards.

Ears
Large, wide at base, with rounded tips

Eyes
Large and luminous green

Coat
Smooth, short and close-lying

Traditional Korat

Along with the Siamese and Burmese, the Korat is an ancient breed that originated in Thailand. While Korat breeders take great pride in claiming that the breed is totally natural and still looks the same as the Korat cat described in a Thai poem written many centuries ago, the TCA describes two different varieties that exist today: the Traditional Korat and the Classic Korat (see page 138). The Traditional has a sturdier body type and more rounded face than the Classic Korat. The Classic is more svelte but not extreme in any way. Traditional Korat breeders, like other traditional breeders, strive to preserve their breed and protect it from developing the problems that selecting for extremes has created in several other breeds.

Legs
Forelegs slightly longer than hindlegs

Paws
Oval-shaped, with five toes in front and four behind

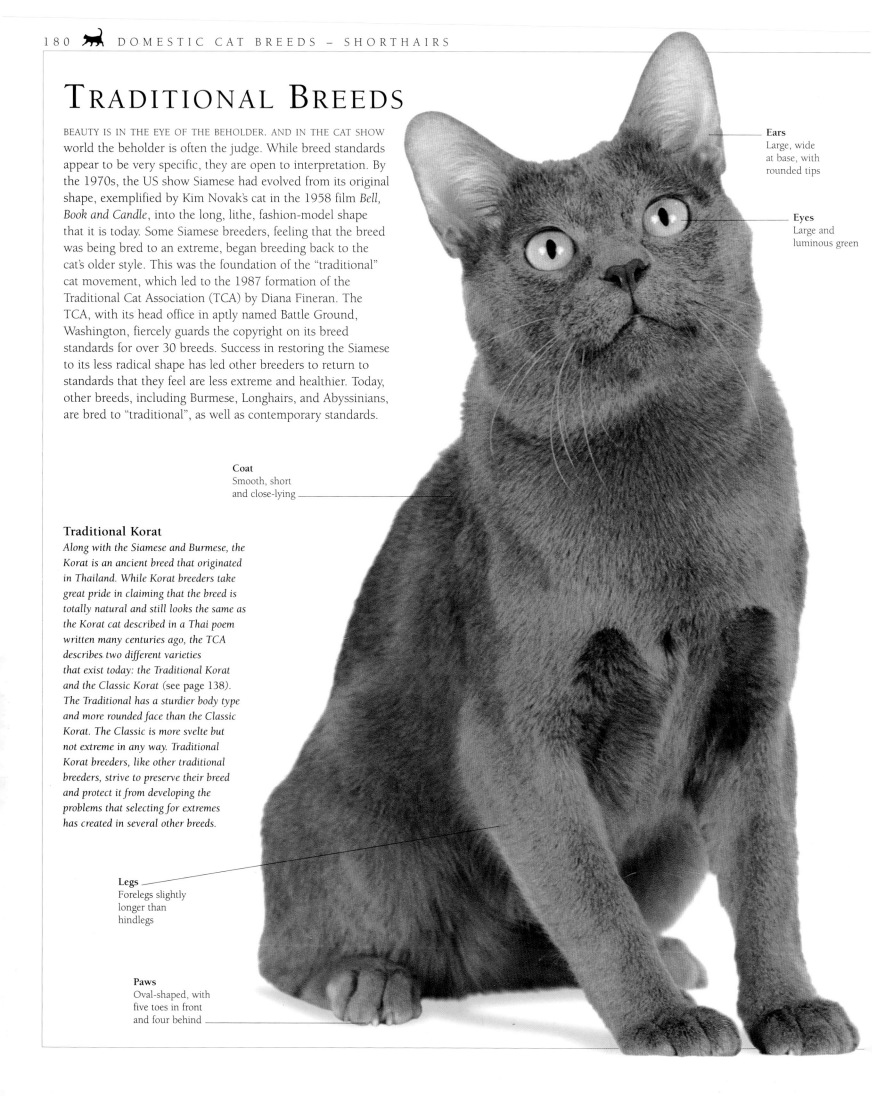

Eyes
Luminous,
vivid blue

Body
Medium to large,
and muscular

Coat
Thick and plush

Traditional Siamese

*This breed preserves the look of the first Siamese cats
imported into the West from Siam (Thailand) over
100 years ago. It has a less elongated body and head
type than the contemporary show cat, and larger,
less slanted eyes. Also known as the Applehead or
Old-Style, the Traditional Siamese, like the
"contemporary", has a longer life expectancy than
most other breeds. Traditionals have short, thick,
velvet-like coats, and are described by breeders
as calm, affectionate cats. They are talkative
individuals and can be extremely loud. Their
vocal traits are apparent from kittenhood.*

Head
Must not be too narrow
or fine-boned

Traditional Abyssinian

*Although the exact origins of the
Abyssinian are unknown, it is believed that
this ancient breed has Egyptian and African
ancestry. It was introduced into the West in the
late 1800s, and in Great Britain there are records
of the Abyssinian dating back as far as 1892. Like
other traditional cat standards, the standard for the
Traditional Abyssinian calls for features that
preserve the look of the original breed. Compared to
the contemporary Abyssinian (see page 134), the
Traditional is a larger cat, with a less extreme body
and head. Unlike the contemporary, which is now
accepted in several colours by most registries, the
Traditional is only accepted in two colours: Ruddy
and Red. This breed was accepted for registration
by the TCA in June 2000.*

Traditional Longhair

*Also known as the Doll Face, the Traditional Longhair
(Persian) is recognized as a separate breed in the United
States by the TCA. The cat is described by breeders as
the ultimate lap-cat, which may explain Queen Victoria's
decision to keep Longhairs. According to the TCA, the
Traditional has a longer, less "pushed in" nose than the
"contemporary" Longhair (see page 188). As a result,
the Traditional Longhair is a healthy cat that does not
suffer from the breathing difficulties experienced by the
"contemporary". Like other longhaired cats, Traditional
Longhairs require daily grooming to prevent their long,
thick coats from matting.*

Coat
Plush and soft

DOMESTIC AND WILDCAT HYBRIDS

MEMBERS OF THE CAT FAMILY MAY APPEAR STRIKINGLY DIFFERENT IN appearance but their genetics are so similar that breeding between species is possible. Accidental matings between lions and tigers in zoos, for example, have produced infertile "ligers" or "tigons". Since the 1960s, domestic cat breeders have both accidentally and intentionally crossed domestic cats with more closely related wildcats, producing fertile hybrids. Within the cat breeding world there is often intense debate over the ethics or the value of creating domestic/wildcat hybrids. The Bengal, a hybrid cross with the Asian leopard cat (*see page 17*), is now recognized by most North American associations, yet it remains unrecognized by the CFA. In Europe, both the GCCF and FIFé now recognize the Bengal, so it is probable that other successful hybrids, such as the Chausie, may also eventually be recognized.

Savannah kitten

Although still a cute kitten, this Savannah will grow to stand 46–61 cm (18–24 in) at the shoulder, making the Savannah one of the largest of the cat breeds. Created from a hybrid cross between domestics and the African Serval, the breed is not recognized by any of the main registries. One of its distinctive features is its thick, spotted coat, which in adulthood closely resembles that of a cheetah.

Coat
Thick and spotted, with soft feel

Ears
Small or medium-sized, with rounded tips

Chausie

Not widely recognized, the Chausie is a hybrid cross between a domestic female and a male Jungle cat (Felis chaus). Breeders claim that Chausies make loyal pets if they are properly socialized from birth. The cat has a large, athletic body and a shorthaired coat.

Body
Large and muscular

Coat
Short, with a satin-like appearance

Coat
Dense and shorthaired

Tail
Three-quarter length, extending just below the hock

Bengal

Rapidly growing in popularity, the Bengal descends from the accidental mating of a male domestic and a female Asian leopard cat in California in 1963. It has since been breed to create a good-natured domestic cat that preserves the beauty of the leopard cat. Although recognized by most registries in Britain and North America for championship status, the Bengal is not, and may never be, accepted by the CFA. This is due to a CFA ruling not to accept any cat breed that has documented non-domestic ancestry.

Chausie (opposite page)

RANDOM BREEDS

THE RANDOM-BRED CAT HAS NO CLUBS TO PROMOTE IT AND NO romantic history or royal connections to beguile the public, yet the "moggie" is still the most popular and widely owned cat worldwide. Type tends to vary from place to place, with stocky, sturdy cats found in cold countries and lighter, more slender cats much in evidence in warmer climates, but the extremes of cobbiness or elongation that have been selectively bred into the pedigree breeds are most unusual in these cats. Typically Oriental colours and pointed patterns remain rare in Western non-pedigrees, although these genes have occasionally filtered from pedigree breeds into the general population through the determined ingenuity of all felines in matters of mating. Occasionally, non-pedigree cats may closely resemble particular breeds but, for the most part, these felines are only reminiscent of the breeds that emerged from their stock.

Cream Tabby
In the first pedigree cats, creams were often regarded as poor reds, because their colour was warm and orange, but too light for a proper red. Over the decades, breeders have bred out the red tones to produce paler, cooler coats in pedigree Creams. However, random-bred individuals will still show the orange hues of those early cats.

Brown Classic Tabby and White
In Europe, the classic tabby came to outnumber the mackerel tabby by the 18th century. It has been suggested that, because it is darker than the mackerel pattern, the classic coat provided better camouflage in the urban environment. Breeds such as the European Shorthair (see page 126) are not so very far from their roots in these cats.

Red Classic Tabby and White

Because the white spotting gene is dominant, bi-colours are common in domestic cats. Reds vary in distribution from area to area within most countries, and can help to indicate whether adjacent cat populations are interbreeding or not. Breeds such as the American Shorthair (see page 120) grew from sturdy domestic cats like this one.

Blue and White

Blue is a common colour in many areas of mainland Europe, and the Chartreux (see page 128) was developed from such stock. Many random-bred cats are a much darker blue than pedigree breeds, and lack the intense and varied eye colours achieved by dedicated selective breeding. The build of this cat is heavier than that of mainland European breeds, but would not disgrace a British Shorthair (see page 112).

Tortoiseshell and White

North European cats tend to have small amounts of white in their coats, while in southern Europe predominantly white patterns, reminiscent of the Turkish Van (see page 216), are more common. Without selective breeding, polygenic factors have a large influence on coat: this cat's plumy tail might indicate that it is genetically a longhair, even though the coat on its body remains short and soft even in winter.

INTRODUCTION TO LONGHAIRS

GENETICALLY, ALL LONGHAIRED CATS SHARE THE recessive allele that causes their coats to grow longer than those of their wildcat ancestors (*see page 75*). The difference between full-coated longhairs and silky semi-longhairs is one of polygenetic factors. Some sources still state that the gene for long hair was introduced into these domestic cats from the wild Pallas' cat of Tibet (*see page 17*), but there is no evidence that this is so: a simple genetic mutation is almost certainly the cause. Although their exact origins are unknown, longhairs occurred naturally centuries ago in Central Asia. Some of these cats reached Europe: the French authority Dr. Fernand Mery reported that specimens were brought to Italy around 1550 and to France shortly after. Early longhairs in Europe were called Russian, French, or Chinese; almost three centuries passed before these cats acquired the classifications we know today.

Balinese

Like its shorthaired cousin the Siamese, the Balinese has no insulating undercoat, and its semi-long hair is more decorative than functional. In summer, these cats may shed so much of their long hair that they look like shorthairs.

LaPerm

Among the newest breeds, the LaPerm is descended from a mutation that occurred in an ordinary farm cat. Most rex mutations lessen the effectiveness of the coat as insulation, but cats living as pets are not unduly disadvantaged by this.

Longhair kittens

Among the first longhair breeds was the Longhair or Persian. Over the years, the coat of this breed has been bred to ever-greater length, and many breeders clip their cats when not showing, as they must otherwise groom the cats several times a day.

Norwegian Forest Cat
Breeds that developed in harsh climates, such as the Norwegian Forest Cat, or the Maine Coon, show their origins in their coats. They tend to have water-repellent topcoats and thick, insulating undercoats.

After the Crystal Palace Cat Show in London in 1871, standards were published for Persians, now called Longhairs (*see page 188*), and Angoras, revived in the Angora (*see page 232*) in Britain, and Turkish Angoras (*see page 220*) elsewhere. The most popular longhair in early American shows was the Maine Coon (*see page 202*). This breed's coat may have come from Central Asia via Britain, but some early settlers' cats may have carried the longhair gene from another mutation. What is certain is that New England's harsh climate favoured the survival of big cats with long, insulating coats. Other longhaired breeds are the result of introducing the longhair gene into shorthaired breeds. New breeds, such as the Tiffanie (*see page 227*) and the Nebelung (*see page 219*) have been developed on both sides of the Atlantic. Perhaps the most extraordinary and striking of all longhaired breeds are the rexed longhairs, with a curl or ripple to the coat, such as the LaPerm (*see page 235*), the Maine "Waves" (*see page 204*), and the obscure Bohemian Rex (*see page 235*).

Turkish Van
The coat of the Turkish Van is extremely fine and, because it lacks an undercoat, it is easy to groom. Although the coat pattern and general type were known for centuries in Turkey, the breed was a relatively late arrival in the West.

Somali
Longhaired kittens occurred in Abyssinian litters for a number of years before an attempt was made to develop them into a separate breed. The resulting Somali is now one of the more popular cats in North America.

LONGHAIR

THIS BREED, USUALLY AN INDOOR DWELLER, IS A LOUNGE LIZARD. The Longhair, also known as the Persian, is a relaxed observer. In statistical surveys of veterinarians, the Longhair is cited as the quietest and least active of cat breeds, and the one most likely to accept other cats into its home. This does not mean that the breed is entirely passive: in Britain and mainland Europe, pedigree cats have more access to the outdoors than they do elsewhere, and the Persian will guard its territory and catch and dispatch prey with surprising ease, given its shortened face. The coat needs daily care: veterinarians are frequently called on to clip densely matted coats. Breed problems include polycystic kidney disease, which can be screened for, and a high incidence of retained testicles.

BREED HISTORY

The first documented ancestors of the Longhair were imported from Persia into Italy in 1620 by Pietro della Valle, and from Turkey into France by Nicholas-Claude Fabri de Peiresc at about the same time. For the next two centuries their descendants, known by a variety of names, were status-symbol pets. In the late 1800s, the Persian was developed within the guidelines of Harrison Weir's first written breed standards. The original stocky build is still an essential mark of today's Longhair breed, although other characteristics have been dramatically altered. This most popular breed was recognized by all registries by the early 1900s.

Brown Classic Tabby
The blotched, or classic, tabby is the traditional pattern for the Longhair, although some North American associations now allow others. The Brown is the original colour, being essentially the natural tabby.

Silver Shaded
Once classified together with the tipped Chinchilla, the darker Silver Shaded is now judged separately. Like almost all silvered Longhairs, it has green eyes. The exception is the Pewter, which has copper eyes but resembles the Silver in all other ways.

Body
Large and cobby, with good muscling

Coat
Long and thick, but not woolly

Tail
Full and short, but not disproportionate

LONG, BUSHY HAIR

COAT IS SOFT

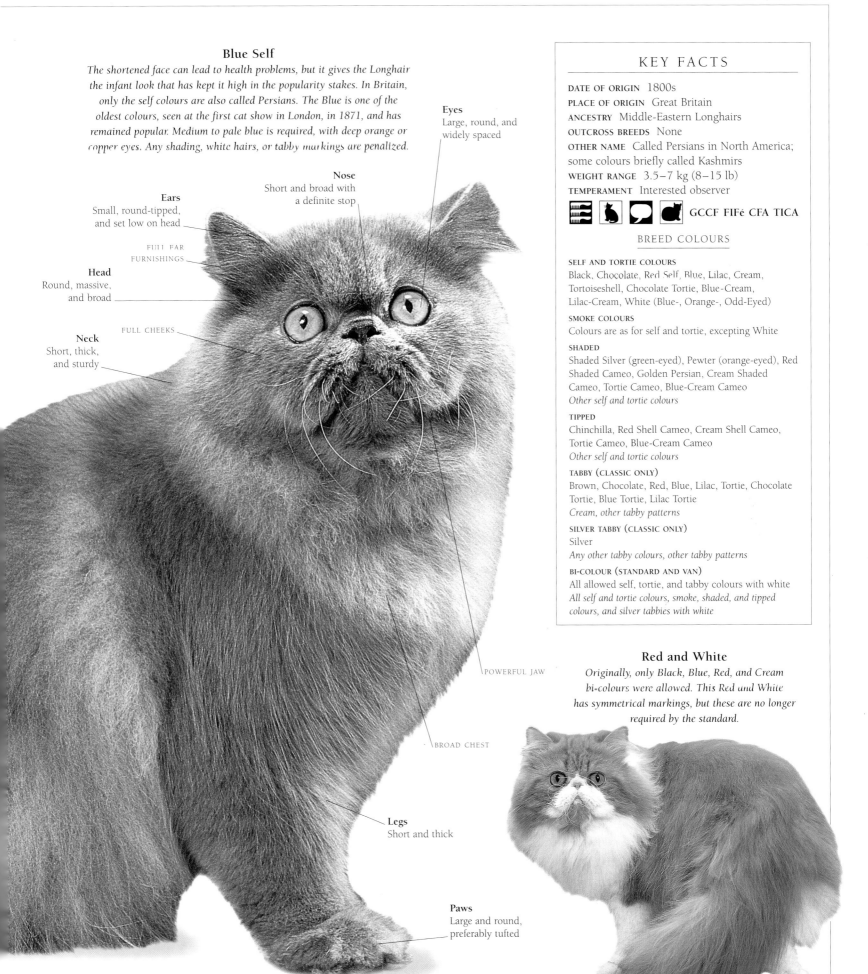

Blue Self

The shortened face can lead to health problems, but it gives the Longhair the infant look that has kept it high in the popularity stakes. In Britain, only the self colours are also called Persians. The Blue is one of the oldest colours, seen at the first cat show in London, in 1871, and has remained popular. Medium to pale blue is required, with deep orange or copper eyes. Any shading, white hairs, or tabby markings are penalized.

Eyes
Large, round, and widely spaced

Nose
Short and broad with a definite stop

Ears
Small, round-tipped, and set low on head

FULL FAR FURNISHINGS

Head
Round, massive, and broad

FULL CHEEKS

Neck
Short, thick, and sturdy

POWERFUL JAW

BROAD CHEST

Legs
Short and thick

Paws
Large and round, preferably tufted

KEY FACTS

DATE OF ORIGIN 1800s
PLACE OF ORIGIN Great Britain
ANCESTRY Middle-Eastern Longhairs
OUTCROSS BREEDS None
OTHER NAME Called Persians in North America; some colours briefly called Kashmirs
WEIGHT RANGE 3.5–7 kg (8–15 lb)
TEMPERAMENT Interested observer

GCCF FIFé CFA TICA

BREED COLOURS

SELF AND TORTIE COLOURS
Black, Chocolate, Red Self, Blue, Lilac, Cream, Tortoiseshell, Chocolate Tortie, Blue-Cream, Lilac-Cream, White (Blue-, Orange-, Odd-Eyed)

SMOKE COLOURS
Colours are as for self and tortie, excepting White

SHADED
Shaded Silver (green-eyed), Pewter (orange-eyed), Red Shaded Cameo, Golden Persian, Cream Shaded Cameo, Tortie Cameo, Blue-Cream Cameo
Other self and tortie colours

TIPPED
Chinchilla, Red Shell Cameo, Cream Shell Cameo, Tortie Cameo, Blue-Cream Cameo
Other self and tortie colours

TABBY (CLASSIC ONLY)
Brown, Chocolate, Red, Blue, Lilac, Tortie, Chocolate Tortie, Blue Tortie, Lilac Tortie
Cream, other tabby patterns

SILVER TABBY (CLASSIC ONLY)
Silver
Any other tabby colours, other tabby patterns

BI-COLOUR (STANDARD AND VAN)
All allowed self, tortie, and tabby colours with white
All self and tortie colours, smoke, shaded, and tipped colours, and silver tabbies with white

Red and White

Originally, only Black, Blue, Red, and Cream bi-colours were allowed. This Red and White has symmetrical markings, but these are no longer required by the standard.

Longhair (overleaf left above, left below, and right) *For breed entry see above*

NEWER LONGHAIR COLOURS

ORIGINALLY RECOGNIZED ONLY IN A LIMITED COLOUR RANGE, the Longhair is now bred in an abundance of new shades. But the colour range seen here is not the only development: the coat, build, and, most dramatically, the face have all changed this century. The coat is now so long that many breeders keep their cats clipped when not showing, rather than groom several times a day. Early Longhairs were less compact, with short, but not flat, faces. While breeders in Europe still select for moderate noses, American breeders and, more importantly, American show judges, have tended to prefer a flatter, or "ultra-type" face. This reached its most extreme in the "Peke-faced" Persian. Due to its narrowed nostrils and tear ducts, the look is no longer desirable. Breeding for large heads has also increased the need for Caesarean deliveries of the Longhair's small litters.

Golden Persian

This colour, with deep tipping on an apricot base, looks like a golden version of the Shaded Silver. There are two versions in America, the Shaded and the Chinchilla; only the first is recognized in Britain. The genetics of the colour are still disputed.

Blue-Cream

Although this colour has existed from the early days of the breed, it was not recognized until 1930; this was mostly because the genetics of coat colours were not understood, and Blue-Creams were only produced by accident.

PASTEL-TONED COAT

Cream Shell Cameo

Chinchillas and Shaded Silvers were recognized early in the Longhair's history, but other tipped Longhairs were not developed until after World War II. This Cream Shell Cameo is essentially a Cream Chinchilla, with warm-toned tipping to the hairs and deep, brilliant, copper-coloured eyes.

COLOURS WELL DISTRIBUTED

Chocolate Tortie

The Chocolate came from a Longhair-Havana cross in the 1950s, and it is now accepted in all the major registries. Chocolate Tabbies and Torties inevitably followed. As in many breeds, the British standard calls for mingled colours, while the North American registries prefer patches.

Lilac Tabby

Together with the Chocolate colours,
Lilac colours were once categorized in
North America as a separate breed,
known as the Kashmir. Coat and type
are now typically Persian: this example
is a young cat, in less heavy coat.

COPPER EYES

BEIGE GROUND

SPOTTED
UNDERPARTS

FADED-LILAC PAW PADS

Chocolate and White

Tortie-and-Whites were recognized in the 1950s, and other colours
in the 1960s. In 1971, markings one-third to half white were
permitted by the standard; this change greatly boosted
popularity. All colours and patterns are
now accepted in both standard
and Van distribution.

TAIL MOSTLY COLOURED

EVEN SHADING

Red Shaded Cameo

The inhibitor gene was present in Longhairs from the
start, but Cameos came to the scene relatively late.
They were developed in a breeding programme in the
United States during the 1950s, and were accepted by
CFA in 1960. They were subsequently taken up
worldwide; breeding began in Europe in 1962.

Cream Shaded Cameo

This dilute version of the Red Shaded Cameo
is a cooler, lighter colour. Only Red, Cream,
Tortie, and Blue-Cream Shaded Cameos are
accepted in Britain, because the original
breeding was done with Tortoiseshells.
The Shaded Silvers and Pewters
have different origins.

COLOURPOINT LONGHAIR

POSSIBLY THE FIRST DELIBERATE HYBRIDIZATION OF TWO BREEDS, this version of the Longhair (*see page 188*) was also the first recognized "export" to another breed of the Siamese pointed pattern (*see page 154*). The resulting cat has the luxurious, thick, long coat of the Longhair, and the exotic colour pattern of the Siamese. Eye colour is less intense than in the Siamese, and the pointing, being heat-related, is softer in the longer coat; masks are generally less dense and body colour is paler. Colourpoint Longhairs also moderate the diverse characters of their parent breeds: while the average Longhair is content to watch the world go by, and the Siamese is always at the heart of the action, the Colourpoint Longhair is an outgoing but relaxed companion. Owners should take care over neutering, however: the sexual precocity of their Siamese forebears may also manifest itself in these cats.

Chocolate Point
An ivory-white body combines with brown points, which should be even in tone and depth.

BREED HISTORY
The first experiments in cross-breeding Siamese and Longhairs were made in the 1920s in Europe: a breed known as the Khmer existed in mainland Europe until the 1950s, and some say that the Birman (*see page 198*) is also a result of these experiments. In the 1930s, American geneticists investigating inherited traits bred a black Longhair with a Siamese. The first generation were all black longhaired cats, but produced a pointed longhair cat when back-crossed. This was later called the Himalayan, after the pointed pattern seen in many Himalayan rabbits, but there was little interest in pursuing it as a breed. In Britain, parallel efforts at creating a pointed cat with good Longhair type progressed through the 1930s and 1940s. The Colourpoint Longhair was accepted in 1955; in mainland Europe, the name Khmer was changed to match. Resurgent interest in Himalayans during the 1950s in North America led to recognition by all major registries by 1961. The status of the type varies internationally: the GCCF and CFA classify them as colours of the Longhair, while others regard them as a separate breed.

Cream Point kitten
The points of all colours take a longer time to develop than in shorthairs, and the points of the dilute colours, such as the Cream Point, take longer to develop fully than those of the darker shades.

KEY FACTS

DATE OF ORIGIN 1950s
PLACE OF ORIGIN Britain and United States
ANCESTRY Longhair/Siamese crosses
OUTCROSS BREEDS Longhair
OTHER NAMES Sometimes called Himalayan in the United States
WEIGHT RANGE 3.5–7 kg (8–15 lb)
TEMPERAMENT Calm and friendly

   GCCF FIFé CFA TICA

BREED COLOURS

SELF AND TORTIE POINTS
Blue, Chocolate, Cream, Lilac, Red, Seal, Blue-Cream, Chocolate Tortie, Lilac-Cream, Seal Tortie

TABBY POINTS
Colours are as for self and tortie points

CREAM TABBY POINT   RED POINT   BLUE POINT   SEAL TABBY POINT

SHADING DEVELOPS ON OLDER CATS

Seal Point
The mask of a mature Colourpoint will cover the face, but should not extend over the rest of the head. Males will have more extensive masks than females.

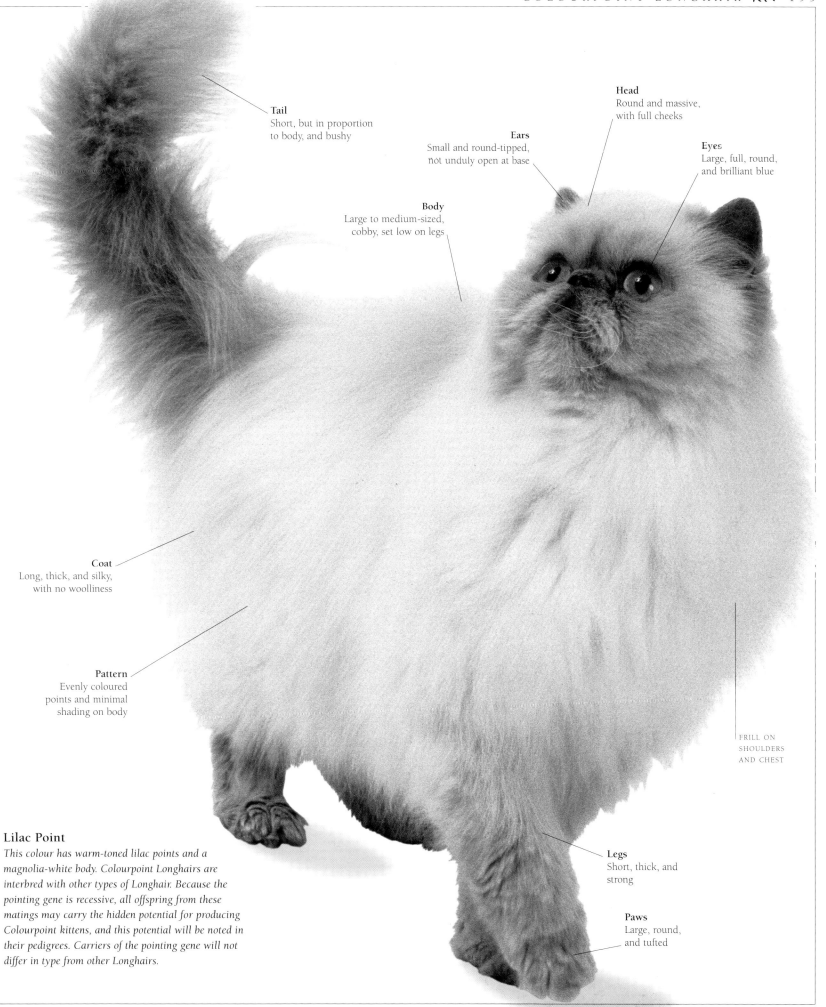

Tail
Short, but in proportion
to body, and bushy

Head
Round and massive,
with full cheeks

Ears
Small and round-tipped,
not unduly open at base

Eyes
Large, full, round,
and brilliant blue

Body
Large to medium-sized,
cobby, set low on legs

Coat
Long, thick, and silky,
with no woolliness

Pattern
Evenly coloured
points and minimal
shading on body

FRILL ON
SHOULDERS
AND CHEST

Legs
Short, thick, and
strong

Paws
Large, round,
and tufted

Lilac Point

*This colour has warm-toned lilac points and a
magnolia-white body. Colourpoint Longhairs are
interbred with other types of Longhair. Because the
pointing gene is recessive, all offspring from these
matings may carry the hidden potential for producing
Colourpoint kittens, and this potential will be noted in
their pedigrees. Carriers of the pointing gene will not
differ in type from other Longhairs.*

Longhair (overleaf left) *For breed entry see page 188* **Birman (overleaf right)** *For breed entry see page 198*

BIRMAN

THIS STRIKINGLY MARKED BREED, WITH ITS MYSTERIOUS HISTORY, is a well-built cat with highlighted paws and large, blue eyes. Its silky hair is not as thick as that of the Longhair (*see page 188*), and is less prone to matting, but daily grooming is still necessary. Breeders say that neutered males demand most attention, while neutered females can be rather bossy. The breed was almost extinct in the West at the end of World War II, when only two individuals remained in France. These were outcrossed with other cats to perpetuate the breed, increasing the genetic base and introducing the potential for a variety of point colours. As with all breeds with a small genetic base, inbreeding may concentrate and increase hereditary problems, but, fortunately, only rare skin and nerve disorders are hereditary in this breed.

BREED HISTORY
According to tradition, the Birman descends from the temple cats of Burma, specifically from Sita, a pregnant female, brought to France in 1919 by August Pavie. Legend links Birmans to a white cat called Sinh who lived in a temple dedicated to Tsun-Kyan-Kse, a golden goddess with sapphire eyes. When the temple was attacked, Sinh took on the goddess's colours and inspired the monks to fight the attackers. The breed may be a distant relation of the similarly patterned Siamese (*see page 154*) and originate in Burma, but a less romantic version holds that Birmans were created by French breeders at the same time as the Colourpoint Longhair (*see page 194*) was developed.

Profile
The Birman profile, seen here on a Red Point, is strong, with a slight dip, but no defined nose stop or break. The chin tapers slightly from the nose, but it should not be receding.

Face
The Blue Point shows a full mask from nose to forehead, connected to the ears by "tracings". In self colours, the mask should be even and dense, and the nose leather should match the coat.

Lilac Point
Together with the Chocolate, this was one of the first "new" colours to be accepted. The points must be pinkish-grey, with nose leather to match, and the body a warm magnolia colour.

Blue Tabby Point
The tabby pattern was one of the earlier additions to the range of Birman points, and now comes in a full range of colours. Tabby Points should show clear frown marks and lighter "spectacles", spotted cheek pads, striped legs, and a ringed tail.

YOUNG CAT HAS INCOMPLETE MASK

Tail
Full and evenly coloured

Point colour
Dense and uniform in self points

KEY FACTS

DATE OF ORIGIN Unknown

PLACE OF ORIGIN Myanmar (formerly Burma) or France

ANCESTRY Disputed

OUTCROSS BREEDS None

OTHER NAME Sacred Cat of Burma

WEIGHT RANGE 4.5–8 kg (10–18 lb)

TEMPERAMENT Friendly and reserved

GCCF FIFé CFA TICA

BREED COLOURS

SELF AND TORTIE POINTS
Seal, Chocolate, Red, Blue, Lilac, Cream, Seal Tortie, Chocolate Tortie, Blue Tortie, Lilac Tortie

TABBY POINTS
Colours are as for self and tortie points

SEAL TORTIE TABBY

CHOCOLATE

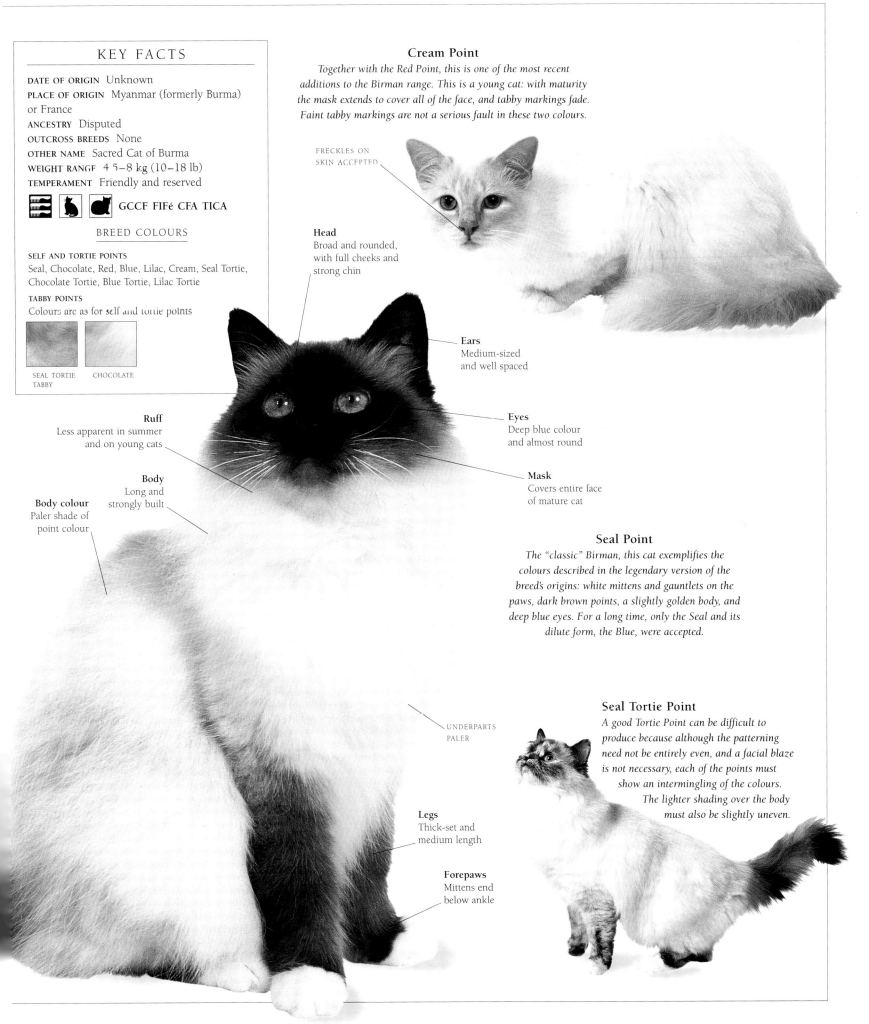

Cream Point

Together with the Red Point, this is one of the most recent additions to the Birman range. This is a young cat: with maturity the mask extends to cover all of the face, and tabby markings fade. Faint tabby markings are not a serious fault in these two colours.

FRECKLES ON SKIN ACCEPTED

Head
Broad and rounded, with full cheeks and strong chin

Ears
Medium-sized and well spaced

Eyes
Deep blue colour and almost round

Mask
Covers entire face of mature cat

Ruff
Less apparent in summer and on young cats

Body
Long and strongly built

Body colour
Paler shade of point colour

UNDERPARTS PALER

Legs
Thick-set and medium length

Forepaws
Mittens end below ankle

Seal Point

The "classic" Birman, this cat exemplifies the colours described in the legendary version of the breed's origins: white mittens and gauntlets on the paws, dark brown points, a slightly golden body, and deep blue eyes. For a long time, only the Seal and its dilute form, the Blue, were accepted.

Seal Tortie Point

A good Tortie Point can be difficult to produce because although the patterning need not be entirely even, and a facial blaze is not necessary, each of the points must show an intermingling of the colours. The lighter shading over the body must also be slightly uneven.

RAGDOLL

Mostly known for its famously placid disposition, the Ragdoll is a big and surprisingly heavy cat. Its medium-long coat has a soft texture and does not mat as readily as that of the Longhair (*see page 188*). Ragdolls are essentially pointed cats, born white and slowly developing colour and pattern over two years. Although well-muscled and with a weight advantage over most other cats, this breed has a gentle disposition. The Ragdoll is open to training with rewards and can easily be induced to use a scratching post. In Australia it is favoured as a breed that shows little enthusiasm for hunting, ensuring popularity in the face of the continent's wildlife problems (*see page 84*). Tales of a high pain threshold, however, have been comprehensively disproved.

BREED HISTORY

Although the Ragdoll is a relatively new breed, its history is confused. In the 1960s, Californian breeder Ann Baker bred the first Ragdolls from Josephine, a white, probably non-pedigree, longhair, and Daddy Warbucks, a Birman-type tom. She claimed that Ragdolls went limp when handled. Baker registered the name Ragdoll as a trademark for her breed, and formed a breed association, but its Ragdolls were not accepted by other registries. Some individuals later bred her Ragdolls, to produce the breed accepted by major registries today. Early in 2000, the breed finally won full CFA acceptance. Similar breeds are being developed today, all with equally cosy names, and some with equally bizarre claims about their traits.

Blue Mitted

There is undoubtedly some Birman in the Ragdoll's ancestry, and it has been passed on in this coat pattern. There is some resistance to the mitted cats, and indeed the whole breed, from those who feel this is a Birman "lookalike". However, there is one difference in pattern: the Ragdoll's gauntlets extend over the hock.

Seal Point

This colour and its dilute, blue, are the most common colours found in Ragdolls: it has proved difficult to produce excellent chocolate and lilac colours, and some breeders believe there may be other unidentified colours in the gene pool. The body colour on Ragdolls is allowed to be deeper than on many other pointed breeds.

Paws
Large and round, tufted

Legs
Medium length, with shorter fur than on body

Neck
Short and heavy set

KEY FACTS

DATE OF ORIGIN 1960s

PLACE OF ORIGIN United States

ANCESTRY Unclear

OUTCROSS BREEDS None

OTHER NAME None

WEIGHT RANGE 4.5–9 kg (10–20 lb)

TEMPERAMENT Genial and relaxed

GCCF FIFé CFA TICA

BREED COLOURS

POINTED
Seal, Chocolate, Blue, Lilac

MITTED
Colours are as for pointed

BI-COLOUR
Colours are as for pointed

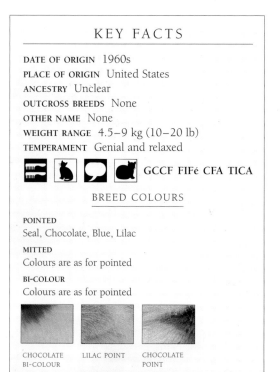

CHOCOLATE BI-COLOUR   LILAC POINT   CHOCOLATE POINT

Ragamuffin

To avoid violating the trademarked Ragdoll name, some breeders used the original Ragdoll stock to develop the more variably coated Ragamuffin. Like the Ragdoll, the Ragamuffin has an affectionate, placid nature. The two related breeds also share many physical features, such as a soft, medium-length coat and large, oval-shaped eyes.

Body
Large, with broad
chest and shoulders

Tail
Long and fully
furred

Ragdoll head

In profile, the face should have a very gentle break for the nose, which is of medium length. This Seal Mitted has a narrow white blaze on the nose, which is allowed in this coat pattern.

Head
Medium to large,
with full cheeks and
a rounded muzzle

Ears
Medium-sized,
wide at the base
with rounded tips

Body
Long and muscular,
with a broad chest

Paws
Large, rounded,
and tufted

COAT FLOWS ON BODY

Tail
Long and bushy,
slightly tapered

Coat
Medium length,
dense and silky

Eyes
Large and oval,
slightly slanted

MAINE COON

STRONG, TRANQUIL, AND LUXURIOUS TO BOTH LOOK AT AND TOUCH, the Maine Coon has recently become an outstandingly popular companion. Maines look their best in winter when the heavy, glossy coat is at its most luxurious. A distinctive characteristic that sets this breed apart from others is its frequent and enchantingly happy chirping trill, which it uses as a greeting to its human or feline family. While the Maine enjoys the company of people, it is not a dependent breed, but is content to pursue its own activities: some owners report that these include swimming. Females retain their dignity more than males, who tend to be slightly goofy, but no Maines are lap cats. These are buddies or pals, not babies, which is perhaps why advertisements for Maine Coons in quality men's magazines outweigh those for all other breeds.

BREED HISTORY

The distant history of the Maine is unknown. Its probable ancestors include British cats that came with early settlers, and longhaired Russian or Scandinavian cats on ships in Maine's ports. The harsh New England winters favour dense coats and a size sufficient for hunting hares. The black-and-white Captain Jenks of the Horse Marines was the "first" Maine, noted at Boston and New York shows in 1861, at the beginning of the breed's early popularity. At the turn of the 20th century, the Maine lost ground to the luxuriously coated Persians. It survived as a breed because farmers recognized its excellent hunting ability. Interest rekindled in the 1950s, and by 2000, it had become the CFA's second most popular longhair breed.

Brown Mackerel Tabby and White
Originally, only brown tabbies were given the name Maine Coons: the coat, together with the massive build and huge tail, made these cats resemble raccoons. Other colours and patterns were generally called Maine Shags in the early years of the breed.

Red Classic Tabby
Only the striped and blotched tabbies are recognized in Maines, in keeping with its traditional image. Reds should show excellent rufousing, setting them apart from the more usual ginger colouring seen on random-bred cats.

Maine face
The preferred look of the Maine Coon, in particular the size and set of the ears, varies between different breed associations. In all, however, the eyes are to be green, gold, or copper, with blue or odd eyes allowed in whites.

Black
The Maine Coon is traditionally associated with tabby colours and patterns in most people's minds, but selfs are widely bred. Dark colours, in particular, show the glossy quality of the coat well: the Maine coat is hard, with a unique texture.

POINTED EAR TIPS

Head
Slightly longer
than it is wide

Ears
Large, upright,
and set high
on head

SQUARE MUZZLE

Eyes
Full, round, and
slightly oblique in set

Neck
Thick in males

LONG BACK

Body
Medium to large,
with solid
musculature

Coat
Long and glossy

Tail
Long, with
flowing fur

Tortie Tabby

*The image of the Maine Coon as a big, shaggy, tabby cat
is so ingrained into public consciousness that any cat
answering to this description is liable to be tagged as a
Maine by the unscrupulous. True Maines comply to a
rigorous standard, and careful breeding is needed to
produce rich colours such as this example consistently.
The size of the Maine Coon has also become a matter of
some myth: claims of breeding lines that tip the scales at
weights of 15 kg (33 lb) remain unsubstantiated.*

KEY FACTS

DATE OF ORIGIN 1860s
PLACE OF ORIGIN United States
ANCESTRY Farm cats
OUTCROSS BREEDS None
OTHER NAME Maine Shag
WEIGHT RANGE 4–10 kg (9–22 lb)
TEMPERAMENT Gentle giant

 **GCCF FIFé CFA TICA**

BREED COLOURS

SELF AND TORTIE
Black, Blue, Cream, Red, Tortoiseshell, Blue Tortie,
White (Blue-, Green-, Odd-, Orange-Eyed)

SMOKE AND SHADED
Colours are as for self and tortie colours with the
exception of White

TABBY (CLASSIC, MACKEREL)
Brown, Red, Blue, Cream, Tortie, Blue Tortie

SILVER TABBY
Colours are as for standard tabbies

BI-COLOURS
All self, tortie, and tabby colours with white

CREAM
SHADED

BLACK SMOKE

BROWN
CLASSIC TABBY

BLUE SILVER
TABBY

MAINE COON VARIATIONS

THE STRUCTURE OF THIS BREED'S COAT IS UNDOUBTEDLY THAT OF A FARM CAT. Despite being long and thick, it requires surprisingly little maintenance, and it is water-repellent, so it is fortunate that washing is rarely necessary. Some of the colours, however, have raised questions among breeders: it has been suggested that Longhairs were used to introduce the smoke and silver colours. This seems unlikely, because many cats in the random-bred cat population of North America, which had nothing to do with Longhairs, carry the inhibitor gene that causes these colours. In Britain, the same colour range is allowed in self, smoke, shaded, tabby, and silver tabby coats. In North America a more complex situation exists.

Blue and White kitten
The sturdy build of this breed is apparent from the earliest age. The Maine Coon can be unpredictable in its maturing, with some excellent breed examples only emerging as such in full adulthood.

Red Shaded Tabby
A good Red Tabby is a rich red tone in both the background colour and the darker markings. The inhibitor gene removes the colour of the background, but the markings should show the same pronounced rufousing as a full-coloured coat.

Maine Wave
These controversial cats are full-pedigree Maine Coons, with a rex mutation. This mutation was once thought to be lethal, but different lines now seem to be breeding healthily. The coat lacks guard hairs, and so is completely unlike the typical Maine coat, and these cats are unlikely ever to be accepted as part of the breed. As the mutation is recessive, they are also unlikely to disappear.

Black Smoke

The heavy topcoat of the Maine conceals the white undercoat over most of the body. Only where the coat breaks – around the legs and neck – does the silvering shine through. The undercoat is much more apparent when the winter coat is shed in the warmer months.

Odd-Eyed White

As in all breeds, Whites can suffer from deafness. This applies mostly, but not entirely, to Blue-Eyed and Odd-Eyed Whites. Some breeders claim that kittens that have large "kitten caps" of darker hair are less prone to this deafness, and are seeking to breed hearing Whites consistently by breeding for larger caps.

Alternative look

This Silver Tabby shows the Maine Coon style preferred by TICA, a primarily North American registry. The face is slightly more angular than the preferred British look, with larger ears set higher on the head, and rounder eyes.

British look

This Silver Tabby is an example of the look that is preferred by the GCCF for the Maine Coon. A broad face with oval eyes is topped by moderately sized ears, angled out from the corners of the head. The body and legs are correspondingly stocky, giving an impresssion of a cat that could hold its own anywhere.

Maine Coon (overleaf left) *For breed entry see page 202* **Norwegian Forest Cat** (overleaf right) *For breed entry see page 208*

NORWEGIAN FOREST CAT

Somewhat reserved with strangers but calmly confident with people it knows, the Norwegian Forest Cat shares attributes with the Maine Coon (*see page 202*) and Siberian Forest Cat (*see page 210*). Large size and long hind legs give the "Wegie" a commanding presence. Norwegian breeders like to think of this "natural cat" as their little lynx. While it makes a gentle household cat, it defends its territory vigorously. It is a superb climber and hunter, and owners who live near streams report that their Wegies fish. The breed is generally sound, although breeding for longer bodies and noses must be guarded against to avoid spinal and dental problems.

BREED HISTORY

Cats arrived in Norway by around AD 1000, at which time Vikings maintained trade routes with the Byzantine East. Proof that cats were traded directly from Byzantium to Norway comes from Norwegian cat populations with coat colours common in Turkey but rare across Europe. It is possible that the Norwegian Forest Cat traces its ancestry back to longhaired Turkish cats. The harsh Scandinavian winter favoured large, longhaired cats, which became popular with farmers. The Wegie was not regarded as a singular breed until the 1930s, and planned breeding did not begin until the 1970s. The first Wegies arrived in the United States in 1979 and in Britain in the 1980s.

Ears
Open, wide at base, and set high on head, continuing lines of face

Head
Triangular, with a long, straight profile and strong chin

MEDIUM FURNISHINGS PREFERRED

Eyes
Large and open but not round, and slightly oblique in set

Body
Large, with solid bone structure and good musculature

Black Smoke and White
The Norwegian type should always remain elegant. While large and robust, the build should not appear stocky, and the facial features should be angular and give an impression of alertness, rather than be rounded or "sweet".

FULL RUFF DESIRABLE

Coat
Smooth, glossy, water-repellent topcoat and woolly undercoat

Legs
Long but not delicate

Paws
Large and round, with tufts between toes

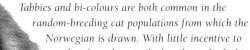

Blue Tabby and White

Tabbies and bi-colours are both common in the random-breeding cat populations from which the Norwegian is drawn. With little incentive to select for colour in the breed standard, this distribution is reflected in the breed.

Silver Tabby and White

In almost all cats, coats will sometimes appear slightly yellowed or "tarnished" due to rufous genes; the effect is most obvious in silver tabbies. Although regarded as a flaw in many breeds, and vigorously selected against, it is not a fault in Wegies.

KEY FACTS

DATE OF ORIGIN 1930s

PLACE OF ORIGIN Norway

ANCESTRY Farm cats

OUTCROSS BREEDS None

OTHER NAME Skogkatt or Skaukatt, Wegie

WEIGHT RANGE 3–9 kg (7–20 lb)

TEMPERAMENT Reserved and contained

  GCCF FIFé CFA TICA

BREED COLOURS

SELF AND TORTIE COLOURS
Black, Red, Blue, Cream, White (Blue-, Green-, Odd-, Orange-Eyed), Tortoiseshell, Blue-Cream

SMOKE COLOURS
Colours are as for self and tortie colours, except White

SHADED, AND TIPPED COLOURS
Colours are as for self and tortie colours, except White

TABBY COLOURS (CLASSIC, MACKEREL, SPOTTED)
Brown, Red, Blue, Cream, Tortie, Blue Tortie

SILVER TABBY COLOURS
Colours and patterns are as for standard tabbies

BI-COLOURS
All allowed colours and patterns with white

Any other colour or pattern except those that indicate hybridization, such as chocolate, lilac, Colourpoint pattern, or these combinations with white

Tail
Long and bushy,
equal in length
to body

Black

Swathed in a thick, black coat for winter, the Norwegian could pass for one of the fierce cats of Nordic legend. Eye colour is unrelated to coat colour, so Black selfs may come with eyes of glowing gold or witch's-cat green. Black coats tend to show a degree of "rusting", unless the cat is rigorously kept out of sunlight.

Silver Tabby

The breed standard for the Norwegian Forest requires that its appearance reflects its natural heritage as a farm cat. The most important features are type and coat quality: there are no points specifically allocated to coat colour in the scoring system. Tabbies are popular, reinforcing the "natural" image.

SIBERIAN FOREST CAT

Famed for its harsh winter, the homeland of this breed favoured large, sturdily built cats with thick, protective coats. It is impossible to be sure of this cat's most ancient ancestry, but what is certain is that the Siberian has been perfected by its environment, just like the Norwegian Forest (*see page 208*). Every aspect of this cat is honed for survival in tough conditions: its topcoat is strong, plentiful, and oily, its undercoat is dense enough to keep out the keenest winds, and its build is large. Sturdy does not mean staid, however: these are active and highly agile cats. The Siberian character is also a product of its past: while sociable, it is disinclined to play the passive lap-cat.

BREED HISTORY

Longhaired cats are found across the northern wastes of Russia. Like many natural breeds, the Siberian was not regarded as noteworthy until fairly recently. Serious breeding to standardize the type began in the 1980s, and the breed is recognized by a wide range of registries in its homeland, including the All-Russian Cat Club. Siberians were imported into the United States in 1990, through the efforts of Elizabeth Terrell. Her cattery name, Starpoint, can be found in the pedigrees of most of the top Siberians there. Among major registries, FIFé and TICA accept the Siberian. Some Russian clubs fear that the cats exported to the West are not always the best. The "TICA face" is different, and the Siberian may develop two distinct looks internationally.

Brown Classic Tabby
The appearance of the Siberian is more reminiscent of a wildcat than any other breed. The face is also unique: great breadth, and oval, slightly tilted eyes, give a distinctly wild and decidedly Asiatic cast. Russian cat clubs wish to see this wild appearance preserved.

Red Shaded Tabby and White
The Siberian is allowed in only black- and red-based colours in its homeland; a wider range is recognized in North America. The inhibitor gene, which produces shaded colours, is naturally present, although not prevalent.

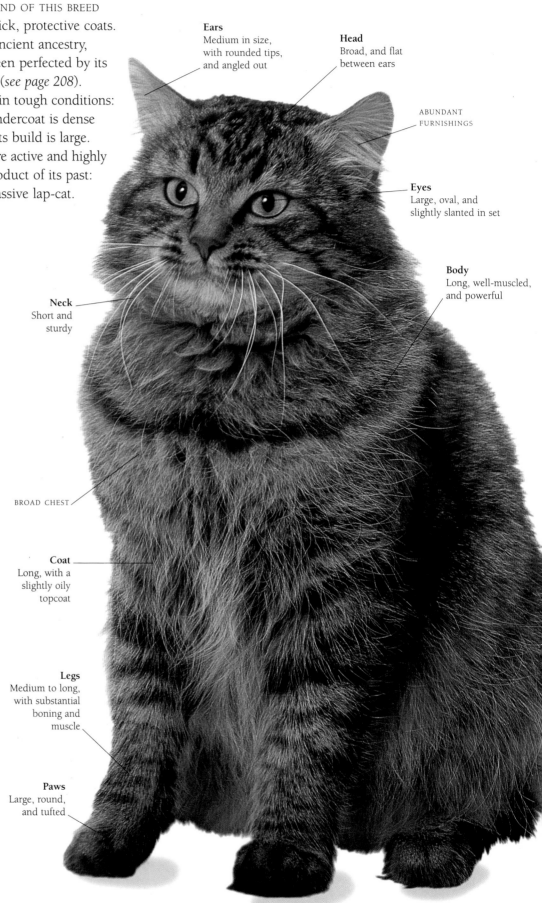

Ears
Medium in size, with rounded tips, and angled out

Head
Broad, and flat between ears

ABUNDANT
FURNISHINGS

Eyes
Large, oval, and slightly slanted in set

Body
Long, well-muscled, and powerful

Neck
Short and sturdy

BROAD CHEST

Coat
Long, with a slightly oily topcoat

Legs
Medium to long, with substantial boning and muscle

Paws
Large, round, and tufted

FEMALES FINER-BONED
THAN MALES

ROUNDED EYES
PREFERRED IN
NORTH AMERICA

Tortie Tabby and White

Siberian females are, as in many breeds, slightly smaller and lighter than the males. In both sexes, the hindlegs are slightly longer than the forelegs when straight, and the body is carried with a slightly arched spine.

KEY FACTS

DATE OF ORIGIN 1980s

PLACE OF ORIGIN Eastern Russia

ANCESTRY Household and farm cats

OUTCROSS BREEDS None

OTHER NAMES None

WEIGHT RANGE 4.5–9 kg (10–20 lb)

TEMPERAMENT Sensible and resourceful

   FIFé TICA

BREED COLOURS

SELF AND TORTIE
Black, Red, Blue, Cream, Tortoiseshell, BLue Tortie
All other self and tortie colours

SMOKE, SHADED, AND TIPPED COLOURS
Colours are as for self and tortie

TABBIES, SILVER TABBIES (CLASSIC, MACKEREL, SPOTTED)
Brown, Red, Blue, Cream, Tortie, Blue Tortie
Ticked pattern, all other self and tortie colours

BI-COLOUR
All allowed self, tortie, and tabby colours
All self, tortie, and tabby colours with white

All other colours and patterns, including
pointed patterns

Brown Mackerel Tabby

The TICA standard for the Siberian head is less "wild" than that preferred by Russian clubs. Although the head should be broad, the impression should be of "roundness and circles", with a sweet expression and eyes that are almost round in shape. North American Siberians have medium to large ears.

Brown Spotted Tabby and White

Originally, there was a great predominance of tabby coats in the Siberian, as is to be expected of a breed that developed outdoors, where natural enemies abound. Breeders are almost certain to wish to develop a wider range of self and shaded colours, but tabbies of all patterns still make up a fairly high proportion of the breed.

Tail
Medium length and
thick, with a
rounded tip

AMERICAN CURL

THIS QUIET AND GENTLE BREED IS SIMPLY THE HOUSEHOLD CAT OF the United States with a single, striking mutation: its ears curl back away from the face towards the back and centre of the head. This distinctive feature gives the Curl a dramatic, pixie-like face, full of astonishment. The trait is dominant, so a Curl bred to any cat should give at least 50 per cent Curls. The rest, called American Curl Straight Ears, are used for breeding programmes or sold as pets. Outcrossing, which has created a shorthaired Curl (*see page 123*), ensures that genetic diversity continues to flourish within the breed. Spontaneous mutations such as this can carry crippling side effects: none are apparent in Curls, but some registries fear that problems may yet emerge.

Seal Point
The pointed pattern, once confined to a single breed, is now found in a vast range of cats. Newer breeds, such as the Curl, usually have this pattern. Longer coats generally soften and lighten the pointing.

BREED HISTORY

In 1981, a stray kitten appeared at Grace and Joe Ruga's home in Lakeland, California. Cats are adept at targeting nurturing individuals: Grace Ruga left food on her porch for the stray, who ate it, liked the ambience of the household, and made it her home. An affectionate black female, she had a long, silky coat and unusual ears. Joe Ruga named her Shulamith, meaning "peaceful one", after the shepherdess in the *Song of Songs*. All Curls trace their origins to Shulamith. In December that year, Shulamith had a litter of four kittens, two of which had the same curly ears. These cats were shown in California in 1983. Fully recognized in North America, the Curl has the distinction of being the first breed to ever win CFA acceptance in two coat lengths. The first Curls to reach Europe arrived in Britain in 1995; they are unlikely to be accepted by the GCCF or FIFé.

Eyes
Walnut-shaped and slightly tilted

Head
Rounded, modified wedge

Ears
Point at back of head

Tortoiseshell and White
This pattern, also called Calico in North America, is almost exclusively female. The Curl's moderate breed standard favours females: in heavier breeds they can lose awards to sturdier males.

Body
Moderately muscled and semi-foreign in build

Coat
Silky and flowing, with minimal undercoat

Legs
Medium length and boning

Tail
Full plume equal to length of body

KEY FACTS

DATE OF ORIGIN 1981
PLACE OF ORIGIN United States
ANCESTRY American household cat
OUTCROSS BREEDS Non-pedigree domestic cats until January 2010
OTHER NAME None
WEIGHT RANGE 3–5 kg (7–11 lb)
TEMPERAMENT Quietly affable

   CFA TICA

BREED COLOURS

SELF AND TORTIE
Black, Chocolate, Red, Blue, Lilac, Cream, White, Tortoiseshell, Blue-Cream
All other self and tortie colours

SMOKE
Colours as for self and tortie, except White and with the addition of Chocolate Tortie
All other self and tortie colours

SHADED AND TIPPED
Shaded Silver, Shaded Golden, Shaded Cameo, Shaded Tortoiseshell, Chinchilla Silver, Chinchilla Golden, Shell Cameo, Shell Tortoiseshell
All other self and tortie colours

TABBIES (ALL PATTERNS)
Brown, Red, Blue, Cream, Brown Patched, Blue Patched
All other self and tortie colours

SILVER TABBIES
Silver, Chocolate Silver, Cameo, Blue Silver, Lavender Silver, Cream Silver, Silver Patched
All other standard tabby colours

BI-COLOURS (CLASSIC AND VAN)
Black, Red, Blue, Cream, Tortoiseshell, Blue-Cream, and tabby colours with white *Other colours with white*

SELF AND TORTIE POINTS
Seal, Chocolate, Flame, Blue, Lilac, Cream, Tortie, Chocolate Tortie, Blue-Cream, Lilac-Cream
All other colours, sepia and mink patterns

LYNX (TABBY) POINTS
As for self and tortie points, excepting Red
All other colours, sepia and mink patterns

MUNCHKIN

Without doubt the most controversial and extraordinary breed to emerge in years, the Munchkin is defined by a single, dominant factor: the long bones in its legs are simply not long. There is no direct effect on other bones, and breeders claim that the short legs have no detrimental side effects; only time will tell. The flexible feline spine may save the breed from the back and hip problems of dwarfed dog breeds, but in all other species dwarfs are prone to arthritis, and dwarfed cats might not escape this. Breeders claim that three sizes of Munchkins exist: standards, "super-shorts" or "rug-huggers", which have very short legs, and an exceptionally small type called "minis". The controversial "mini" is provisionally recognized by the UFO under the name Mei Toi.

BREED HISTORY

Dwarfed individuals occur in many species: the cat is no exception. The Munchkin originated in a mutation in Louisiana in 1983. As breeders began to work with the mutation, outcrossing to non-pedigree cats, controversy grew along with popularity. TICA granted the Munchkin "new breed" status in 1995; as yet, it is the only major registry to recognize the breed, and the standard is still very general. Breeders of other pedigree breeds have expressed fears that dwarfed "editions" of their breeds will emerge. Although the lure of novelty cannot be denied, and some breeders may choose to pursue such a course, the TICA standard specifically bars any other pedigree breeds as outcrosses, and no registry would be likely to look favourably on dwarfed versions of established breeds.

Munchkin head
The shape is between an almost equilateral triangle and a modified wedge: "medium" and "moderate" are the words most used in the standard. This is likely to be reassessed as the breed is developed. Eye and coat colours are not related to each other.

Tail
Medium thickness, tapering to rounded tip

Body
Medium-sized, with level spine or slight rise from shoulder to rump

Head
Medium-sized, neither round nor wedge-shaped

Ears
Triangular, moderately large in size

Eyes
Large and walnut-shaped, with open expression

Legs
Shortened but straight, with paws turned out slightly

Red and White
The Munchkin is recognized in all possible colours and patterns; it would also be almost impossible to limit the range in a breed that is outcrossed to random-bred cats. Tabbies and bi-colours are more common than Oriental shades and patterns.

Black and White kitten
It is apparent at birth which kittens in a litter are Munchkins and which are not. Proponents of the breed claim that one of the breed's most attractive traits is not in its physique, but its personality: Munchkins are said to keep their kittenish curiosity and comical behaviour into adulthood.

KEY FACTS

DATE OF ORIGIN 1983
PLACE OF ORIGIN United States
ANCESTRY Household cats
OUTCROSS BREEDS Non-pedigree cats
OTHER NAME None
WEIGHT RANGE 2.25–4 kg (5–9 lb)
TEMPERAMENT Appealing and inquisitive

BREED COLOURS

All colours and patterns, including pointed, mink, and sepia

SCOTTISH FOLD

WITH THE SAME DISTINCTIVE EARS AS ITS SHORTHAIRED COUSINS (*see page 119*), the longhaired Scottish Fold's coat gives it a lush, warm look. Like all longhairs, the breed is best seen in winter, when it sports an imposing ruff, elegant breeches, and a huge, fluffy tail. All kittens are born with straight ears, which begin to fold at about three weeks of age. The joint problems that result from breeding Fold-to-Fold appear at four to six months: a short, thickened tail is a sign that might be missed in a longhaired kitten, so tails should be checked carefully – and always gently.

Head
Well rounded, with prominent cheeks and whisker pads

Ears
Small, with rounded tips, tightly folded to head

Body
Medium, rounded, and firm

Eyes
Large, rounded, and sweet in expression

Coat
Medium to long, soft, and standing away from body

Legs
Medium length, sturdy, but not coarse

BREED HISTORY

Cats with folded or pendulous ears have been recorded for over two centuries. All Scottish Folds, however, can be traced back to Susie, a white farm cat born in 1961 in Scotland. Two geneticists, Pat Turner and Peter Dyte, oversaw the early development of the breed, and found that Susie carried the longhair gene, which could be carried in shorthaired offspring and appear in later generations. The Fold is still rare: the absence of any longhaired outcross breed makes the longhaired version even rarer.

Tortie Tabby and White

This combination of tortoiseshell and tabby is also called "patched tabby" in CFA or "torbie" in TICA. The tabby pattern should show clearly in both brown and red patches, which are larger and more defined in bi-colours such as this one. The required eye colour, which should be as brilliant as possible, is gold.

Blue Smoke and White

This individual falls short of the show standards for the Fold in two respects: the ears are not pressed tightly to the head, and the face shows slight tabby markings rather than the perfectly even blue that the standard requires. Such cosmetic quirks, however, make no difference to the cat's character or qualities as a pet.

KEY FACTS

DATE OF ORIGIN 1961
PLACE OF ORIGIN Scotland
ANCESTRY Farm cat, British and American Shorthairs
OUTCROSS BREEDS British and American Shorthairs
OTHER NAME Highland Fold, Longhair Fold
WEIGHT RANGE 2.4–6 kg (6–13 lb)
TEMPERAMENT Quietly confident

    CFA TICA

BREED COLOURS

All colours and patterns, including pointed, sepia, and mink

BROWN TABBY   RED TABBY   LILAC   WHITE

SELKIRK REX

THIS IS POSSIBLY THE MOST STRIKING OF ALL OF THE REXED BREEDS. It shares the distinction of being longhaired with the LaPerm (*see page 235*), but its appearance is quite unique. The long, thick coat is at its best in heterozygous cats, with one rexing and one straight-haired gene: this combination gives a loose, ringleted effect. All three hair types are present in the coat, and a longhaired Selkirk in moult may shed as much hair as a Longhair (*see page 188*). The longhaired Selkirk has the same relaxed outlook as its shorthaired counterpart (*see page 118*).

BREED HISTORY

The very first Selkirk Rex, Miss DePesto of NoFace, or Pest, was a shorthaired kitten born at a pet rescue service in Montana. She came into the household of Jeri Newman, who was a breeder of Longhairs, or Persians. Jeri mated the new arrival to her black Longhair champion, Photo Finish of Deekay. The resulting litter included longhaired and shorthaired kittens, with a mixture of straight and curly coats. This variety showed not only that Pest's rexing mutation was a simple dominant, but that she, like many random-bred cats, was carrying the recessive longhair gene. From the start, therefore, the Selkirk Rex has had both longhaired and shorthaired classes: the two are not formally separated, and the allowed outcrosses for the breed continue to include the Longhair.

Tortoiseshell Shaded
Although kittens appear very curly at birth, they may lose their curl almost completely, and young cats will go through an untidy-looking stage before they become curly again. The degree of curl on the body varies with climate, season, and hormonal factors.

Red Shaded Tabby
The ringlets of the coat reveal the white undercoat, making this pattern far less dramatic than it is in a longhair with a straight coat. Tabby markings on the body are softened by the curl: the less obscured they are, the better. Frown lines and spectacles remain clearly visible on the face.

KEY FACTS

DATE OF ORIGIN 1987
PLACE OF ORIGIN United States
ANCESTRY Rescued cat, Longhair, Exotic, British and American Shorthairs
OUTCROSS BREEDS Pedigree parent breeds
OTHER NAME None
WEIGHT RANGE 3–5 kg (7–11 lb)
TEMPERAMENT Patiently tolerant

   CFA TICA

BREED COLOURS

All colours and patterns, including pointed, sepia, and mink

CREAM   WHITE   BLUE

Head
Rounded, with short, squared-off muzzle

Ears
Medium, pointed, and set well apart

Eyes
Round and widely spaced

CURLY WHISKERS

Body
Muscular and rectangular, with slight rise to hindquarters

Coat
Soft, falling in loose, individual curls

Legs
Medium and substantially boned, with large paws

Tail
Thick, tapering slightly to rounded tip

Selkirk head
Unlike the other rexed breeds, the Selkirk Rex has a sturdy, rounded look. The muzzle is short and broad, with a distinct nose stop, and both the cheeks and the whisker pads are full. The colour of the eyes is not related to that of the coat.

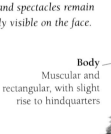

TURKISH VAN

With its incredibly soft coat, and large, rounded eyes, this breed might appear to be the ideal, soft-centred lap-cat. Turkish Vans descended, however, from rural cats in an area where life was far from easy, and retain minds and spirits of their own. This is a cat to be respected and won over, not taken for granted. The breed is distinguished for two reasons: the restricted colour of its coat, so distinctive that the pattern is called Van even in other breeds, and its reputation for enjoying a dip in hot summer weather, which has earned it the alternative name of "Swimming Cat" in its homeland.

BREED HISTORY

The modern history of this breed began when two cats were brought to Britain in 1955. The breed spread across Europe, but acceptance by registries was slow. In the 1970s, the first Van kittens reached the United States, where the breed is now accepted by CFA and TICA. In GCCF, only the original red and the more recently accepted cream are allowed. Some registries also allow black-based colours. FIFé has altered its rules to include these colours.

Tortie and White

Tortie Vans appeared when black was introduced to the breed. The "thumbprints" of colour above this cat's tail make it less than perfect according to the breed standard: the difficulty of meeting the standard means that although the Van is scarce, there are relatively high numbers of pet kittens available.

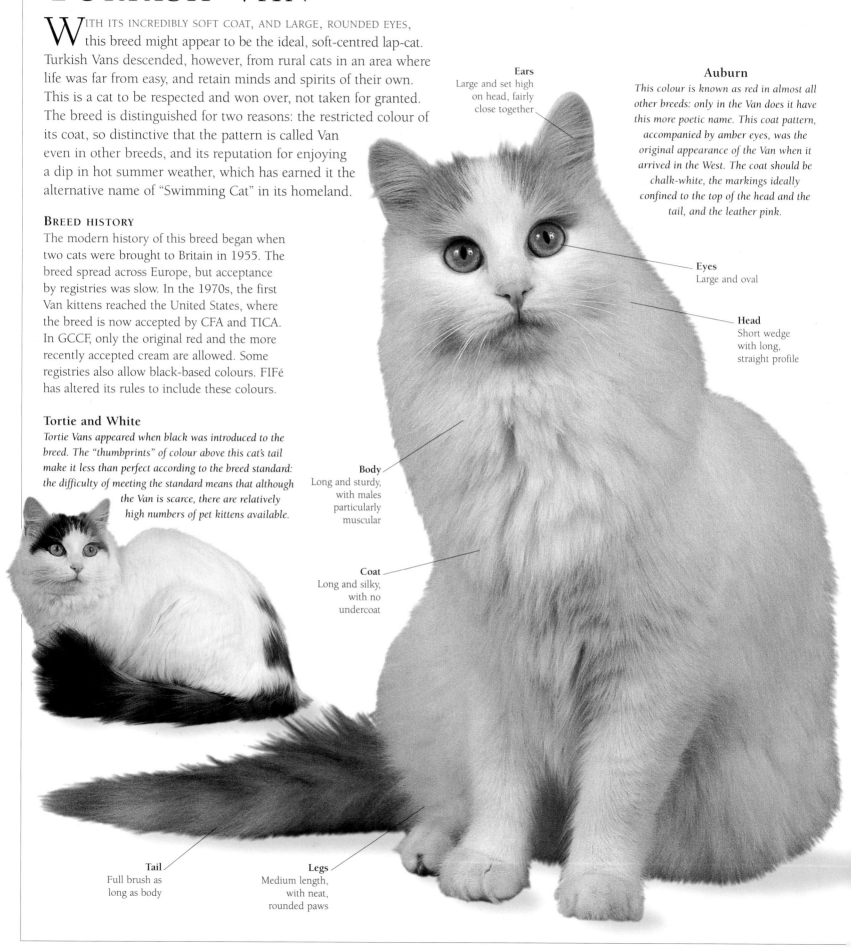

Ears
Large and set high on head, fairly close together

Auburn
This colour is known as red in almost all other breeds: only in the Van does it have this more poetic name. This coat pattern, accompanied by amber eyes, was the original appearance of the Van when it arrived in the West. The coat should be chalk-white, the markings ideally confined to the top of the head and the tail, and the leather pink.

Eyes
Large and oval

Head
Short wedge with long, straight profile

Body
Long and sturdy, with males particularly muscular

Coat
Long and silky, with no undercoat

Tail
Full brush as long as body

Legs
Medium length, with neat, rounded paws

Van face

The colour markings on a Van's head should not extend below the level of the eyes, or beyond the base of the ears at the back. Ideally, there should be a white blaze on the forehead, dividing the colour into two separate areas, and the ears should be coloured.

Local detail

The Van pattern occurs in naturally high numbers in the cat population of the Near East, but is rare elsewhere. Its unusualness impressed visiting artists in the 19th century, and in many paintings of bazaar and camp scenes a Van-patterned cat is to be seen tucked in a corner between the silks and the rugs. This cat is in The Arab Scribe, Cairo, by John Frederic Lewis (1805–1876).

Blue

Blue can vary widely in depth, and this cat, an early Blue, is darker than most breed standards allow. Introducing new colours into the Van inevitably also brought in some undesirable traits. The rich, golden eye colour is a difficult quality to fix, and some green eyes have appeared.

COAT
"BREAKS"
READILY

Cream

The dilute of the Auburn and White, this colour resembles it in all aspects of pattern and eye colour. The Van's soft coat is water resistant, and tends to "break" open over any curves, rearranging itself with every move.

KEY FACTS

DATE OF ORIGIN Pre-18th century
PLACE OF ORIGIN Lake Van region, Turkey
ANCESTRY Household cats
OUTCROSS BREEDS None
OTHER NAME Turkish Swimming Cat
WEIGHT RANGE 3–8.5 kg (7–19 lb)
TEMPERAMENT Self-possessed

   GCCF FIFé CFA TICA

BREED COLOURS

BI-COLOURS (AMBER-, BLUE-, ODD-EYED)
Auburn, Cream, with White
Black, Blue, Tortoiseshell, Blue-Cream, with White

CYMRIC

THIS SOLID BREED MATCHES THE ORIGINAL SHORTHAIRED MANX IN ALL BUT COAT, which is semi-long and double. Like its shorthaired cousin the Manx (*see page 116*), the Cymric produces variant "stumpies" and "longies" with some degree of tail, as well as the showable, tailless "rumpies". Although the breed originated in North America, its name comes from "Cymru", the Welsh word for Wales; there is a belief that Wales had its own strains of tailless cats. The breed is also known as the Longhaired Manx, but it is the more poetic name that has gained the wider currency.

BREED HISTORY

Although its name suggests a Welsh ancestry, this is an exclusively North American breed. Manx cats have always produced the occasional longhaired kittens and, in the 1960s, breeders, including Blair Wright in Canada and Leslie Falteisek in the United States worked to gain these longhaired variants recognition. By the 1980s, CFA and TICA both recognized the cats as a separate breed with the name Cymric, but CFA has now reclassified them as Longhaired Manx. The breed is not recognized in Britain.

Orange-Eyed White
White Cymrics' eyes may be deep blue, brilliant copper, or one of each colour. The coat should be pure white, with no hint of any yellowing, and no stray coloured hairs should be present.

Brown Mackerel Tabby and White
The coat colour and markings are relatively unimportant in the Cymric's points system. Most important are complete taillessness, a cobby appearance, and a rounded head with prominent cheeks and whisker pads. Cymrics can take two years to develop the required deep flanks and rounded body.

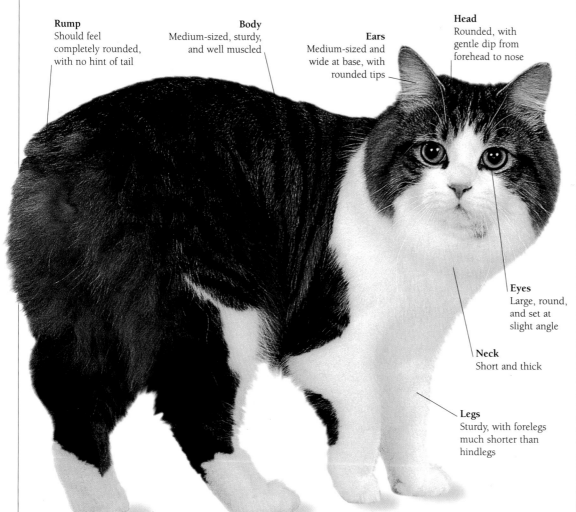

Rump
Should feel completely rounded, with no hint of tail

Body
Medium-sized, sturdy, and well muscled

Ears
Medium-sized and wide at base, with rounded tips

Head
Rounded, with gentle dip from forehead to nose

Eyes
Large, round, and set at slight angle

Neck
Short and thick

Legs
Sturdy, with forelegs much shorter than hindlegs

KEY FACTS

DATE OF ORIGIN 1960s
PLACE OF ORIGIN North America
ANCESTRY Manx
OUTCROSS BREEDS Manx
OTHER NAME Longhaired Manx
WEIGHT RANGE 3.5–5.5 kg (8–12 lb)
TEMPERAMENT Friendly and even-tempered

CFA TICA

BREED COLOURS

SELF AND TORTIE
Black, Red, Blue, Cream, White, Tortoiseshell, Blue-Cream
All other self and tortie colours

SMOKE
Black, Blue
All other self and tortie colours

SHADED AND TIPPED
Shaded Silver, Chinchilla Silver
All other self and tortie colours

TABBIES (CLASSIC, MACKEREL)
Brown, Red, Blue, Cream, Brown Patched, Blue Patched
Spotted and ticked patterns, all self and tortie colours

SILVER TABBIES
Silver, Patched Silver
All other standard tabby colours

BI-COLOURS (STANDARD AND VAN)
All self and tortie colours with white
All colours and patterns with white

POINTED
All colours and patterns in pointed, sepia, and mink

BLACK AND WHITE | CHOCOLATE (NOT CFA) | RED TABBY | BLUE

NEBELUNG

THE SILVER-TIPPED, BLUE HAIR GIVES THIS BREED A LUMINOUS
elegance. Light reflects off the Nebelung's guard hairs, creating
a misty incandescence. Only when you stroke the semi-long hair
against the grain do you notice the solid-blue ground colour to
the shafts of both the guard hair and the down. This rare breed,
the name of which is German for "mist-creature", is based upon a
"lost" longhaired strain of the Russian Blue (*see page 132*).

BREED HISTORY

Blue shorthairs and longhairs from Russia
were exhibited over 100 years ago. Shorthairs
became known as the Russian Blue, but the
longhairs lost their separate identity. In 1986,
Siegfried, the founding father of this revived
breed, was mated to his longhaired sister, who
produced longhaired blue kittens. In 1987,
TICA approved the Nebelung as a breed, and
it was also recognized by TCA in 1993, but
other major registries have not accepted it.

Nebelung head
*A modified wedge of seven flat
planes, the face has a slight smile.
The green eyes take time to
develop, being yellow in kittens
and younger cats. A green ring
should appear around the pupil by
four months, and the eyes should
turn green with maturity.*

Blue
*The Nebelung standard is very similar to that of the
Russian Blue, calling for the same lithe appearance,
the same silver tipping, and a semi-long version of the
characteristic double coat. Whether this breed will
develop to be identical to the Russian in all but coat
length, or diverge in other ways, remains to be seen.*

KEY FACTS

DATE OF ORIGIN 1984
PLACE OF ORIGIN United States
ANCESTRY Russian Blue
OUTCROSS BREEDS Russian Blue
OTHER NAME None
WEIGHT RANGE 2.5–5 kg (6–11 lb)
TEMPERAMENT Retiring
TICA

BREED COLOURS

SELF COLOURS
Blue

Eyes
Very slightly
oval, green, and
widely spaced

Ears
Wide at base, with
slightly rounded tips

Neck
Long and slender
under ruff

Body
Lithe and slender,
but not tubular

Coat
Fine, double, and
medium length,
with silver-tipped
guard hairs

Head
Modified wedge,
with flat
forehead and
straight profile

Paws
Small and round,
with an impression
of being on tiptoes

Tail
Long and fluffy, tapering
from thick base to fine tip

TURKISH ANGORA

GRACEFUL AND ATHLETIC, WITH FINE BONES AND A SILKY COAT, this small to medium-sized cat is covered by a single coat that shimmers when it moves, and comes in all colours other than the Oriental shades. They are vivacious cats, considered quick-witted and quick-moving by breeders. Some published information still repeats the theory that the breed descended from the wild Pallas' cat (*see page 17*). The unlikely story is that Tartars domesticated the Pallas' cat and took it into Turkey. The Turkish Angora's medium-long coat is probably the result of a mutation that occurred centuries ago in an isolated population of domestic cats in central Asia. As with other breeds, white blue-eyed cats may be born partially or totally deaf. This is due to a colour gene defect.

National cat
Turkish Angoras often appear on stamps: this Odd-Eyed White commemorates 25 years of the Turkish veterinary service.

BREED HISTORY

Angoras from Turkey first reached France and Britain in the 17th century, but it was not until the mid-1800s that breeders chronicled the differences between this breed and longhairs from Persia and Russia. By the beginning of the 1900s, cross-breeding with other longhaired cats had led to the virtual extinction of the breed outside Turkey. The breed is said to have been saved through a breeding programme at Ankara Zoo, although this may be a rather romanticized tale. Certainly, breeders in Sweden, Britain, and the United States imported Angoras after World War II from Turkey, where the breed is now protected, and established a successful breed revival programme. Turkish Angoras are not recognized in Britain, where there is an unrelated breed called the Angora (*see page 232*).

Ears
Large and high set, slightly pointed

Eyes
Large and oval, slightly slanted

Head
Modified wedge of small to medium size

Neck
Slim and graceful

Body
Long and slender, but muscular

Coat
Fine and silky with negligible undercoat

Legs
Long, with hindlegs longer than forelegs

Tail
Tapering from wide base to fine tip

FULL PLUME OF HAIR

Red Shaded
The inhibitor gene has been present in the Turkish Angora for centuries. Combined with the agouti gene it gives silvered, or shaded, colours; these are not recognized by CFA, but are by TICA and FIFé. When the coat is shed in the summer, the silver becomes more apparent, especially on the face, as here.

Tortie and White
The predominantly white Van bi-colour pattern is prevalent in Turkish cats. In the Turkish Angora bi-colours, the underparts should be uniformly white, and the black and red patches should be solid and even in colour.

RUMP HIGHER THAN SHOULDERS

Black
The coat of a Black Turkish Angora must be coal black, and solid from the roots to the tip. As in all cats, the colour tends to take on a rusty tinge with prolonged exposure to sunlight, recovering its solid tone when the summer coat is shed and the fuller winter coat comes in. All longhairs are at their best in winter, but especially black cats.

Tortoiseshell Smoke
Smoke Angoras were first documented in Britain in the latter part of the 19th century. Smoke Turkish Angoras should appear to be full-coloured cats in repose, the undercoat only apparent with movement. The loss of the long coat in summer decreases the effect.

Turkish Angora head
The head is a smooth wedge, with the narrow muzzle continuing the lines without a pronounced pinch at the whisker pads. The eyes may be any colour from copper through gold and green to blue.

KEY FACTS

DATE OF ORIGIN 1400s
PLACE OF ORIGIN Turkey
ANCESTRY Household cats
OTHER NAME None
OUTCROSS BREEDS None
WEIGHT RANGE 2.5–5 kg (6–11 lb)
TEMPERAMENT Energetic exhibitionist

FIFé CFA TICA

BREED COLOURS

SELF AND TORTIE COLOURS
Black, Red, Blue, Cream, Tortie, Blue-Cream, White
All other self and tortie colours

SMOKE COLOURS
As for self and tortie colours, except White
As for self and tortie colours, except White

SHADED COLOURS
As for self and tortie colours, except White

TABBY COLOURS (CLASSIC, MACKEREL)
Brown, Red, Blue, Cream
Spotted and ticked, all other self and tortie colours

SILVER TABBY COLOURS (CLASSIC MACKEREL)
Silver
Spotted and ticked, all other self and tortie colours

BI-COLOURS
All self and tortie colours with white
All other colours and patterns with white

RED	BLUE TABBY	BLUE-CREAM	SHADED SILVER

Turkish Van (overleaf left) *For breed entry see page 216* Somali (overleaf right) *For breed entry see page 224*

SOMALI

WITH ITS BUSHY TAIL, ARCHED BACK, AND APPEARANCE OF walking on tiptoes, this striking cat is one of the world's most popular newer breeds. Like its shorthaired Abyssinian forebear (*see page 134*), the Somali has a ticked coat: each hair on its body has three to twelve bands of colour. The bands are darker than the ground colour and produce a glossy, vibrant shimmer when the cat is in full coat. The facial markings are very striking, resembling theatrical eyeliner. This is an active breed: living up to its slightly wild looks, it is a natural hunter and thrives on outdoor activities. The Somali will only accept confinement if introduced to this lifestyle at an early age.

BREED HISTORY

The genetic roots of this breed go back to founder stock in Britain. Longhaired kittens appeared occasionally in Abyssinian litters, and in the 1940s, breeder Janet Robertson exported Abys to North America, Australia, and New Zealand. Descendants of these Abys sometimes produced fuzzy, dark kittens and in 1963, Canadian breeder Mary Mailing entered one of her longhaired cats in a local show. The judge, Ken McGill, asked her for one to breed from, and the official first Somali was McGill's May-Ling Tutsuta. Evelyn Mague, an American Aby breeder, was also developing longhairs, which she named Somalis. Using McGill's stock, Don Richings, a Canadian breeder, began working with Mague, and by the late 1970s the breed was fully accepted in North America. Somalis appeared in Europe in the 1980s, and by 1991 had worldwide recognition. More colours are accepted in Europe than in North America.

Fawn

In summer, the Somali loses much of its coat and may appear almost shorthaired, but for the plumed tail. The Fawn is the dilute of the Sorrel, and has a pale mushroom or oatmeal undercoat and fawn ticking. In the pale colours, good contrast between base colour and ticking is hard to achieve.

KEY FACTS

DATE OF ORIGIN   1963
PLACE OF ORIGIN   Canada and United States
ANCESTRY   Abyssinian
OUTCROSS BREEDS   None
OTHER NAME   Longhaired Abyssinian
WEIGHT RANGE   3.5–5.5 kg (8–12 lb)
TEMPERAMENT   Quiet but extrovert

     GCCF FIFé CFA TICA

BREED COLOURS

TABBIES (TICKED)
Usual, Chocolate, Sorrel, Blue, Lilac, Fawn, Cream, Usual Tortie, Chocolate Tortie, Sorrel Tortie, Blue Tortie, Lilac Tortie, Fawn Tortie

SILVER TABBIES (TICKED)
Colours are as for self and tortie

BLUE   SORREL   CREAM

Ticking
At least three dark bands on each hair

Tail
Long, with full brush of hair

DARK TIP

Lilac

The Lilac's warm-toned coat has an oatmeal base with lilac-toned ticking; the paw pads and nose leather should be a matching mauvish-pink. Like all self and tortie Somali colours, it may be white only around the lips and chin.

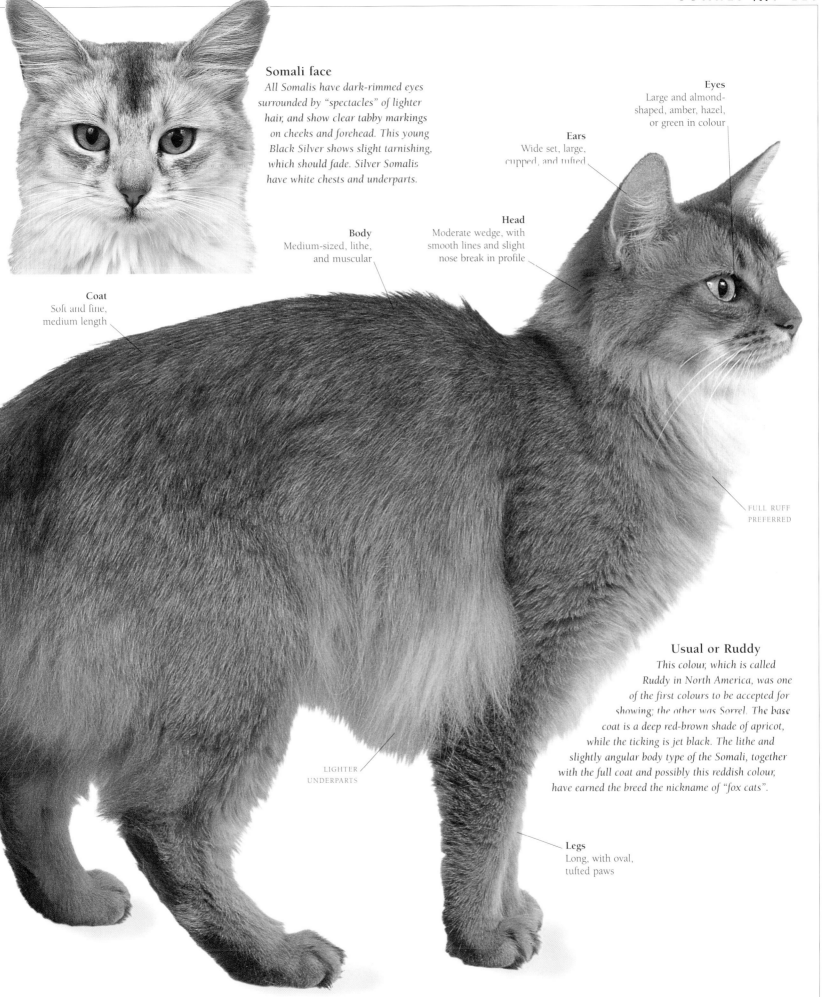

Somali face
All Somalis have dark-rimmed eyes surrounded by "spectacles" of lighter hair, and show clear tabby markings on cheeks and forehead. This young Black Silver shows slight tarnishing, which should fade. Silver Somalis have white chests and underparts.

Eyes
Large and almond-shaped, amber, hazel, or green in colour

Ears
Wide set, large, cupped, and tufted

Head
Moderate wedge, with smooth lines and slight nose break in profile

Body
Medium-sized, lithe, and muscular

Coat
Soft and fine, medium length

FULL RUFF PREFERRED

LIGHTER UNDERPARTS

Usual or Ruddy
This colour, which is called Ruddy in North America, was one of the first colours to be accepted for showing; the other was Sorrel. The base coat is a deep red-brown shade of apricot, while the ticking is jet black. The lithe and slightly angular body type of the Somali, together with the full coat and possibly this reddish colour, have earned the breed the nickname of "fox cats".

Legs
Long, with oval, tufted paws

CHANTILLY/TIFFANY

STILL EXTREMELY RARE, THIS BREED HAS A MODERATE DISPOSITION, neither as quiet as a Longhair nor as active as Oriental-type longhairs. The Chantilly will communicate happiness with an endearing "chirp", that sounds like a pigeon cooing; once you have been chirped at, you become addicted to this pleasurable sound. Although the first examples of the breed were deep chocolate in colour, the Chantilly comes in the whole range of Oriental colours, as well as tabby patterns. This is a late bloomer: the medium length, single coat is not fully mature until two to three years of age.

Chocolate Tabby kitten
Although this breed is usually thought of in terms of self colours, tabbies are also bred. Kittens can take some time to show their potential: eye colour in particular can take years to reach its full intensity.

BREED HISTORY

This companionable and undemanding breed is not as new as its recent fame suggests, but its history is beset by confusion. In 1967, in New York, Jennie Robinson purchased a pair of chocolate-coloured, gold-eyed, longhaired cats of unknown background, although the colour implied a Burmese parentage. Signe Lund, a Florida breeder, bought the cats, and coined the name Tiffany. As she also bred Burmese, the association with that breed was inadvertently perpetuated. In the 1980s, confusion led to the breed almost vanishing. In 1988, in Alberta, Canada, Tracy Oraas re-established the breed and researched their history, concluding that they were probably an offshoot of the breeding programme that created the Angora (*see page 232*). The name Tiffanie was now being used by British breeders for the longhaired variety of the shaded Asian or Burmilla (*see opposite*), causing yet more confusion, so the Canadian breeders have altered the name to Chantilly/Tiffany.

Chocolate

This is the original colour, which has led to the breed being tagged "the chocoholic's delight". Deep golden eyes appear to glow against the coat's even shade of warm, rich brown. Nose leather and paw pads match, and chocolate-brown whiskers complete the look.

KEY FACTS

DATE OF ORIGIN 1970s
PLACE OF ORIGIN Canada and United States
ANCESTRY Uncertain
OUTCROSS BREEDS Angora, Havana Brown, Nebelung, Somali
OTHER NAME Tiffany, Foreign Longhair
WEIGHT RANGE 2.5–5.5 kg (6–12 lb)
TEMPERAMENT Gentle and conservative

BREED COLOURS

SELF AND COLOURS
Chocolate, Cinnamon, Blue, Lilac, Fawn
Black

TABBY COLOURS (MACKEREL, SPOTTED, AND TICKED)
Colours are as for self colours

Ears
Medium-sized, wide at base, with rounded tips

Head
Triangular, with gently curved profile, slightly indented at eye level

Eyes
Golden or copper, oval, and slightly slanted

Legs
Medium length, well muscled but not stocky

Body
Medium length, slender, and elegant

Tail
Plumed, equal in length to body

TIFFANIE

ALTHOUGH SOMETIMES CONFUSED WITH THEIR NAMESAKES FROM
North America (*see opposite*), in fact Tiffanies have nothing
to do with that breed. Essentially longhaired Asians (*see page 142*),
they are descended, in well-recorded breeding programmes, from
Chinchilla Longhairs (*see page 188*) and Burmese (*see page 148*).
Tiffanies take after their longhaired forbears only in coat; the
conformation is Burmese. In temperament, they combine the traits
of their parent breeds to great advantage, with more liveliness than
the average Longhair and more restraint than the Burmese: the
breed standard lays stress on good temperament. An easygoing,
easy to care for longhair, the Tiffanie deserves wider popularity.

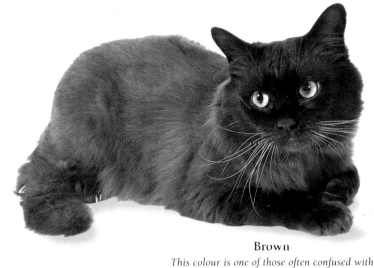

Brown
*This colour is one of those often confused with
the Chantilly/Tiffany. Although it looks like a
very dark chocolate, the colour is in fact black,
slightly degraded by the sepia pointing pattern; it
is called Sable in the Burmese. Sepia pointing is
allowed in self Tiffanies; the long coat prevents
them being mistaken for Burmese.*

BREED HISTORY

The Tiffanie, essentially a longhaired Burmese,
is the only longhaired member of the Asian
breed group. This group's origins can be traced
back to an accidental mating, in London in
1981, of a Chinchilla Longhair and a Lilac
Burmese belonging to Baroness Miranda von
Kirchberg. The first-generation offspring were
shorthaired, shaded Burmillas, but subsequent
breedings inevitably brought the recessive
longhair and sepia pointing genes back to the
surface. The breed group was developed with
the help and support of Burmese breeders,
and remains quite distinct from this breed.
There are two distinct threads in the Asian
group, however: FIFé cats come from some
of the same lines as GCCF cats, but more
diverse lines are now used in Britain.

Ears
Medium to
large, continuing
lines of face

Head
Short wedge, with
distinct nose-
break in profile

SHADES FROM
YELLOW TO
GREEN ALLOWED

Eyes
Neither almond-
shaped nor round,
slightly oblique
in set

Blue Silver Shaded
*Shaded Tiffanies are the longhaired equivalent of
the Burmilla, the original Asian coat pattern. All
members of the Asian group aim for the same
physical conformation, differing in coat colours,
patterns, or length. The semi-long coat of the
Tiffanie is longest in the ruff and on the tail.*

Body
Medium build, with
straight back and
good musculature

Coat
Semi-long,
fine and silky

Tail
Medium to long,
elegantly plumed

Legs
Medium length,
with round paws

KEY FACTS

DATE OF ORIGIN 1970s
PLACE OF ORIGIN Great Britain
ANCESTRY Burmese/Chinchilla crosses
OUTCROSS BREEDS Burmese/Chinchilla
OTHER NAME None
WEIGHT RANGE 3.5–6.5 kg (8–14 lb)
TEMPERAMENT Lively and affectionate

    GCCF

BREED COLOURS

SELF COLOURS (SOLID, SEPIA)
Black, Chocolate, Red, Blue, Lilac, Cream, Caramel,
Apricot, Black Tortie, Chocolate Tortie, Blue Tortie,
Lilac Tortie, Caramel Tortie

SHADED COLOURS (SOLID, SEPIA)
Colours are as for self colours

TABBIES (SOLID, SEPIA, ALL PATTERNS)
Brown, Chocolate, Red, Blue, Lilac, Cream, Caramel,
Apricot, Black Tortie, Chocolate Tortie, Blue Tortie,
Lilac Tortie, Caramel Tortie

BALINESE

SLENDER, FINE-BONED, AND REFINED AS ROYALTY IN APPEARANCE, the Balinese is in fact an intensely social breed, happiest when it is underfoot or at the centre of activity. Highly inquisitive, it is indefatigable in its investigation of vacuum cleaners, cupboards, and shopping bags. Its tubular body permits it to accomplish Houdini-like acts of contortion: veterinarians know that, like the Siamese (*see page 154*), this is a superb escape artist, seemingly able to pick locks. It is a talkative cat, sometimes appearing, like its shorthaired cousin, to talk to itself. A typical pointed cat, it does not have dramatically long hair, and from a distance some might be mistaken for a Siamese except for the graceful plume of a tail.

Striking namesake
Delicate movements, unshakeable poise, and a lithe physique are all prerequisites for the Balinese temple dancers who were the inspiration for this breed's name.

BREED HISTORY
Siamese have produced semi-longhaired kittens for a long time: whether they historically carried the longhair gene or were outcrossed is not clear. A longhaired Siamese was actually registered with the CFA in Britain, in 1928. Breeders usually passed these sports on as pets, until after World War II, when Marion Dorsey, in California, began to breed to produce longhairs. Longhair Siamese were exhibited in 1955, and were recognized in 1961. Breeders of Siamese protested at the use of the name, and a breeder suggested Balinese because the cats reminded her of Balinese temple dancers. In the mid-1970s, the breed arrived in Europe. Here, all colours are called Balinese: in North America, some are more usually known as Javanese (*see page 230*). Recently, the stockier Traditional Balinese has been recognized by the TCA.

Chocolate Point
As in all pointed cats, the coat darkens with age: a cat's show "career" may be quite short. This cat's shading is too heavy to show now, but he was a winner in younger days.

Lilac Point
The dilute version of the Chocolate, this colour is a study in delicacy. The warm, magnolia coat may bear soft lilac shading, while the nose leather and paw pads are pinkish or faded lilac to tone with the points. As in all the colours, the eyes should be a clear, vivid, and brilliant blue. This colour is also generally called Lilac Point in the various North American associations, although it has also been called Lavender Point and even Frost Point.

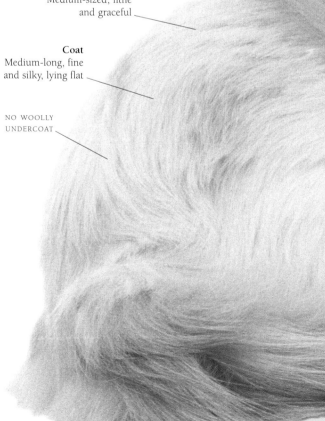

Body
Medium-sized, lithe and graceful

Coat
Medium-long, fine and silky, lying flat

NO WOOLLY UNDERCOAT

Blue Point
The standard for the Blue calls for a body of glacial white, with points and any shading on the back to be a cold blue. The nose leather must also be blue. It should not be possible to mistake a Blue Point for a Lilac Point in any respect.

Seal Point
Points of an even, deep brown and a soft shading of fawn on the body characterize this colour. The Seal is less warm than the Chocolate, just as in the dilutes of these two colours the Blue is cooler than the Lilac.

Balinese head
Seen full face (left), the Balinese face is wide between the ears, narrowing in straight lines to a fine muzzle. This Blue Tabby Point also shows clear facial markings in the mask. Seen in profile (below), the nose should be straight and the chin strong. This Seal Tortie Point has a fully developed mask.

Ears
Large, wide at base, and pricked

Mask
Connected to ears by faint tracings

DILUTE COLOURS SHOW VESTIGIAL FROWN LINES

Head
Long and wedge shaped, with elegant lines

Eyes
Widely spaced, and Oriental in shape and set

ABSENCE OF FRILL PREFERRED

Tail
Long and plumed

Points
Clearly contrasting with pale body

Paws
Small and oval, with pads to match points

KEY FACTS

DATE OF ORIGIN 1950s

PLACE OF ORIGIN United States

ANCESTRY Longhaired Siamese

OUTCROSS BREEDS Angora in Britain, Siamese in the United States

OTHER NAME Some colours called Javanese in U.S.

WEIGHT RANGE 2.5–5 kg (6–11 lb)

TEMPERAMENT Energetic and exhibitionist

GCCF FIFé CFA TICA

BREED COLOURS

BALINESE POINT COLOURS
Seal, Chocolate, Blue, Lilac

JAVANESE POINT COLOURS (IN CFA)
Red, Cream, Blue-Cream, Lilac-Cream, Seal Tortie, Chocolate Tortie, and tabby versions in all colours

Cinnamon, Fawn, Smoke, Silver and parti-colour versions

RED TABBY   BLUE TORTIE

NEWER BALINESE

ORIGINALLY, ONLY SEAL, BLUE, CHOCOLATE, AND LILAC POINTS WERE recognized in the Balinese and Siamese: breeders have worked to create the other colours and patterned points we know today. In Britain and Australasia, the name Balinese encompasses all colours and patterns, just as the name Siamese covers all of their shorthaired cousins. However, CFA in North America and some smaller associations in other countries still recognize only the four "traditional" colours in these breeds. Other colours, such as red and cream, and the tabby and tortie patterns, are classed as separate breeds. The longhairs are called Javanese, while the shorthairs are known as Colourpoint Shorthairs. To complicate matters further, many European cat organizations use the name Javanese for the Angora (*see page 232*), a different breed entirely, but one to which the Siamese did contribute. The Balinese can be thought of as a longhaired Siamese, although its fine coat is not as long as that of most Longhairs, and does not form a ruff.

Seal Tabby Point
As with all pointed breeds, kittens are born white and young cats are paler than older individuals. However, some shading is inevitable on all tabby points, and the Seal Tabby Point in particular is noticeably shaded, with a cream body shading to a warm fawn colour over the back. There should be no vestige of tabby markings on the body. The legs, tail, and face should show clear tabby markings; the whisker pads in particular should be heavily spotted.

BRIGHT, INTENSE BLUE EYES

FACIAL BLAZES ALLOWED

Chocolate Tortie Point
Tortie Point Balinese must show some mingling of colours on each point, no matter how small or uneven, and any shading on the ivory body colour should be uneven. Chocolate Tortie Point patches are a mixture of light chocolate and varying shades of red; larger red areas may show faint tabby marks. The nose leather and paw pads are a mixture of chocolate and pink.

BODY FREE OF MARKINGS

MOTTLED SHADING ON BODY

Chocolate Tabby Point
Tabby points must show clear facial markings, with distinctly spotted whisker pads and dark-rimmed eyes. The legs and tail should also be barred or ringed, but it is not possible to distinguish which tabby pattern is present on a pointed cat. Tabby points are called lynx points in North America.

Seal Tortie Point

With points of deep, seal brown mixed with shades of red, this is the darkest of the tortie points. As in all other tortie colours, each point must show a mixture of colours, but the mixture need not be even: a point may be almost entirely one colour or another, as long as it is not completely that colour. This includes not just the mask, but the backs of the ears, which should be mottled.

Blue Tabby Point

Unlike self-pointed Balinese, tabby points may have noses that do not match their points. A Blue Point must have a blue nose, but a Blue Tabby Point may have a pink nose outlined in blue. The eyes, however, must follow the same standard as self points. The "thumb" marks on the backs of the ears may be less apparent in a Blue Tabby Point than in other tabby points.

SPOTTED
WHISKER PADS

Seal Tortie Tabby Point

All points of a Tortie Tabby Point must show both tabby marks and mingled colours, and body shading should be uneven, as in tabby points. In tabby points and tortie tabby points, the colour is not judged quite as strictly as it is in the self points: variances in the exact tone are allowed, for some body shading is also to be expected in these patterns.

ANGORA

THIS BREED IS SIMILAR IN BOTH TEMPERAMENT AND TYPE TO THE Oriental breeds – lively and inquisitive, and long and lean, with a tail that forms a delightfully elegant plume. The fine, silky, medium-length coat has no woolly undercoat, so this cat is fairly easy to groom. Balinese (*see page 228*), Siamese (*see page 154*), and Oriental Shorthairs (*see page 160*), are all permitted outcrosses in Angora pedigrees. The Angora breed suffers from a profusion of confusing names. Not related to the Turkish Angora (*see page 220*), in mainland Europe it is called the Javanese to avoid confusion with it, but some North American associations use Javanese for some colours of Balinese. In North America, the British Angora has been called the Oriental Longhair, with the misleading implication that it is directly descended from the Oriental Shorthair. There is now an Oriental Longhair (*see page 234*) with that descent, and it is to be hoped that Angoras can simply remain Angoras from now on.

BREED HISTORY

The Angora was developed in Britain by Maureen Silson, who mated a Sorrel Abyssinian (*see page 134*) to a Seal Point Siamese in the mid-1960s, attempting to produce a Siamese with ticked points. The descendants inherited both the cinnamon trait, producing cinnamon Oriental Shorthairs, and also the gene for long hair, which eventually led to the creation of the Angora. Descendants of this mating are behind the majority of today's British Angoras. The breed is not related to the 19th-century Angora, or to the revived Turkish Angora, or to the recently created Oriental Longhair.

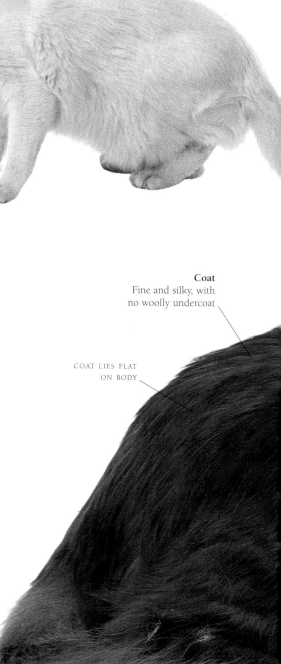

Blue-Eyed White
To many people, this is the classic colour of the historical Angora, or French cat. The blue eyes are bright and vivid, similar to those of the Siamese, rather than the paler baby-blue of western breeds.

Coat
Fine and silky, with no woolly undercoat

COAT LIES FLAT ON BODY

KEY FACTS

DATE OF ORIGIN 1970s
PLACE OF ORIGIN Britain
ANCESTRY Siamese/Abyssinian crosses
OTHER NAMES Javanese (Europe), previously Oriental Longhair (United States), Mandarin
OUTCROSS BREEDS Siamese, Balinese, Oriental Shorthair
WEIGHT RANGE 2.25–5 kg (5–11 lb)
TEMPERAMENT Energetic exhibitionist

    GCCF

BREED COLOURS

SELF AND TORTIE COLOURS
Black, Chocolate, Cinnamon, Red, Blue, Lilac, Fawn, Cream, Caramel, Apricot, White (Blue-, Green-, Odd-Eyed), Tortoiseshell, Chocolate Tortoiseshell, Cinnamon Tortoiseshell, Blue Tortoiseshell, Lilac Tortoiseshell, Fawn Tortoiseshell, Caramel Tortoiseshell

SMOKE, SHADED, SILVER SHADED, AND TIPPED
Colours are as for self and tortie colours, with the exception of White

TABBIES (ALL PATTERNS)
Brown, Chocolate, Cinnamon, Red, Blue, Lilac, Fawn, Cream, Caramel, Tortie, Chocolate Tortie, Cinnamon Tortie, Blue Tortie, Lilac Tortie, Fawn Tortie, Caramel Tortie

SILVER TABBY COLOURS (ALL PATTERNS)
Colours are as for standard tabbies

CHOCOLATE TORTIE TABBY   CINNAMON TORTIE TABBY   CARAMEL TORTIE

Chocolate
The rich, warm tones of this colour create a truly luxurious-looking cat. The colour should be even all over, but all hair tends to lighten in the sun, so longer areas of the coat, such as the tail, may become paler than the rest.

Tail
Long, tapering to fine end

PLUME-LIKE TAIL

Red Silver Shaded

In shaded Angoras, the undercoat is a very pale, but warm, shade. The silver shaded colours, like this Red, have a pure silvery white undercoat, which provides a more dramatic contrast to the tipping. The Angora coat is slow to mature: this young cat has a short coat.

Historical model

The Angora was bred to recreate the elegant longhairs recorded in paintings such as this (1813) portrait by Louis Léopold Boilly (1761–1845).

Body
Medium-sized, svelte, and muscular

Ears
Large and following lines of wedge

WIDE BASE

Head
Moderate, triangular wedge

Neck
Long and slender

Eyes
Green in all colours of Angora except white

SLANTED EYES

FINE MUZZLE

Paws
Small and oval

HINDLEGS LONGER THAN FORELEGS

Legs
Long, slim, and well muscled

Cinnamon

The first Angora, Cuckoo, was a Cinnamon. The gene for this colour came from the Abyssinian parentage, where it is known as Sorrel. The tone should be warm, and eye rims and nose leather should match the coat.

ORIENTAL LONGHAIR

THIS BEAUTIFUL, FULLY COLOURED, SEMI-LONGHAIRED VERSION of the Oriental Shorthair completes the quartet of Oriental breeds. Just as the Siamese (*see page 154*) has a semi-longhair counterpart in the Balinese (*see page 228*), the Oriental Shorthair (*see page 160*) has its own silky alternative in the Oriental Longhair. The coat of the Oriental Longhair lacks an undercoat and tends to lie flat against the body. In summer, but for the plumed tail, it may look similar to shorthairs. This breed reflects its family in all respects: it bears the colours of the Oriental, and the soft coat and plumed tail of the Balinese.

BREED HISTORY

In spite of the efforts of breeders to control matings, cats still manage to spring some surprises. In 1985, an Oriental Shorthair and a Balinese at Sheryl Ann Boyle's Sholine cattery mated to produce a litter of silky, semi-longhaired Orientals. The breed was then developed, and is now recognized by TICA and FIFé. The CFA controversially combined Oriental Longhairs and Shorthairs into one breed group in 1995. As with Oriental Shorthairs, breed associations differ as to the status of pointed kittens. There may be some confusion in name between this cat and the British Angora (*see page 232*), which has been called an Oriental Longhair in North America: visually and historically, the two are distinct.

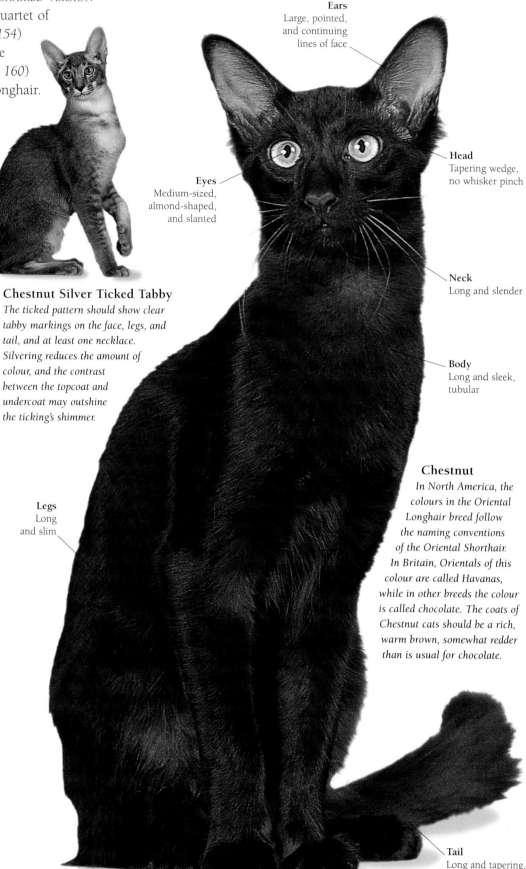

Ears
Large, pointed, and continuing lines of face

Head
Tapering wedge, no whisker pinch

Eyes
Medium-sized, almond-shaped, and slanted

Neck
Long and slender

Body
Long and sleek, tubular

Legs
Long and slim

Tail
Long and tapering, softly plumed

Chestnut Silver Ticked Tabby
The ticked pattern should show clear tabby markings on the face, legs, and tail, and at least one necklace. Silvering reduces the amount of colour, and the contrast between the topcoat and undercoat may outshine the ticking's shimmer.

Chestnut
In North America, the colours in the Oriental Longhair breed follow the naming conventions of the Oriental Shorthair. In Britain, Orientals of this colour are called Havanas, while in other breeds the colour is called chocolate. The coats of Chestnut cats should be a rich, warm brown, somewhat redder than is usual for chocolate.

KEY FACTS

DATE OF ORIGIN 1985
PLACE OF ORIGIN North America
ANCESTRY Oriental Shorthair, Balinese
OUTCROSS BREEDS Siamese, Balinese, Oriental Shorthair
OTHER NAME None
WEIGHT RANGE 4.5–6 kg (10–13 lb)
TEMPERAMENT Friendly and inquisitive

    FIFé CFA TICA

BREED COLOURS

All colours and patterns, except pointed, sepia, and mink
All colours and patterns, including pointed, sepia, and mink

BLUE TORTIE TABBY | FAWN AND WHITE | WHITE | BLACK

LAPERM

MOST OF THE REXED BREEDS IN EXISTENCE HAVE ORIGINATED, AND therefore been developed in, shorthairs. The longhaired LaPerm and the Selkirk Rex (*see page 215*) are the only longhaired curly cats accepted by major registries. The Bohemian Rex, which is a proposed rexed Longhair, has never achieved recognition, and the rexed Maine Coon (*see page 204*) is also highly controversial. Although LaPerms are descended from random-bred American stock, they have a foreign appearance, with a wedge-shaped head and lean build. They are very active, inquisitive cats, and the breed description reflects their farm-cat ancestry, noting that they are "excellent hunters".

BREED HISTORY

In 1982, a farm cat in The Dalles, Oregon, produced a litter of six kittens that included a single bald kitten. In spite of this disadvantage, the kitten survived, and she eventually grew a coat. Unexpectedly, her coat, unlike that of her littermates, was curly and soft to the touch. Linda Koehl, the owner and founder of the breed, named this kitten Curly. Over the next five years, Koehl bred a number of curly coated kittens, which were to become the basis of the LaPerm breed. The gene is dominant, so wide outcrossing to increase the gene pool can be done while still producing reasonable numbers of rexed kittens. TICA has granted the LaPerm "new breed" status.

KEY FACTS

DATE OF ORIGIN 1982
PLACE OF ORIGIN United States
ANCESTRY Farm cats
OUTCROSS BREEDS Non-pedigree cats
OTHER NAME Also called Dalles LaPerm
WEIGHT RANGE 3.5–5.5 kg (8–12 lb)
TEMPERAMENT Affectionate and inquisitive

BREED COLOURS

All colours and patterns, including sepia, pointed, and mink

WHITE

CURLY HAIR
AT EAR BASE

Eyes
Large and expressive, slightly slanted in set

Ears
Wide set and continuing lines of face

Head
Medium-sized, modified wedge, with rounded contours

Body
Medium size and build, with good musculature

Legs
Hindlegs longer than forelegs

Coat
Medium length and ringleted, with heavy undercoat

Tail
Long, tapering plume

Red Tabby
The proportion of red cats in the random-bred population varies geographically. However, it is predictable that there are few red selfs because dedicated breeding is usually needed to eliminate tabby markings from this colour. In any new breed, especially one outcrossing to non-pedigree cats, this work has to begin from scratch.

Blue Mackerel Tabby kitten
The first LaPerm was born bald and then grew a rexed coat, but most LaPerms are born with a slightly curly coat. After a bald stage in their first year, their coats grow back with more curl.

KURILE ISLAND BOBTAIL

BOTH THIS BREED AND ITS HOMELAND ARE OF UNCERTAIN ownership. The Kurile Island chain, running from the easternmost point of Russia to the tip of Japan's Hokkaido Island, is disputed between the two nations. The Kurile Island Bobtail is of quite a different type to the Japanese Bobtail (*see opposite*), although it has the same short tail. Its coat, conditioned by the harsh winters of its northern home, is longer and thicker than that of its more southerly relation, and its build is sturdier. A relatively small range of colours is recognized by the breed standard. A friendly breed, it nonetheless retains its independence.

BREED HISTORY

Until recently, there was only one widely known bobtailed breed, the Japanese. With the advent of a more open attitude in the countries of the former Soviet Union, new breeds are emerging, among them some surprises such as this hitherto obscure breed. The Kurile Island Bobtail represents the same mutation as the Japanese Bobtail, and has almost certainly been present on the Kurile Islands for centuries. While this genetic similarity causes no problems to the Russian bodies that register the Kurile, the shared mutation that causes its bobtail might be a barrier to the breed's acceptance in Europe.

Red Self
Sex-linked red is common in cats from this area, and is ideally accompanied by copper-coloured eyes. The coat is longest over the throat and in "breeches" on the hindlegs. Males often develop a full face, with clear jowls.

Tortie and White
Even female Kuriles like this one develop an impressive build by maturity, with strong legs and broad shoulders. The hindlegs are longer than the forelegs, and the spine is carried with a slight rise from shoulders to rump. The tail should be a soft ball of longer hair, carried high.

KEY FACTS

DATE OF ORIGIN Pre-18th century
PLACE OF ORIGIN Kurile Islands
ANCESTRY Domestic cats
OUTCROSS BREEDS None
OTHER NAME None
WEIGHT RANGE 3–4.5 kg (7–10 lb)
TEMPERAMENT Busy and friendly

BREED COLOURS

SELF AND TORTIE
Black, Red, Blue, Cream, Tortoiseshell, Blue-Cream, White

SMOKE, SHADED, AND TIPPED
Colours are as for self and tortie, except White

TABBIES (CLASSIC, MACKEREL, SPOTTED)
Brown, Red, Blue, Cream, Brown Tortie, Blue Tortie

SILVER TABBIES
Colours are as for standard tabbies

BI-COLOURS
Any allowed colour with White

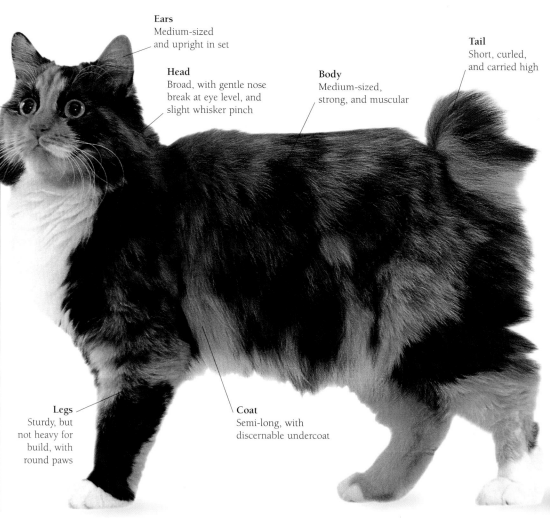

Ears
Medium-sized and upright in set

Eyes
Oval, slightly tilted, wide range of colours

Head
Broad, with gentle nose break at eye level, and slight whisker pinch

Body
Medium-sized, strong, and muscular

Tail
Short, curled, and carried high

Legs
Sturdy, but not heavy for build, with round paws

Coat
Semi-long, with discernable undercoat

JAPANESE BOBTAIL

SOCIABLE AND INQUISITIVE, THIS LONGHAIRED BREED STILL EXISTS only in small numbers. The Japanese Bobtail is bred by very few breeders worldwide. This is in part because the longhair gene is masked in longhair-to-shorthair breedings, and the low numbers of longhairs means that mating only longhair-to-longhair could lead to serious inbreeding. In this semi-longhaired version of the Bobtail, the short tail makes a full, fluffy pom-pom: this trait does not carry spinal or bone deformities with it. The Bobtail is highly gregarious but easily bored and, when bored, can be mischievously destructive.

BREED HISTORY

Although it has only recently received much publicity, this cat is a natural variant of the shorthaired Bobtail (*see page 164*). Examples of both types can be found in Japanese prints and paintings going back over the last three centuries (*see page 38*), and the longhaired cats would have had an advantage in the cold climate of Japan's northernmost areas. The documented breeding history, however, only dates from 1968, when shorthaired Bobtails were brought to the United States, carrying with them the longhair gene, which was noticed in the early 1970s. Now that the shorthaired version is firmly established in North America, the less common longhaired cat is gaining some ground there. It has yet to gain recognition in Britain.

Bobtail face
An almost equilateral triangle, the face has gentle curves and high cheekbones. Odd eyes are prized in Japanese Bobtails, especially in the tortie-and-white coat pattern, known as Mi-ke.

KEY FACTS

DATE OF ORIGIN 700s
PLACE OF ORIGIN Japan
ANCESTRY Household cats
OUTCROSS BREEDS None
OTHER NAME None
WEIGHT RANGE 2.5–4 kg (6–9 lb)
TEMPERAMENT Vibrantly alert

   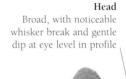 **FIFé CFA TICA**

BREED COLOURS

SELF AND TORTIE COLOURS
Black, Red, Tortoiseshell, White
All other self and tortie colours, including pointed, mink, and sepia

TABBY COLOURS
All colours in all four tabby patterns

BI-COLOURS
Black, Red, Tortoiseshell, with White
All other colours and patterns with white

Head
Broad, with noticeable whisker break and gentle dip at eye level in profile

Ears
Large, set wide apart and upright

Eyes
Large and oval, with definite slant when viewed in profile

Red and White

The overall look of the Japanese Bobtail is one of long, clean lines. Although well muscled, it is athletic rather than massive in build. The legs, correspondingly, are slender but not fragile. The hindlegs are longer than the forelegs, but they are carried bent so that the torso is almost level when the cat is standing or walking.

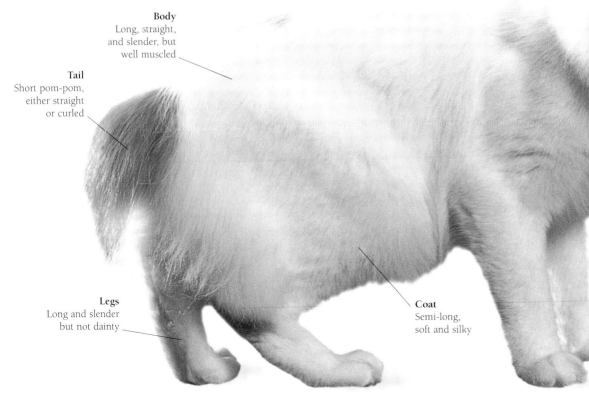

Body
Long, straight, and slender, but well muscled

Tail
Short pom-pom, either straight or curled

Legs
Long and slender but not dainty

Coat
Semi-long, soft and silky

EXOTIC

THE EXOTIC LONGHAIR IS IDENTICAL IN LOOKS TO A LONGHAIR. Although there are genetic differences between the breeds, it is impossible to tell them apart by appearance. Similarly, the Exotic Longhair is accepted in the same colours and patterns as the Longhair (see page 188). According to breeders, however, the Exotic Longhair is more outgoing and adventurous than the typical Longhair. This is a playful, inquisitive breed, and most individuals get on well with other cats and dogs.

BREED HISTORY

This is curious and requires concentration. The Exotic Shorthair (see page 108) is a shorthaired Longhair. It developed from breeding Longhairs with American Shorthairs, which produced exotic-looking shorthaired cats, with cobby, compact Longhair-type bodies. The CFA recognized the Exotic Shorthair in 1967. In order to retain the Longhair's body and head type, the CFA's breeding policy still allows Exotic Shorthairs to be bred periodically to Longhairs. This means that longhaired kittens appear in Exotic Shorthair litters but these offspring cannot be registered, either as Longhairs or Exotics. To overcome this dilemma, cat breeders created the Exotic Longhair, the longhaired version of the shorthaired version of the longhaired Longhair.

Blue and White Bi-colour
As a preferred minimum, the standard requires the cat to have white feet, legs, underparts, chest, and muzzle. In the show ring, less white than this minimum is penalized. An inverted "V" blaze on the face is desirable.

Paws
Large and round

Tail
Short, must not be carried high

Ears
Small, tilted forward, rounded at tips

Dilute Calico
The standard calls for white with patches of blue and cream. The eyes may be copper, as seen in this cat, blue, or odd-eyed. An odd-eyed cat must have one blue and one copper eye, both with equal colour depth.

Coat
Long and glossy

KEY FACTS

DATE OF ORIGIN 1960s
PLACE OF ORIGIN United States
ANCESTRY Longhairs/American Shorthairs
OUTCROSS BREEDS Longhair
OTHER NAMES None
WEIGHT RANGE 3–6.5 kg (7–14 lb)
TEMPERAMENT Gently inquisitive

BREED COLOURS

All Exotic Shorthair/Longhair colours and patterns

Turkish Angora kittens (next page) For breed entry see page 220

RANDOM BREEDS

BY FAR THE MOST COMMONLY OWNED DOMESTIC CAT IS THE HUMBLE, random-bred "moggie". Even in countries with high populations of pedigree cats, these self-selected pets outnumber them four to one. While some people are set on the looks and personality traits of a certain breed, random-bred cats usually satisfy those of us with less definite requirements. A cat's personality depends on its early experiences, so a random-bred cat is as friendly as one makes it, although it may not achieve the chattiness of Oriental breeds or the extreme placidity of some longhaired breeds. Only a few of these "moggies" have long hair, because it is a recessive trait, but cats in the style of the Angora (*see page 232*) or the Maine Coon (*see page 202*) do turn up: after all, these breeds sprang from such non-pedigree cats.

Blue
Blue, the defining colour of several naturally developed breeds, is often found in the feline populations of mainland Europe. The semi-foreign build of a cat like this indicates a heritage in common with southern European breeds, such as the Turkish Angora (see page 220); cats like this are often seen on Greek islands.

Green-Eyed White
This cat, which has often been mistaken for an Angora, has a white coat as pure as that of any pedigreed cat. The eyes may not be the brilliant green achieved by careful selective breeding of top show cats, but many pet-quality pedigrees also fail to meet such rigorous standards.

Black

Many longhairs with northern origins, such as the Maine Coon (see page 202) or the Norwegian Forest Cat (see page 208) have broad faces, a sturdy build, and thick, almost shaggy coats. These breeds were developed from random-breeding farm cats, which still produce beautiful individuals.

Cream and White

Creams are less common than reds among random-bred cat populations. In both colours, even selfs almost always show some ghost tabby markings. Breeders of pedigree cats select carefully to minimize these markings and produce consistently clear self coats, but even in random-bred cats the occasional individual shows a suprisingly solid colour.

Shaded Silver

The shaded and tipped Longhairs (see page 188) epitomize elegance. The extraordinarily long coat of that breed is the result of a century of breeding, and is highly unlikely to turn up in non-pedigree domestic cats. However, the shading, the dramatic black eyeliner, and the outlined nose all appear, and many "moggies" have coats that would not have disgraced the breed in its earlier days.

CHAPTER SIX

CARING FOR YOUR CAT

Cats that are kept for companionship depend upon us for food, shelter, and freedom from illness or suffering. Choose a cat that fits into your family life. Some require extensive grooming, others demand considerable attention, while all need balanced nutrition and both mental and physical stimulation. Take extreme care if your cat goes outdoors: natural predators or cars are likely to be the greatest dangers. Understand the principles of first aid and train your cat to miaow when you call its name. For most cat owners, these are satisfying responsibilities, forming the basis of our bond with this fascinating creature.

ABOVE CAT PLAYING WITH A CATNIP TOY

LEFT KITTENS NEED A PLACE TO CALL HOME

YOU AND YOUR CAT

IF YOU LIVE WITH A CAT ALREADY OR ARE PLANNING TO DO SO, IT is important to remember that the best relationship is based upon understanding its emotional, as well as physical, needs and desires. It takes a little imagination to think like a cat. Remember, left alone, a cat is a self-sufficient individual: it does not wither emotionally if it lives in the absence of others of its own kind. Be true and honest with yourself. Understand why you want to live with feline companionship. By understanding your motives as well as your cat's you are in the best position to provide and receive lifelong, high-quality care and companionship.

The role of intimate pet is one that cats have only recently begun to play. For centuries cats simply used us as a source of food and shelter, while we used them as vermin eliminators. The relationship was "commensal", both parties benefiting, neither being adversely affected. In the last half of the 20th century, this changed radically. In 1970, cats only made up 10 per cent of my patients; by 2000, this had risen to over 60 per cent. Today we live with cats primarily for social and psychological reasons. Studies over the last 30 years help to explain why this is so.

CATS ARE GOOD FOR US

In the 1970s, Dr. Aaron Katcher observed that when cat owners stroke their own cats their blood pressure dropped. By the early 1980s, Dr. Ericka Friedmann reported that cat owners are more likely to survive one year after a major heart attack than people without pets. In the early 1990s, Dr. Warwick Anderson found that cat owners are less at risk of heart disease than non-pet owners. The rewards of living with a cat seem straightforward, but the relationship is complex. Outwardly, we appear to be the care-givers: we nurture our cats, providing food,

Beneficial modification
When cats live in colonies, they do enjoy social contact such as licking and rubbing. Neutering pet cats reduces their inclination to dominant behaviour, and this allows them to enjoy more physical contact with us without needing to become aggressive.

accommodation, health care, contact comfort, and affection. We mother them. But, in a subtle way, they too are care-givers, offering unexpected physiological and emotional returns. Cultural anthropologist Constance Perin was the first to hypothesize about why our blood pressure drops when we stroke our own cats or why, for that matter, we deny any antisocial behaviour on their part even when we know it to be true. Perin theorized that the physiological benefits of stroking a cat evolved from the rewards we got from physical contact with our mothers as infants. A baby is reassured and relaxes when there is physical contact, and stroking your cat stimulates the same chemical pathways in your body. Skin temperature, blood pressure, and heart rate all drop. Stress diminishes. In a physiological sense, perhaps even in a deeply buried psychological way, our cats "parent" us; they too are care-givers.

Young companions
Combining children and pets always needs careful consideration. Very young children can be careless or overly inquisitive, and may frighten a nervous cat or provoke a defensive reaction. Older children, however, can make ideal companions, happy to play with cats and willing to devote a great deal of time to them.

KITTEN KINDERGARTEN

In the 1990s, Australian veterinarian Dr. Kersti Seksel devised a two-week programme, "kitty kindergarten", for cats under 14 weeks of age. Concerned that an increasingly indoor lifestyle denied cats natural outlets for normal behaviour, Dr. Seksel used this programme to train owners in how to enrich their cats' environment and extend the social sphere to include other cats, dogs, and people. Owners learn: to adopt variations of the methods used by mother cats to carry and groom kittens; how to use play to enhance a cat's mental and physical abilities; and how to use a cat carrier so that a cat enjoys going in it. Perhaps most importantly, they learn never to ask of a cat more than nature intended it to do.

Prolonging infant dependency
If fed and handled by people in the early weeks of their lives, cats will come to regard us as substitute mothers. This forms a good basis for rewarding companionship.

PROVIDING FOR YOUR CAT
Of course, we are good for our cats in obvious physical ways. We provide them with safe and secure territories, although the size of a typical human living space is not necessarily large enough for a typical cat. We offer them a constant supply of food, although we may sometimes mistakenly think that taste is the most important factor. We ensure their good health, firstly by providing preventative inoculation and parasite control, but also by organizing medical treatment when they are injured or ill.

Family provider

The basis of the cat's relationship with us has always been food: from the beginning, cats found rich pickings around human dwellings, and the least nervous moved in with us. Even today, our willingness to provide shelter and a supply of food enables them to overcome innate territorial instincts and live with us and with other pets.

We also provide emotional support: a leg to rub against, a hand to tickle a chin or stroke a back, or someone to snuggle up to. It is with this emotional support that we sometimes fail.

THINK LIKE A CAT

It is difficult to think like any other animal, especially if that animal is emotionally quite different from us. Our other popular pet, the dog, is easier to understand. Like us, the dog is a gregariously sociable species, enjoying the company of its own kind. We make excellent companions because we share so many needs, such as physical contact or a social hierarchy.

Cats are different. Cats raised by their own mothers, living in a colony of related cats in the absence of human influence, want contact neither with other unrelated cats nor with us. If forced to live together in a small living space, two unrelated cats will probably fight: another cat is a threat to the supply of food, to territory, and mating success. If forced to live with us, the relationship is slightly easier, because we are sufficiently different from cats to be seen as a different threat: a human is simply a potential predator. Once a cat learns that its physical well-being is not at risk from a specific human, it relaxes in that person's presence. Cats that have little early contact with us like the security of our homes and the ready availability of food but they do not, as dogs do, want our affection.

If, however, a cat learns early in life to look upon humans as an extended family, it will grow to see us not only as non-threatening, but also as potential mother substitutes. This is the objective of most cat owners: they want cats to be family members, there to cuddle and play with. This is realistic for cats raised by people, but less so for cats denied early social integration with us.

LEARNED HELPLESSNESS

It is a blunt truth that for most owners the ideal feline companion is one who has learned to be helpless, to be dependent on humans. This is not as malign as it sounds, for these cats are not truly helpless. They have learned new ways to be independent, and we are instrumental in these behaviours. If our cats go outdoors, we want them neither to roam nor to get in fights: the ideal cat stays in its own garden, uses only body language to keep other cats away, and retreats indoors when harm is imminent. The ideal cat does not prey upon garden songbirds, but rather only bats playfully at moths and eats commercial cat food. Through combinations of careful selective breeding and early learning we have partly succeeded in producing cats that behave this way.

Also, while pet cats appear to be dependent, their behaviour can often be found to involve subtly successful dominance when examined more closely. Most cat owners are perfectly sensible individuals, yet they can become slaves to a cat's culinary or social whims. People are brow-beaten into submission because it is the cat's mealtime or playtime, and most accede to these feline demands. To the ever adaptable cat, we are cat an easy touch, providing rich pickings for a modified and highly successful lifestyle. It is important to acknowledge this truth before choosing to live with a cat.

Communal living

Early learning is crucial in cat behaviour. Two kittens from the same litter are the most likely to make good companions as adults, provided there is shelter and food for both.

CHOOSING A CAT

YOUR CHOICE OF CAT DEPENDS UPON AESTHETIC AS WELL AS MORAL decisions. A cat's looks are important: the choice of size, colour, coat length, even sex, is highly personal. Equally important are your moral decisions. Do you want to offer a home to a lost or unwanted feline? Do you want your cat to live safely indoors or have a more risky but potentially more exciting outdoor life? Do you want your cat neutered or do you feel that this is "unnatural"? Owning a cat is a considerable commitment, and it is important to answer these surprisingly difficult questions before making your decision on who to choose as your feline companion.

Perhaps the most common way for cats to enter our homes and lives is on their own terms: arriving at our doorsteps, willing acceptors of both food and affection. If, however, you have decided to acquire a cat, you have the luxury of making your own personal choice.

INDOORS OR OUTDOORS
Your first and perhaps hardest decision is whether to keep your cat indoors or allow it to roam outdoors. Until the 1950s, when cat litter became widely available, owners had little choice but to allow their cats outdoors. Surveys in the United States indicate that approximately 50 per cent of cats now live exclusively indoors, a figure that surprises even the experts. Indoor living is partly a cultural phenomenon: American pedigree cat associations strongly advise cat owners to keep their cats indoors. Dr. Nicholas Dodman, director of the Behaviour Clinic at Tufts University School of Veterinary Medicine,

Feral cats and strays
If you acquire a stray from a shelter, find out about its background. Ferals, born and raised away from humans, will usually be apprehensive. Strays, which were initially raised with people, will be much more companionable.

Pedigree cats
Pedigree varieties of cat, such as this Korat, offer considerable advantages: known looks, size, and personality. To a limited extent, however, they have a higher incidence of inherited medical problems.

Non-pedigree cats
Non-pedigrees, such as this red spotted tabby, form the genetic root stock for all pedigrees and exhibit most of the varieties of coat colours and textures.

notes that American cats have a much higher incidence of anxiety-related problems, such as urine marking, than British cats. He suspects that the reason for this is that many more British cats, pedigree or not, are allowed outdoors.

Your choice of cat should be made on the foundation of its lifestyle. Raised from kittenhood, most cats willingly and contentedly live indoors, but cats that have experienced the thrills of outdoor life often find them difficult to give up.

CATS AND OUR HEALTH

Many more people are likely to be allergic to cats than to dogs, because of their sensitivity to a protein in cats' saliva and skin. Cats are naturally hygienic animals, and cover their hair in saliva as part of their fastidious grooming routine. People with an allergy to cats should always keep them out of the bedroom, and wipe them at least daily with a clean, damp sponge to reduce the amount of saliva left on the hair.

Ritual cleansing
Cats wash themselves in a ritual fashion. First they lick their body, then paws, then use their licked paws to cleanse each side of their head.

RESCUED CATS
Cat rescue shelters often have feral cats needing homes. Mature ferals, raised outside human contact, are apprehensive. If you crave contact and feline affection, these individuals are not suitable pets for you. If, on the other hand, you enjoy the satisfaction of offering an independent creature a secure home and a regular supply of nourishment, it is highly rewarding to take on an adult feral. Initially, feral kittens under seven weeks old may hiss and spit, but will still have sufficient adaptability to accept and eventually enjoy some human companionship.

Once they are beyond this age, it is more difficult for them to overcome their ingrained fears. Rescue shelters also have strays: cats that are familiar with people, born in other homes but now needing new ones. These cats can make extremely satisfying but often nomadic companions.

Happy families
Cats raised together from kittenhood make the best companions. If you are planning to introduce a kitten to your cat, find one that has stayed with its littermates until it is 12 weeks old.

NEW ADDITIONS TO THE FELINE FAMILY

If you think your resident cat will benefit from a feline friend, it is often advisable to think again. Not all cats want friends. A resident cat can be disconcerted when a new person or dog walks through the door, but indoor cats in particular are likely to be severely affronted if they find their territory being invaded by another feline.

When you bring your new cat home, restrict it to one room, leaving your resident cat to roam the rest of the home. Leave food near the closed door of the room for your resident cat to find; it will eventually associate the pleasure of food treats with the scent of the new cat. After a few days, allow your resident feline to enter the new cat's room, only when the new cat is in a deep sleep, and only for a look. Increase these exposures, leaving nothing to chance. These gradual introductions reduce the likelihood of fights, or of one cat simply leaving for pastures new.

MALE OR FEMALE

Most pet cats are neutered (called spaying in a female), but if you choose an unneutered male, remember that cats use urine to mark their territory. Male cat urine is particularly pungent. Some males are content to leave their urine only in their toileting sites, but others use it to spray objects such as fences, trees, or if indoors, curtains and walls. Unneutered males can be very affectionate with people but are most likely to fight with other cats. This is a serious problem, because fighting is the most successful way for potentially lethal viruses such as feline leukemia virus (FeLV) and feline immune deficiency virus (FIV) to spread from one adult to another. Unneutered males are more likely to be carriers of these viruses than neutered individuals. Neutering has only a moderate effect on temperament and behaviour, making both males and females slightly less excitable and more affectionate.

GROOMING REQUIREMENTS

The cat is "self-cleaning": its barbed tongue is ideal for standard coats, fairly good for longer coats of heavy guard hair, but unfortunately inadequate when long coats consist of fine down. Cats with Persian-type coats are high-maintenance pets, and require careful, daily grooming. They are also prone to hairballs and need preventative attention for this problem.

Thin coats
Thin coats evolved in hot climates. Consequently, breeds such as the Singapura retain their thin coats only in temperate or warm climates.

Long coats
Some longhaired breeds need moderate grooming, but this Longhair requires daily, whole-body grooming.

Short coats
When allowed outdoors in cold climates, the British Shorthair naturally develops a thick, insulating down.

Neutered cats
The neutered cat, male or female, is more easygoing, friendly, hygienic, playful, and affectionate than unneutered individuals.

Unneutered cats
Unneutered males develop visible secondary sex characteristics such as cheek jowls, thick neck skin, and heavier muscles. Unspayed females are much more excitable in temperament than spayed females.

HOUSEHOLD HARMONY

When choosing a cat, consider the reaction of the whole household to a new feline and vice versa. If you have a dog and acquire a new cat, plan its introduction carefully. Train your dog to accept the cat's presence. Regardless of size, your dog is likely to be dominated by the cat but remember that even the best-trained dog will sometimes give chase to its "own" cat.

HEALTH CHECK

Ask your vet to examine your chosen feline's eyes, ears, nose, mouth, and body for any signs of disease or parasites. If your cat is a pedigree, ensure the viral health status of its parents by asking the breeder for proof of their virus-free condition. Most pedigree cats will have been vaccinated before leaving the breeder; ask for a vaccination record and show it to your vet.

Living with other animals
Most dogs accept cats into their homes, but some, especially terriers, can be troublesome. It is wise to supervise all initial contact. Restrict your cat to one room, and when it is asleep, allow your dog to see and sniff it.

DOG ACCEPTS KITTEN

A LIFE INDOORS

CATS HAVE BECOME OUR MOST POPULAR COMPANIONS FOR GOOD reason, being self-sufficient, self-cleaning, quiet, and independent. This does not mean, however, that they can or should be ignored. Most urban cats now spend their entire lives in the safety of their owners' homes; to help your cat make the most of an indoor life, it is vital to provide outlets for its emotional and physical needs. This means active play between you and your cat, something that should be appealing to both of you. With compensation for the loss of the physical and sensory challenges of the outdoor world, cats do come to enjoy this secure and comfortable life.

All cats have certain basic needs, such as their own food and water bowls, preferably shallow and ceramic, and hygienic bedding. In cold climates, beds suspended from radiators are favourites. Put comfortable bedding in the cat's carrier; if a cat feels secure in its own bed, it is less frightened when taken to the vet or cattery.

Even if your cat lives indoors, ensure it is vaccinated against infectious diseases. Some of these, especially flu viruses, can be transmitted on shoes and clothing, and your cat should also be protected in case it strays or needs urgent kennelling or hospitalization.

Compatible living patterns
Cats may sleep for 16 hours a day, and are most likely to sleep through the middle of the day when owners may be out of the house. But they can wake in an instant and are most active and playful in the early morning and evening.

Check your home for potential dangers. Keep the doors of washing machines and dryers closed, and always check inside before turning them on; cats enjoy sleeping in these warm places. Spray electric flexes with bitter non-toxic spray, to prevent kittens chewing on them. Never use products licensed only for dogs on your cat; some are toxic to felines.

LITTER AND LITTER TRAYS
Cats are naturally fastidious, using a specific toileting site and covering their waste. Only unneutered or otherwise dominant cats use unburied waste as territory markers. Outdoor cats favour earth or sand for burying waste; indoor cats need similarly "diggable" substances.

When choosing litter, remember that the odour and texture must appeal to your cat; if introducing a new cat, set up two trays with

Early learning
Kittens will continue to prefer the litter they first learn to use, but supervision is needed: some vets feel kittens are particularly likely to try to eat clumping litter.

different types of litter and let the cat choose. Early learning is pivotal, so if possible continue with the litter your cat first used. Supervise any kitten starting to use a litter tray until you see that it understands what the litter is for. Place trays in easily accessed but secluded places,

away from busy areas, because cats enjoy privacy. They also do not naturally share their toileting sites, so if you have more than one cat, provide each with its own tray. Clean litter trays at least every other day.

PLAYING WITH YOUR CAT

Play teaches cats about their abilities and the world around them, but for an indoor cat it has added importance. It is a way to release pent-up energy, which would normally be vented by stalking and capturing meals outdoors. Indoor cats may release energy by indulging in a "mad half hour" or even performing a "wall of death", racing around the room until they are literally running on the walls.

Play "capture-and-release" games with your indoor cat: dangle an object in front of the cat, perhaps through the rails of stairs. Mimic the movement of natural prey, and try to vary the circumstances each time. This type of play provides stimulation that outdoor cats get from grabbing at moths and butterflies.

Traditional toy
A fluffy object on the end of a string is ideal for the "grab-and-hold" games favoured by cats. A ball of wool is fine, but ensure that the end cannot be swallowed.

TYPES OF CAT LITTER

Cat litter should be an absorbent substance with a pleasant odour, but cats and owners are likely to have different opinions on smell.

The best way to control odour is by frequent cleaning. If cats are used to one type of litter, they often prefer its continued use.

Wood-based
These pellets expand as they absorb moisture. They can be composted or dried out and burned, but longhaired cats may trail them out of the tray.

Earth or sand
These are usually cats' favourites, but are bulky and nonbiodegradable. They are fine if disposal sites are available, but not ideal for urban use.

Clay
This is the most common type of litter. It clumps moderately well and often contains odour-absorbing granules. Do not flush clay litter; it clogs drains.

Non-absorbent
Made from wax-coated pelleted corn cobs, this is used in special trays that drain into a collecting tray below. The litter is rinsed and re-used.

THE IMPORTANCE OF NEUTERING

The urge to mate is hormonally controlled. To keep a hormonally active cat indoors without an outlet for this basic need verges on inhumane.

A female cat ovulates only after mating (*see page 70*). If she is denied access to males, her seasons are prolonged, and unspayed indoor female cats eventually reach a stage where they are more often in season than not, a state that imposes unnecessary stresses on the cat's body and behaviour. Females benefit from being spayed before puberty; early spaying also eliminates the later risk of mammary cancer.

Neutering male cats benefits you and your cat. It helps to eliminate the sour smell of male cat urine and diminishes the male's need to mark his territory with urine. Perhaps more importantly, neutering reduces the risk of fights with other cats. Should your cat stray outside, fights, specifically bites, are the most common route of transmission of potentially lethal viruses, such as feline leukemia virus (FeLV) and feline immune deficiency virus (FIV).

CAT CLAWS

Cats' claws are used to grasp, to climb, and also to scratch, leaving visible territorial markers. These natural needs can be problematic, but if you choose to live with a cat you choose to live with natural cat behaviour, including climbing and scratching. From the start, provide your cat with a scratching post and climbing frame, and clip the 10 front nails routinely, at least every three weeks.

TOYS FOR MENTAL STIMULATION

Cats are marvellously inventive at creating their own toys out of virtually anything small and lightweight they chance upon, from balls of string to butterfly wings.

Just like outdoor cats, indoor cats need to stalk and climb, chase and pounce, and bite and bat at objects. Creative toys, from simple dangling items to elaborate cat gymnasiums, offer an enjoyable release for these behavioural needs. Invest in a selection of toys that stimulate your cat's natural inclination to stalk, chase, grab, and bite in a safe context.

GLOVE TOY

BRIGHTLY COLOURED BALLS

"SNAIL" ON ROD

CAT TRIES TO CATCH MOVING MOUSE

"FISH" ON ROD

CAT PLAYS WITH CATNIP TOY

INDOOR PROBLEMS

THE CAT'S ADAPTABILITY IS MARVELLOUS, SO MUCH SO THAT MOST live contented lives as indoor individuals. Even in this relatively new environment, quite radically different from their original home, cats find ways to do what cats need to do: climb, mark territory, hunt, even find their own food. Climbing the Christmas tree, attacking our ankles, or raiding the kitchen is fun for cats, but can be problematic for us. Through simple training it is possible to direct your cat's instincts in safe directions. A sedentary existence may appear comfortable, but overweight, lazy, indoor cats are more prone to certain medical conditions.

You may think that your indoor cat has problems but remember, what is a problem to you may simply be natural for a cat. Cats lead three-dimensional lives. They explore and climb. They keep their weapons sharp and mark their territory. They stalk and chase small things that move quickly, and they eat vegetation. These are natural cat activities, and so climbing curtains, clawing the sofa, stalking your ankles, or eating houseplants are not "wrong" in a cat's mind; they are normal activities. Unfortunately, most of us find them unacceptable; fortunately there are simple ways to alter them.

EXPLORING AND CLIMBING

An indoor cat naturally assumes it is permitted anywhere. If you want to keep your cat out of particular places, for example the bedroom of a family member, simply close the door to that area. If this is not practical, use small heat or motion sensors, available from hardware or security shops. A cat crossing the beam of the sensor activates a high-pitched sound, scaring it off. These are inexpensive and excellent for training an indoor cat in your absence. For specific locations, such as kitchen work surfaces, double-sided adhesive tape is marvellous: cats hate the sticky feeling under their paws.

Defining boundaries
Combine discouragement of unwanted behaviour with provision of an acceptable outlet for your cat's natural curiosity or playfulness. Make these distinctions early: what may seem "cute" in a kitten is annoying or even destructive in an adult cat, but hard to unlearn.

Climbing is a natural activity: satisfy your cat's need by providing acceptable indoor locations. For example, placing warm bedding on top of a cupboard acts as an inducement to climb in that area. Use mild indirect punishment to discourage climbing in inappropriate places. Indirect punishment means your cat does not know it is being reprimanded by you: for example, as your cat starts to climb in a no-go area, make a loud noise by dropping an aluminium drink can with some coins in it, or squirt your cat with a water pistol. Sensible cats soon associate their activity with these mildly unpleasant consequences. Of course, provide suitable rewards when your cat climbs where you want it to climb.

SCRATCHING FURNITURE
To prevent damage to your furniture, provide your cat with furniture of its own. Do not hide the scratching post in a corner or distant room. One of the functions of scratching is to leave a highly visible marker of a cat's presence, so to start with, place the post in the middle of your cat's favourite room. Once your cat is using it, move it very gradually into a less obtrusive position. Aim to eventually position the scratching post near your cat's favourite sleeping area. Catnip rubbed on the post, and lavish verbal praise, stroking, and food rewards when the post is scratched all promote further use of it. If your cat is already scratching furniture, temporarily cover the damaged area with heavy plastic or double-sided adhesive tape to make the surface texture unappealing.

ATTACKING PEOPLE AND PLANTS
Channel your cat's need to stalk and pounce by providing well-designed toys to pounce on. There are a variety of excellent battery-powered toys that stimulate stalking, capturing, and holding "prey" (*see page 249*). If your cat has already decided that stalking your ankles is more fun, arm yourself with a water pistol and shoot when you see the glint in your cat's eyes.

Cats naturally often graze on grass, to help digestion and for the satisfaction of chewing on items with fibre. Provide your housebound cat

Preferred post
Cats choose sturdy, stable, ideally tall furniture to use as their scratching posts. The scratching post should be covered in fabric with a vertical weave. Sisal and cut (non-looped) carpet are attractive textures to cats. A little trial and error on your part will help you discover the textures your indoor cat prefers.

with its own "cat grass" to chew on. You can grow it from seed or buy it at pet shops. Any fast-growing, thin grass is suitable. Check all your houseplants to ensure you do not have any that are potentially dangerous if chewed, such as *Dieffenbachia*. If your cat is already attracted to your existing houseplants, put them out of reach or temporarily surround them with double-sided adhesive tape.

LIVING WITH OTHER CATS

When allowed to roam freely, cats stake out their own territory. It is estimated that on US and European farms there are about five cats per square kilometre. The densest population of free-living cats ever investigated, centred around a Japanese fish-processing plant, where the area contained the equivalent of 2,000 cats per square kilometre. But this is still less crowded than two cats sharing an average human home. As comfortable as it may appear to be, living indoors with us is unusual, and crowded if it is a multiple-cat household. What we think of as problems are bound to occur.

If two outdoor cats do not like each other, one bullies the other until it leaves. Locked in our homes, this simple solution becomes impossible. One potential outcome is not simply tension between the individuals but also behaviour such as urine marking and increased irritability, sometimes directed at the humans of the household. If you plan to be a multiple-cat household, acquire two kittens at the same time. If your cat has kittens, keeping one is also a simple way to increase the resident cat population with a minimal risk of serious cat fights – although it must be recognized that feline family disputes are not uncommon.

Eating houseplants

Most cats will want to chew on houseplants, so avoid toxic ones. Provide grass in pots or non-toxic plants such as the spider plant for safe chewing. Plants with toxic or irritant effects (often, but not always, noted on the label) and those with sharp spikes are best avoided. If cats tip plants over when exploring them, or scratch in the top of the pot, try placing large, heavy pebbles on the surface of the potting mixture.

CAT SEEKING VEGETATION TO CHEW FOR FIBRE

TOILETING AND MEDICAL PROBLEMS

A cat may refuse to use its litter tray and toilet elsewhere for a variety of reasons, some minor, other potentially serious. The location of the tray or the texture of the litter may be at fault. You may not be cleaning the tray often enough, or you may be cleaning it so fastidiously that you leave cleanser smells in it rather than natural ones. A cat may avoid a litter tray used by another household cat. Loss of toilet training may also be a sign of medical problems. Your cat may associate pain when defecating, for example from blocked anal sacs, with its litter tray, and continue to urinate in the box but defecate outside it. Similarly, pain on urinating may induce a cat to urinate elsewhere while continuing to defecate in the box. Painful urinating is caused by crystal formation in the urine, a bladder or urethra infection (bacterial cystitis) or even emotion (interstitial cystitis). Affected cats often urinate in unusual locations such as sinks or bathtubs. Your veterinarian will need a urine sample to help determine the cause of the problem.

One of the little-recognized problems of living indoors is the tendency for a cat to become overweight. The reason is obvious: excess calories being eaten in the form of tasty food, without the chance of regular exercise to burn them off. Indoor cats are liable to become slovenly, or even slothful, so watch weight and calories (*see page 262*). There is a direct relationship between being underactive and overweight, and serious urinary tract conditions.

Fatal attraction

Some pets combine better than others. Cats can be kept together with animals that would naturally be their prey, but they will need plenty of distractions to prevent their natural instincts from taking over.

THE PRINCIPLES OF BASIC TRAINING

The basic strategy for "positive reinforcement" training is simple: stimulate your cat to do what you want it to do and, as it does so, give a reward, often accompanied by a verbal command. Reward training is really quite simple (*see page 275*), but there will be times when your cat does something you do not want it to do, such as scratching furniture. Saying "No" is not sufficient. Cats need to climb, scratch, and stalk. The basis of cat training is to channel these activities into acceptable alternatives.

Limited value

Do not overuse direct reprimands. A cat is more likely to stop an activity if it prompts an unpleasant consequence not related to you.

A LIFE OUTDOORS

IF YOUR CAT GOES OUTDOORS, SAFETY IS A PRIME CONSIDERATION: both your cat's safety and yours. Examine the risks before deciding whether your cat can lead an indoor-outdoor existence. Perfect safety cannot be guaranteed, but be reasonable: are there dangers from road traffic, heights, or open water? Are there natural cat predators in your vicinity? Be scrupulous in protection against parasites and diseases, and provide suitable identification. It may also be important to be a good neighbour and discourage your cat from visiting other gardens. Cats can be the glue or, all too easily, the solvent in neighbourly relationships.

Plan your cat's introduction to the garden with care. If you have recently moved, give your cat time to adjust to its new indoor environment and find its own personal places: this can take from one to three weeks. Once your cat feels secure in its new home, allow it to have access to the garden just before a mealtime – the lure of tasty food is likely to bring it back in from the exciting outdoors. Assume that your garden is already "owned" and operated by a resident cat, and be vigilant. A water pistol can be a useful weapon. Squirting a cat with water will discourage it from returning to your garden.

Living the high life
A well-planned garden provides a safe but stimulating environment for a cat. There should be a variety of sights and sounds, and a choice of sunny and shady places to snooze. Ensure there are different levels to climb and explore, with high points for watching the territory.

CAT-FLAPS
Indoor-outdoor felines need their own exit and entrance. A partly opened window fitted with security locks to prevent further opening is one simple option, as long as it is low enough for your cat to reach. A more efficient alternative is a cat-flap: the most sophisticated cat-flaps can even be "personalized", activated by a small magnet on your cat's collar.

PLANNING THE GARDEN
The more creative you are with your garden, the more time your cat will spend there, rather than roaming. Elevated structures provide areas in which to soak up the sun, while wooden posts are ideal for overseeing territory as well as to use as scratching posts. Shrubs create cool shade in the heat, and grassy or large-leaved plants make a natural jungle. Instead of bird feeders, include flowering plants that attract butterflies, such as buddleia (*Buddleja davidii*). Keep planting dense all year to avoid bare soil, which is attractive for toileting. If your cat still digs, lay chicken wire, plastic mesh, or gravel between plants.

Rather than leaving the toileting decisions wholly to your cat, provide a toilet area exactly where you want it, preferably camouflaged, in a place where toddlers have no access but is easy for you to reach for routine cleaning. Dig a small sandpit and scent it with litter from your cat's own indoor litter tray. Regularly sift the site, and bury, dry and burn, or flush the faeces; do not put them on a compost heap.

Easy access
You will need to show your cat how a cat-flap works. Hold it fully open at first, and have a reward ready on the other side. Graduate to partially opening the flap, and your cat should quickly get used to the idea.

PREVENTATIVE MEDICINE
If your cat frequents areas accessible to other animals, it will almost certainly be exposed to parasites and diseases (*see page 270*). All cats should be inoculated against the most common viral infections, enteritis and flu. Outdoor cats face greater hazards. Feline leukemia virus (FeLV) is transmitted in saliva, through bites or even

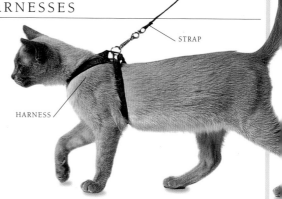

CAT HARNESSES

It is possible to take an indoor cat outdoors, especially if you plan ahead from kittenhood and introduce your feline to a body harness at an early age. If your cat is older, introduce the harness indoors and allow plenty of time for your cat to become accustomed to the feel of the harness. Never force your cat to go outdoors just because you feel it should. If it is fearful of the outdoors, allow it to remain in the security of its own home.

STRAP

HARNESS

PROPER IDENTIFICATION

All cats, even those that live indoors, should carry identification. An engraved disc or a canister containing your cat's information can be attached to a safe collar, which has a section that breaks away if caught on an obstruction. A tiny microchip transponder inserted under the skin on your cat's neck will provide a permanent ID, even if your cat slips its collar. If frightened, cats are likely to hide rather than return home. Training your cat to meow when it hears its name called (*see page 275*) is an effective way of finding a frightened or injured cat.

IDENTITY TAG

IDENTITY CANISTER

mutual grooming and licking. Ensure that your cat is protected against this potentially fatal infection, and also against rabies in areas where it is a risk. Fleas and other skin parasites are an additional hazard, and cat fleas enjoy meals from a variety of mammals, especially dogs and people. Use the modern, fast-acting treatments, which are highly effective. The most common tapeworm is ingested by eating fleas, while roundworms are passed from mother to kittens and through contaminated faeces. Both can be controlled with effective wormers.

BEING A GOOD NEIGHBOUR

It is a fact that on average about 20 per cent of people actively dislike cats, so do take your neighbours feelings into consideration. Long before there are any problems, discuss with them the fact that offensive, foul-smelling urine is produced not by neutered pet cats but by unneutered, often feral tom cats. Neutered cats almost invariably use their own toilet site (especially if you have provided one and encouraged them to use it) and cover their droppings. Only confident, territory-owning cats – once more usually unneutered tom cats – leave their droppings visible and unburied as territory markers. Hopefully, your neighbours enjoy seeing your cat sunning itself in the garden. But if they do not, respect their wishes.

RESTRICTED ACCESS

You may decide you do not want to let your cat roam freely, for any of the considerations listed here or for other reasons. One survey found that over 50 per cent of outdoor cats have a

serious accident at some time; another found that 42 per cent of cat fatalities were caused by road-traffic accidents. The outdoors may be a cat's natural habitat, but cats evolved in the spare surroundings of North Africa, not today's urban jungles. You may also find legal limitations imposed upon you: in some places local laws restrict the free movement of cats. Usually these laws are implemented to protect local bird or other wildlife populations.

If there are any such considerations where you live, but you still want your cat to have access to the outdoors, be creative with large enclosures for patios or balconies. Patios are easy to enclose, using wooden fencing and mesh for cover. Climbers quickly make such a cover attractive, and your cat can bask safely.

Innate hunter
Long grass provides perfect cover for cats to stalk through, either in play or in earnest hunting. If other small prey are available, many cats will ignore birdlife in the garden, but some cannot resist the challenge. Always site any feeders in the middle of an open area, well away from cover, or simply forgo them altogether.

Natural high
A clump of catnip (Nepeta cataria) planted in your garden may be attractive to your cat, but bear in mind that it could also act as a magnet to other neighbourhood felines. If this is the case, it may be best grown in a conservatory or enclosed space.

OUTDOOR PROBLEMS

THE GREATEST CONSIDERATIONS FOR OUTDOOR CATS ARE DANGERS. Rural cats face dangers from their natural predators, but urban and suburban cats face a greater variety of risks. There are dangers from falling out of windows, road traffic, toxic substances, diseases that thrive in dense cat populations, or simply from getting lost.

While you cannot remove these risks, you can minimize them. Ensure your cat is neutered, vaccinated against infectious diseases, and routinely treated for internal and external parasites. Equip it with ID and a reflective collar. Prevent scavenging by keeping rubbish secure, and take measures to prevent unwanted predation.

Like us, cats willingly choose to live in urban environments. They seem to thrive on the adrenalin surges that come from this type of environment; lawns where it is easy to see birds, rubbish that attracts rodents, ready-made human territory markers, such as fences, that can be adopted as their own feline perimeters. But you should always anticipate outdoor problems. Do your own homework, wherever you live: assess where there are risks, to your cat, to you and your family, and also to the environment if your cat goes outdoors, and reduce those risks as much as possible.

THE OPEN ROAD
More young outdoor cats die from trauma than from disease, and vehicles are the most common cause of serious or fatal cat injuries. Never expect your cat to develop road sense. Some do, but only those that survive through trial and error, a hazardous method. The glare of oncoming headlights at night can virtually paralyze a cat. If your cat goes outdoors, even if it has only minimal access to roads, equip it with a reflective collar to reduce night-time risks. This gives a greater chance that drivers will see your cat either by the roadside or on the road itself.

BOUNDARY DISPUTES
In natural circumstances, cats space themselves out in their own territories, meeting only when they chance upon each other at their territorial perimeters, or during mating. Generally, a natural territory is large enough to provide a constant food supply for one feline stomach. In the artificial circumstances of the urban environment, cats find themselves living in denser feline populations than ever before. This crowding leads to unique social pressures. In congested urban environments, cats often use human territory markers, such as fences and hedges, as their territory markers, and fight with intruders on their turf. Cats use ritual displays as far as possible to avoid fights: staring, hissing, spitting, and swatting are all rituals used to intimidate other cats. These displays, which evolved to avoid injury, are usually effective, with the intruder leaving the resident's territory. Fights only occur when such rituals have not been effective.

Neutered cats content themselves with relatively small territories, while unneutered cats of both sexes need to patrol larger areas. An unneutered tom cat may need a territory seven times larger than that of a neutered female, and will have to defend it all; this is one reason why it is vital to neuter your cat.

It is difficult to prevent unneutered feral tom cats entering your garden, even invading your

Finding a lost cat
If your cat goes missing, contact your local veterinary clinics, cat shelters, the police, your microchip or tag registry, and internet lost-and-found sites. Put up posters in the area with a photograph of your cat, a reward offer, and appropriate information. Make sure that you can be contacted at all times. When your cat has returned, pass on the news to all organizations that have offered help.

home through your cat's flap. Some feral cats liberally spray their urine to stake a territorial claim. If this is a problem, arm yourself with a noise-maker or high-powered water pistol for when social encounters become problems, and install a magnet-operated cat-flap.

DISEASES AND PARASITES

City living has given feline viruses that had been uncommon until relatively recently an unparalleled opportunity to multiply. The higher the concentration of cats in an area, the greater the likelihood of border disputes. More disputes mean more fights; more fights mean more bites; and more bites mean more opportunities for previously rare viruses such as feline immune deficiency virus (FIV) and feline leukemia virus (FeLV), which are spread via saliva, to be transmitted.

There are also diseases that pose a risk to people. Rabies is increasingly diagnosed in cats living in rabies-endemic areas. If your cat goes outdoors in a locality where rabies exists, or will be visiting such an area, ensure that it is vaccinated against this invariably fatal infection.

Parasites are also a problem, and again not just for your cat. While most animals can carry toxoplasmosis parasites in their muscles, cats uniquely transmit the disease to us or other animals via their faeces. Because it is contracted by eating contaminated wildlife, this is primarily a problem in outdoor cats, although it can also affect indoor cats that are successful mousers. The greatest danger is damage to a baby during pregnancy: any pregnant woman whose cat goes outdoors, and especially if it preys on birds or mice, should wear protective gloves when handling a cat's litter.

Outdoor cats can also contract the intestinal parasite *Giardia* by eating affected wildlife or drinking contaminated water. *Giardia* affects most mammals, including us. Routine worming with an effective medication controls *Giardia* and is a must for all outdoor cats.

HUNTING AND SCAVENGING

While the human environment carries risks for a cat, it has to be admitted that cats can also make trouble for other animals or humans.

Cats are natural born killers: we might find this distasteful, but it is a relevant fact. Hungry cats will kill only to eat; well-fed cats kill more frequently, and simply for the excitement of the hunt. Cats will prey on whatever is available. Indoor cats are restricted to flies, moths, maybe mice, and occasionally our ankles. Outdoor cats have real birds, small mammals, and frogs. This is potentially catastrophic to resident wildlife: there is at least one instance, off the coast of New Zealand, where one single cat drove a species of bird, the Stephen's Island wren, to extinction. There may be no threat to the bird population where you live, but it is still your responsibility to ensure that your cat causes as little harm as possible. Wherever you live, if your cat shows an inclination or ability to stalk and kill birds, do what you can to stop it. For example, do not set up a bird-feeding table in your garden, and attach a bell, or two, or three, to your cat's collar. There are devices available for use in your garden to deter birds from landing. Investigate whether any of these is suitable in your circumstances.

Scent messages
Cats will adopt high points to view their territory and high boundaries to define it. A garden wall or fence may help your cat to define an acceptable territory. Boundaries will be patrolled regularly, as scents left by other cats carry valuable social information.

Feline graffiti
While you may be relieved that your outdoor cat sharpens its claws on a tree rather than on your furniture, remember that the scratch marks are also there to declare ownership of a territory. Be aware of other cats in the neighbourhood and whether your cat runs the risk of getting into territorial fights.

POISONS AROUND YOUR HOME

People are often less careful of poisons outside their home than inside it. Make sure your cat does not have access to your garage or garden shed, where you may store toxic substances such as paint remover, paint, motor oil, or rodent poison. Danger usually comes not from consuming these substances, but from walking through them or licking fur that has become contaminated by chemicals. Antifreeze is a particular problem, because it is actually tasty to some cats. Standard antifreeze, ethylene glycol, is toxic to cats, initially causing a loss of balance, leading to convulsions, coma, and death if untreated. A safer antifreeze, propylene glycol, is available. Use this if your cat has access to your garage or might lick up drips that are on your drive. In your garden, avoid using toxic metaldehyde slug bait or other dangerous substances.

FOXGLOVE

Poisonous plants
Many garden plants, such as these foxgloves (Digitalis), contain toxins. If your cat tends to nibble on plants, find out which plants in your area are dangerous to small animals, and remove them from your garden.

FOOD FOR LIFE

OUTDOOR CATS ARE HIGHLY SKILLED AT FINDING POTENTIAL sources of food, but your indoor cat is wholly dependent on you to provide balanced nutrition. It is our responsibility to ensure that our cats' diet is as risk-free as possible. In any diet there is a balance between benefits and risks: a freshly caught bird provides a wholly natural diet, but carries the risks of intestinal damage from bone and feathers, or infection from bacteria and parasites. Similarly, poor home-cooking or commercial cat food may be deficient in the essentials of life. Cat nutrition is common sense; science can explain why some foods are not only good, but vital.

The right diet is even more important for cats than it is for us. As omnivores, we obtain the essential nutrients from a vast array of foods, converting both meat and vegetables into the essential amino acids and fatty acids upon which life depends. The cat cannot convert vegetable fat and protein into these building blocks. It is an "obligate carnivore", which must eat other animals to survive. This need is a consequence of the cat's original diet, which consisted of wildlife just large enough to fill a single stomach. In a sense, the cat's digestive system has become lazy, depending upon the consumption of other animals for essential substances. According to the circumstances, cats will consume a variety of foods, including insects, small mammals, birds and their eggs, reptiles, amphibians, and fish. While well-fed pet cats can be fussy eaters, a hungry cat will eat almost anything.

WATER IN THE CAT'S DIET

Most animals have a high water content, about 80 per cent, so a cat can often survive on the liquid it obtains from prey. This is a reminder of the cat's evolutionary origins in the relatively dry climate of North Africa. Some pet cats, especially those fed on wet foods, may seldom drink, giving the impression that they do not need water, but of course water is an essential at all times of life. It absorbs certain water-soluble vitamins, and is absorbed by fibre to add bulk to the diet (*see page 259*). A healthy cat has taut skin that snaps back into place when gently pinched over the top of the neck. If this skin does not snap back, a cat may be dehydrated, either not consuming enough water or losing too much. Dehydration is dangerous, reducing the efficiency of the body and increasing risk from disease.

Unfair competition
Although cats will eat together, it is best to feed them separately. This reduces the inclination to gobble food and ensures that even the smallest gets a fair share.

A natural diet

If your cat hunts and eats its prey, take this into account when calculating its food requirements. All parts of live prey provide nutritional benefits. Muscle is a source of fat and protein, the bones and organs contain vitamins and minerals, and feathers and hair provide valuable fibre. Eating the bones also helps to keep the cat's teeth clean.

VERSATILE BACTERIA

All cats digest their food the same way, with the help of "good" bacteria in the intestines. You can think of these bacteria as guests: the unwanted guests are asked to leave while the best guests keep their environment safe and tidy. They quickly create a stable environment, a "homeostatic" balance, where they survive and reproduce. This is not only good for digesting food. It also protects your cat from harmful bacteria and is an active component of an efficient immune system.

At birth, a kitten's intestines are colonized by bacteria in its mother's milk. Some of these are particularly efficient at producing enzymes necessary to digest milk. Later these bacteria dwindle in numbers, making milk digestion more difficult. If a cat were to consume milk all of its life, there would never be a need for this change, but dietary needs evolve not only from infant dependence to self reliance, but also according to physical and emotional health, hormones, activity, and age (*see pages 262–63*).

FROM KITTEN TO ADULT

An average kitten weighs little more than 100 g (4 oz) at birth and typically around 5 kg (11 lb) a year later. To attain this 50-fold increase in body mass, good nutrients are essential. Poorly nourished mothers raise kittens that are slow at learning to walk, crawl, and climb and go on to be poor mothers themselves; the effects of malnutrition can be perpetuated for generations.

When cats reach full size, nutrient needs drop to "maintenance" levels, varying according to the cat's health and activity levels.

Later in life, older cats do not need drastic changes in their food, but subtle modifications are valuable; increased vitamins and certain minerals are beneficial.

APPETITE AND WEIGHT CHANGES

Regardless of age, steady weight and appetite are signs of good health. Weight increases or losses mean that balances are upset. Almost invariably, your veterinarian will recommend dietary changes as part of the solution to the diagnosed problem. A healthy cat's appetite is constant. For housebound cats with little

Food treats
Cats, like humans, enjoy snacks. Major cat-food manufacturers produce a variety of dry or semi-moist treats. It is claimed that some of these complementary food treats help to keep a cat's teeth clean and tartar-free.

Mother's milk
Cat's milk has almost twice the nutritional value of either cow's or goat's milk. The high quality of the milk is important, because for the first three weeks of life kittens depend solely on their mother for their nutrition, and they continue to need her milk for at least two or three weeks more after this. Lactating mothers need much larger amounts of food than normal to satisfy the high demands placed upon them.

mental or physical activity, eating becomes the most exciting daily activity, and an increased appetite may simply be a sign of boredom.

For some cats, increased appetite may also indicate illness, such as an overactive thyroid gland (hyperthyroidism) or diabetes. A cat that is eating less or picking at food is often a greater cause for concern. This may indicate gum or tooth disease, or diminished taste or smell, and is an important sign for a variety of illnesses. If your cat is eating less, seek veterinary advice immediately. After illness, encourage a cat to eat once more by warming the food to about 35°C (95°F). This releases the natural aromas, which are the best trigger to appetite.

MILK AND CREAM

Milk is vital for all young mammals, but dairy products may upset digestion in some adult cats, causing diarrhoea. This happens because the adult cats' intestines may not contain enough of the bacteria that produces an enzyme to digest lactose, the sugar in milk. Feeding lactose-free milk, available for lactose-sensitive people and also sold as "cat milk", overcomes this problem for cats that have a lifelong love of milk.

BALANCED NUTRITION

ACHIEVING A BALANCED DIET IS MORE COMPLICATED FOR A CAT, than it is for ourselves or a dog. While all animals need protein, fat, carbohydrates, vitamins, and minerals in a healthy balance, a cat's needs are precariously fine-tuned. For example, cats need high levels of protein to grow, maintain, and repair body tissues.

Cats have unique needs for animal-derived protein and fat, but are also susceptible to medical conditions caused by excess amounts of certain foods. Each component in a cat's diet has a specific purpose and all, readily available in whole rodents or birds, are necessary in any food prepared either commercially or at home.

Special kitten needs
Although kittens are small, they need food that is high in energy, both to grow and to maintain good health. Their digestive systems differ from those of adult cats, and they should be given special kitten foods. For both nutritional and social reasons, kittens should not be weaned until their mother decides the time is right.

ESSENTIAL ANIMAL PROTEIN AND FATS
Protein provides amino acids, the building blocks of all body tissues, and enzymes, which support the body's chemical reactions. Cats can manufacture most amino acids, with the vital exception of taurine: this is acquired only by eating meat. Taurine deficiency causes blindness and heart disease.

Fat is the cat's most energy-dense nutrient, with more than twice as much energy per gram than protein. It also carries the fat-soluble vitamins. Animal fat contains essential fatty acids (EFAs) that cats cannot make, such as linoleic acid, which is vital for growth, wound healing, and liver function, and arachidonic acid, which is needed for blood clotting and coat condition. Recent research has also found that EFAs play vital roles in controlling skin disorders, allergies, and damage associated with heart disease and arthritis.

Cat owners may think that their pets are fussy eaters, but often there is logic in a cat's eating habits. Evolution has taught the cat that fresh is best, as older meat may be tainted by bacterial contamination. Evolution has also ensured that a cat's nutritional needs are met by its natural prey. Birds and mice provide excellent balanced nutrition, including a satisfying variety of textures to chew on, and no nutrients are supplied in excess. In our desire to be kind to our cats we can upset this balance unwittingly, feeding a poorly balanced diet and causing unexpected problems.

ENERGY REQUIREMENTS AND SOURCES
The energy in food is calculated in kilocalories, commonly called calories. If a cat is to maintain a stable weight, the energy consumed should be equal to the energy spent in activity. Cats' energy needs vary enormously with their inherent metabolic rate, age, activity level, and environment. A small, sedentary, indoor adult may need only 100 kcal a day, but a large, active, heavily lactating mother cat may need over 1,500 kcal. As an approximate guide, an indoor, easy-going cat needs about 50 kcal per kilogram of body weight (25 kcal per pound of body weight). Unfortunately, few cat-food manufacturers list energy content on the labels.

Protein, fat, and carbohydrates provide energy. Protein is broken down into amino acids, carried by the blood to help the body grow and repair. Fat is broken down into fatty acids, essential for healthy cells and an efficient immune system. Soluble carbohydrates are converted into sugar to provide energy, while carbohydrates in the form of fibre aid digestion.

Essential oil
Oily fish are a good component in a cat's diet, as they contain vitamin A, which the cat cannot manufacture itself from other sources. Not all cats relish fresh fish, however, and bones present a hazard.

Ring the changes
Whatever you feed, vary the diet to prevent your cat from becoming addicted to a single food, which can be a considerable problem if that food is not nutritionally complete.

Occasional salads
Plants are part of a cat's diet; they are a satisfying foodstuff and may also be used to induce vomiting. Indoor cats should be provided with boxes of fresh grass to chew; in its absence, felines might chew on your household plants.

THE VALUE OF FIBRE

Broadly speaking there are two forms of fibre: insoluble and soluble. Soluble fibre keeps food in the stomach for longer; insoluble fibre stimulates the bowel. Fibre is a natural part of a cat's diet, coming from the fur, feathers, and viscera of its prey. Until recently it was not thought to be as important for cats as it is for other animals, but many veterinarians and nutritionists feel that it may be valuable in treating conditions such as obesity, sugar diabetes, constipation, inflammatory bowel disease, and excess fat in the bloodstream. The amount of fibre a cat needs varies with age and lifestyle. Beet pulp, chicory, rice bran, and breakfast bran are sources, and psyllium is an excellent form of soluble fibre.

VITAMINS AND MINERALS

The vitamins A, D, E, and K, are soluble in fat. Vitamin A maintains good eye health; vitamin D is needed for bone development; vitamin E is an antioxidant; and vitamin K is necessary for blood clotting. We can manufacture all of these, but cats can only make vitamins D, E, and K; they cannot make their own vitamin A and must therefore eat parts of animals that are high in vitamin A, such as liver.

The water-soluble vitamins include vitamin C and the B-group. Unlike us, cats can make their own vitamin C. Take care if you give your cat a vitamin C supplement: excess is excreted in the urine as a substance called oxalate, and in the last 20 years there has been an increase in the incidence of oxalate bladder stones in cats. The B-group vitamins have diverse roles to play in good health and metabolism.

Natural balance

The cat's natural diet provides it with a perfect balance of nutrients. For example, a mouse's liver is a tasty and nutritious feline treat, but the naturally high level of vitamin A in animal liver can be toxic to cats if too much is eaten over an extended time period. A cat cannot eat excessive amounts of liver in its natural diet, because to do so it would have to eat a very large number of mice, which is simply a physical impossibility for a species with a small stomach.

Recently, deficiency in folic acid, a member of the B-group, has been linked to human heart disease, and the same may be true in cats.

Like vitamins, minerals act at the cellular level. Calcium and phosphorus are necessary to build bone and for cell membrane and nerve function. Meat is low in calcium, and a meat-only diet may lead to serious bone and joint conditions, while feeding excess calcium supplement may trigger a zinc deficiency. Iron is vital for red blood cell production. Excess magnesium was once considered a cause of lower urinary tract conditions, but this is not so; magnesium deficiency may be associated with increasingly common heart conditions. Selenium, essential for a healthy enzyme system is, like vitamin E, a natural antioxidant. It may also play a role in the immune system. Sodium is necessary in transporting nutrients across the cell membranes.

As advances in nutrition have highlighted the importance of trace vitamins and minerals, nutrients formulated to pharmaceutical

ANTIOXIDANTS

An antioxidant is a substance that destroys chemicals called free radicals. Free radicals are atoms in the body that damage cell membranes. Cats can manufacture some of their own antioxidants, such as vitamins E and C, but also need the building blocks of vitamins and minerals such as selenium and zinc in their diet to maintain a healthy, active antioxidant system.

YEAST TABLETS   VITAMIN POWDER   CAT SWEETS

Healthy fibre
Nutritionists have discovered that fibre from vegetable sources promotes the growth of benign bacteria and suppresses the growth of unpleasant bacteria that give rise to intestinal gas.

standards have been marketed for therapeutic use as "nutraceuticals". The North American Veterinary Nutraceutical Council is a self-appointed body that monitors the manufacture of these supplements: not all products conform to the acceptable high quality standards.

COMMERCIAL AND HOME-MADE FOODS

IF WE HAD THE TIME, MANY OF US WOULD PREPARE FRESH MEALS for our cats. To do so takes not only time for preparation but also time for learning about the cat's nutritional needs. Alternatively, there is a vast array of commercially produced cat foods, many of them based upon our ever-increasing scientific understanding of the cat's unique nutritional needs. The choice of feeding regimes, be it home-cooking, dry or wet commercial food, a regular or special diet, is yours. Through understanding and interpreting labels and knowing your cat's energy needs, it is possible to provide healthy, tasty, and nutritious variety for your cat.

Eating dried food
Dry food is not a natural texture for cats, and if they do not eat it as kittens they may refuse to touch it as adults. If you feed dry foods, be sure to provide your cat with plenty of fresh water at all times.

The greatest cost, and also one of the greatest considerations in keeping a cat, is what to feed it. You are making decisions that affect not only your cat's health but also the quality of its life. Early learning is vital to future eating habits: for example, if a cat does not eat bones as a kitten it may never learn how to do so. If it eats only one texture of food, for example dry food, it may be unwilling to change to a different texture later on. Talk with your veterinarian about what is best for your cat. When making any changes in your cat's feeding routine, remember that cats dislike sudden change, and shift from one food to another gradually, over a one- to three-week period. This gives the bacteria in the cat's gut time to modify to meet the new digestion demands.

VARIABLE OR FIXED FORMULAS
Reputable cat-food manufacturers only use surplus nutrients from foods produced for the human food sector. Other manufacturers may use products that have been deemed unfit for human consumption. Whatever the texture of commercial food, the recipes follow either a fixed or a variable formula. Fixed formulas remain constant and form the "premium" or "super-premium" end of the market. These are the most expensive foods, because in them the manufacturers always use the same ingredients, making no substitutions. A variable formula product is not necessarily inferior, as long as the varying ingredients remain of high quality and nutritional value. Variable formula foods maintain the same energy value.

READING FOOD LABELS
Interpreting labels is more complicated than it should be, because the information provided is often vague. For example, a label might give the following information:
- Crude protein.................................8%
- Crude fat.....................................6%
- Fibre..1%
- Moisture....................................78%

This typical or "guaranteed analysis" gives the protein, fat, fibre, and moisture in percentages. Because moisture content is so variable, it is impossible to compare these percentages directly with those of other cat foods. To do so you must convert the percentages into a dry-matter basis, in which all moisture is removed (*see Formula to calculate nutrition, right*). Converting information to a dry-matter basis allows you to compare different foods. This is important if you want to compare levels or sources of energy: cats need at least 26 per cent protein on a dry-matter basis, and the quality of the protein becomes important in older cats.

FORMULA TO CALCULATE NUTRITION IN COMMERCIAL FOODS

DRY-MATTER NUTRIENT CONTENT =

$$\frac{\text{THE LABEL'S NUTRIENT PERCENTAGE} \times 100}{\text{DRY-MATTER CONTENT PERCENTAGE}}$$

EXAMPLE:

- CRUDE PROTEIN $= \dfrac{8 \times 100}{22} = 36.4\%$

- CRUDE FAT $= \dfrac{6 \times 100}{22} = 22.7\%$

- FIBRE $= \dfrac{1 \times 100}{22} = 4.5\%$

Calorie (kcal) content is rarely stated on a label. As a guideline, a standard 200 g (7 oz) can of cat food contains 150–200 kcals of energy.

An ingredients list shows the constituents in descending order of weight, and very specific wording is used for this listing. "Meat" means muscle; "meat byproducts" or "meat derivatives" mean viscera, bone, and marrow, unappetizing

TYPES OF COMMERCIAL FOODS

Commercial cat foods are available in three different textures: moist or wet, semi-moist, and dry. The choice of what to feed is yours, but remember that texture is important to cats, and their natural prey consists of a satisfying combination of soft, chewy, and crunchy bits.

Dry foods are most convenient for you and the best taste delicious to cats, while moist foods in cans or sachets contain more water, usually around 80 per cent, beneficial for some cats. (*see Urinary problems, opposite*). Semi-moist foods make up only a small part of the market.

TYPES OF DRY FOODS

TYPES OF SEMI-MOIST FOODS

CANNED WHITE FISH

CANNED POULTRY

CANNED TUNA

CANNED LAMB

BASIC HOME-COOKED DIET

This recipe with medium-fat chicken contains about 800 kcal of energy, enough to feed a typical house cat for three days. For greater weight control, use lean chicken.

HOME-COOKED CHICKEN

- CHICKEN .... 140 g
- LIVER .... 30 g
- UNCOOKED RICE .... 70 g
- STERILIZED BONE MEAL .... 10 g
- IODIZED SALT .... 2 g
- SUNFLOWER OR CORN OIL .... 5 ml

Cook the rice, bone meal, salt, and oil in twice their combined volume of water for 20 minutes. Stir in the chicken and liver, and leave to simmer for another 10 minutes. Blend thoroughly and keep refrigerated.

Feeding trials

Look for the words "feeding trials" on the label of your cat food. This indicates the manufacturer conducted real trials with real cats, and the food should be appetizing to them. The words "formulated to meet the nutritional needs" mean the manufacturer is relying on laboratory analysis, not feeding trials.

to us, but all natural components of a cat's diet. "Meat meal" means dry products that have been rendered from animal tissues.

Feeding guidelines on packaging suggest the quantity to feed each day. These guidelines are often generous, being based on the needs of an active, young cat. If your cat does not fit this profile, especially if it is an indoor cat, talk to your veterinarian about energy requirements.

SHELF LIFE

Canned and vacuum-packed wet foods are pasteurized before sealing and have extensive shelf lives. They have no added preservatives, come in single-meal or single-day quantities, and are meant to be eaten quickly once open.

Dry foods are cooked under pressure and then dried. Fat is applied to the surface of the dry particles for odour and palatability, but fat spoils when exposed to heat, humidity, light, or even oxygen, so dry food needs preservatives to enhance shelf life. Antioxidants (*see page 259*) such as vitamins C and E are preservatives. Vitamins can come from natural sources or be synthesized: curiously, natural antioxidants do not preserve food for as long as synthetic ones. These dry foods are produced in multi-meal packages. For added safety, transfer your cat's dry food into a sealed container and store it in a cool, dry location.

URINARY PROBLEMS

Some cats produce a sticky substance in their urine that is associated with lower urinary tract complications. This substance "captures" urinary crystals, causing a partial or complete blockage to passing urine. Recent studies at Ohio State University suggest that cats with this condition benefit from eating wet rather than dry foods, as this dilutes the cat's urine. If changing from one food type to the other, do so over several weeks.

HOME-COOKING

Home-cooking is excellent if you understand the unique nutritional needs of cats. Because of modern abattoir methods, it is best not to feed raw meat: there is always risk of bacterial contamination. Any raw meat can also harbour the protozoal parasite *Toxoplasma gondii*, for which the cat is the intermediate host, and to which we are susceptible. Avoid single-protein diets: to take just two examples, a diet that is restricted only to tuna can cause severe liver problems, and a diet solely of muscle meat will lead to decalcification of the bones. Of course, a vegetarian diet is impossible, as cats must eat animal protein and fat to survive. When you select ingredients for your cat's diet, always use products fit for human consumption.

Chewing it over

Bones contain the perfect ratio of calcium and phosphorus for the maintenance of a strong skeleton. They also provide interesting texture. A cat can swallow a mouse whole and will usually consume prey in its entirety, including the bones.

COOKED MEAT AND VEGETABLES

COOKED POULTRY

SCRAMBLED EGG

Varied protein

Protein is of prime importance to cats and available from a range of sources. Always keep some variety of ingredients in the diet to prevent imbalances, and do not indulge your cat if it becomes fatty.

GIVING BONES

Bones, such as those in well-cooked chicken necks, can make an excellent and satisfying source of nourishment if introduced early in life. They can be dangerous, however, both for cats that never learned how to eat bones and for cats that wolf down their food in a dog-like manner. Avoid uncooked bones, as they may be contaminated with harmful salmonella bacteria.

ADJUSTING FOOD INTAKE

EVERY CAT IS A UNIQUE INDIVIDUAL WITH ITS OWN FOOD NEEDS. These needs are affected by age, sex, activity level, metabolic rate, local climate, and a host of other considerations. No particular quantity of a specific food is correct for all cats, but it is possible to develop guidelines for the energy intake that your cat needs.

This is vital, because whether your cat maintains a healthy weight depends upon you. Just like us, pet cats are increasingly suffering from unnecessary and worrying obesity. The sooner this is realized, the easier it is to modify your cat's diet. Fat cats may look cute, but excess weight is detrimental to good health.

Different breeds have differing basic builds, and also carry any excess weight in different places. The compact or "cobby" British Shorthair is naturally rotund, while the modern elongated, slim-shouldered Siamese and Orientals look like fashion models. Naturally stocky breeds show their weight gain all over and so appear to have a greater tendency to gain weight, but the fashion-model breeds can be overweight without showing it: they "hide" their excess baggage in a fat flap between the hind legs.

THIN CATS

With the exception of Siamese and Orientals, cats are rarely thin: by far the most likely causes of thinness in a pet cat are illness or starvation. Mothers can become thin because of the huge demands on them to feed their kittens. In older cats, over 10 years of age, who are still eating well, weight loss is always an important sign of a serious medical condition such as diabetes, kidney failure, an overactive thyroid gland, or other illnesses. Use the assessment chart opposite to assess your cat's body condition.

FAT CATS

According to veterinary statistics, around 30 per cent of cats are overweight today. The figure is probably higher for indoor cats and lower for active outdoor cats. Obesity is the most common dietary problem seen in cats, and it is increasing in incidence. The future can be seen even in young cats:

Knowing their limits
Many manufacturers produce foods specially formulated for kittens. Most kittens can be fed on demand, because they will stop eating when their stomachs are full.

when spaying females at five to six months of age, one can find that although the cat looks elegant on the outside, inside they have already laid down copious amounts of excess fat. This happens because, in simple terms, the cats are fed such nutritious and tasty food that they consume more energy than they expend in activity. The excess is converted to fat and stored. This is excellent if a cat needs to call on such reserves in the future, but few do. Instead, the excess weight is associated with skin and joint problems and an increased risk of developing sugar diabetes.

Excess weight creeps up on cats; because they are small, we often miss this surreptitious weight gain. There may only be 500 g (1 lb) difference from one year to the next, but in reality this represents a dramatic 10 per cent increase in weight, which is equivalent to a person putting on perhaps 8 kg (18 lb) in a year. If we did that we would be concerned,

Breed differences
If you choose to have a pedigree cat, find out when you acquire it what you should expect in terms of build and weight throughout its life. Even the stockier breeds should never be fat, but rather densely muscular. It is relatively easy to see weight gain on the shorthaired breeds such as these: with longhaired breeds it is even more important to weigh the cat at frequent intervals.

and you should be equally concerned if your cat either gains or loses this amount of weight between annual visits to your veterinarian.

ATTACKING FAT

Controlling feline obesity is as much a problem for you as it is for your cat. A satisfactory and lasting solution depends upon your willingness to change the status quo: the answer is not to put your cat on a crash diet but to change your pet's eating and exercise habits permanently.

With regard to food, write down everything your cat eats: meals, treats, and any scavenging. Making a list makes you more aware of the extras. Either gradually reduce the quantity of food you are offering over three weeks or, on a similar time scale, gradually switch to a less energy-dense diet. Low-fat, high-fibre foods offer fewer calories while maintaining bulk.

LOW-CALORIE DIETS

Manufacturers seldom state the calorie (kcal) content of their foods, although they will give you this information if you contact them. It is difficult to compare the energy levels of "low-calorie" or "lite" cat foods. Even if a package states how much less fat it contains than other foods, this percentage is not equivalent to a similar reduction in calories. As a rule of thumb, assume that a "low-calorie" food has 15–25 per cent fewer calories than an average product from the same manufacturer.

Cuddly cats
Although some breeds have developed to look soft and rounded, they should still maintain a healthy body weight; no breed standard calls for a fat cat.

Eating for seven

Pregnancy and motherhood place high demands on a cat. She provides all the energy that fuels her kittens' growth, both in the womb and for the first weeks of life, and her food intake throughout this time should be increased accordingly.

ENERGY REQUIREMENTS DURING LACTATION (KCAL)				
ADULT WEIGHT	2 KG	4 KG	6 KG	8 KG
LACTATION WEEKS 1–2	220	440	660	880
LACTATION WEEKS 3–4	300	600	900	1200
LACTATION WEEKS 5–6	420	840	1260	1680

The fibre effectively "dilutes" the calories in food. Your cat may sulk or harass you for more food: try not to give in to this psychological warfare. Talk to your veterinarian for reassurance and to be sure of your pet's real needs rather than its desires (*see page 258*).

For indoor cats, increasing levels of exercise can be difficult. It depends upon you stimulating more activity. Increased games or even, in the right circumstances, the introduction of another pet can increase energy consumption and halt excessive weight gain.

CHANGING ENERGY NEEDS

There will be times in a cat's life when its needs vary from the average dietary requirements. For example, cats benefit from increased energy consumption to help them recover from weight loss, whether this is a result of illness or simply due to the absence of food.

The greatest need for increased energy occurs when mothers are feeding kittens. Cat's milk is almost twice as nutritious as cow's milk, and to produce such an energy-dense milk requires greatly increased energy consumption. Even after her kittens have been weaned, a mother still needs 50 per cent more energy than usual, to replace the normal reserves she used during lactation.

Needs change again in later years. All the reputable cat-food manufacturers now produce special diets for older cats. These foods, which may be wet or dry, usually contain about 10–20 per cent fewer calories, increased vitamins and minerals, and more easily digested sources of protein and fat.

ASSESSING YOUR CAT

Use this chart to assess your cat's physical condition, comparing top and side views with the pictures below. If your cat is at either of the extremes shown here, you should consult your veterinarian promptly for advice on the cause and treatment of its condition.

Emaciated
- Ribs showing, no fat cover
- Bones at base of tail raised, with no tissue between skin and bone
- No palpable abdominal fat
- Severe abdominal tuck

Thin
- Ribs easily felt, with minimal fat cover
- Bones at base of tail raised, and covered in minimal fat
- Minimal abdominal fat
- Waist obvious behind ribs

Ideal
- Slight fat cover on ribs
- Bones at base of tail smooth, covered in thin layer of fat
- Minimal abdominal fat
- Waist can be seen behind ribs

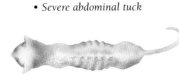

Overweight
- Ribs not easily felt, with moderate fat cover
- Bones at base of tail felt under moderate layer of fat
- Moderate abdominal fat
- Waist hardly discernable

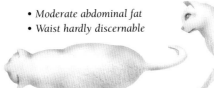

Obese
- Ribs not felt, due to thick fat cover
- Bones at base of tail difficult to feel through fat
- Extensive abdominal fat
- No waist, abdomen distended

Weighing a cat

Weigh your cat regularly and keep a note of its weight so that you will be aware of any gradual changes. If your cat simply will not sit still for this procedure, weigh yourself first holding the cat and then without it, and calculate the difference – but you must have accurate and clear scales to use this method reliably.

MAINTAINING GOOD HEALTH

HEALTH DEPENDS ON THE BODY'S ACTIVE DEFENCES AND ITS ABILITY to recognize damage and carry out repairs. All living creatures are capable of this defence and repair, and all living creatures use similar methods. The cat is particularly adept at maintaining its good health and well-being, although the pressures of living in a human environment have created new and unexpected problems. It is not only road traffic that threatens the urban cat. Microbes that caused little problem in the cat's evolution have benefited from the pet cat's rapidly evolving lifestyle, and now pose threats to good health that are particularly difficult to counter.

Natural good health
The cat evolved to operate at peak efficiency in its natural environment and keep itself in a state of good repair throughout its expected lifespan. Almost all of its excellent design is still to be seen in the domestic cat today, despite the vastly different environment in which it now lives. It is easy to read and admire the signs of good health: bright eyes, glossy coat, supple movement, and clean skin. Any variance from these high standards deserves our care and attention as responsible owners.

A carapace-like blood clot, the scab, forms a temporary repair to the damaged skin. At the same time, new skin cells begin to grow at the edges of the wound; these eventually grow into a new layer of skin to fill in the hole. New blood vessels sprout from the nearest supply, ensuring that the new skin has a healthy level of nourishment.

PERPETUAL REPAIR
Just as this self-repairing skin covers and protects the cat's body, every living cell inside the body has a covering wall that protects the contents inside. Just like the skin, these cell walls are not static structures: sections of wall are constantly being dragged inside the cell for examination and necessary repairs. Scavengers inside each cell, called lysosomes, recognize and eliminate any defective parts of the cell wall. At the same time, chemical regulators called cytokines, proteins so small and scarce

Like all animals, cats heal themselves. They do so every moment of the day, although we may never see it, because damage is constantly occurring and being repaired at the molecular or cellular level. We become aware that there is a problem only when damage is so extensive that organs are injured. In these circumstances veterinarians and cat owners help their cats by providing health care and treatment, but in fact such treatment simply creates the best possible environment for self-repair to take place. Whenever there is damage, the body strives to regain its former harmonious balance.

THE NEED FOR PERFECTION
Everything about your cat is geared towards maintaining the status quo, where all systems are in good order and working properly. If damage occurs, extraordinary repairs begin almost instantly. For example, skin punctures from cat fights are common injuries, and they are quickly repaired, but most of the repair work is under the visible scab and goes unseen.

As soon as a tooth or claw penetrates the skin, monitoring white blood cells, circulating in the bloodstream, note the damage and a whole army of defensive immune system cells is called into action to kill invading infections and clear up the debris left behind (*see page 266*).

HEREDITARY CONDITIONS

Selective breeding from small gene pools enhances desired characteristics, such as coat colour or texture, but can also unwittingly concentrate potentially damaging genetic material. Hereditary health problems, while still uncommon, are increasing in frequency in pedigree cats. These include:
- Hip dysplasia in Maine Coons
- Patellar luxation in Devon Rexes
- Deafness in blue-eyed white cats
- Cataracts in Himalayans
- Progressive retinal atrophy in Abyssinians

Concentrating the problem
Hip dysplasia is not unique to Maine Coons; it occurs in the general feline population, but it does have a higher than normal incidence in this breed.

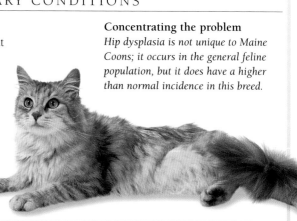

All in a day's work
Cats evolved in a tough environment where they could naturally expect to gain injuries from hunting and occasional fights. Their immune response to such injuries is swift and effective, although they are now vulnerable to more subtle diseases that can be transmitted in fights.

that they are almost impossible to detect, work to either stimulate or inhibit new cell growth according to what is best for that organ. For example, when a cat's skin is torn through fighting or some other injury, it is cytokines that stimulate the growth of new cells to cover the skin defect. This kind of repair is happening continually in all organs.

MOLECULAR REPAIR

Even more fascinating is the repair that occurs at the molecular level, in the coding of genetic information, or DNA. As old cells die and new cells are created to replace them, DNA passes genetic data from one generation to the next (*see page 74*). But in this constantly repeating process of copying and recopying, mistakes will happen. These mistakes occur in part because of the law of averages, but also as a result of external factors ranging from excessive exposure to natural threats, such as ultra-violet light or radiation, to interference from carcinogenic (cancer-causing) substances, either natural or manmade. When a DNA molecule is damaged or fails to copy itself correctly, special proteins called enzymes see

A helping hand
The cat's extraordinary resources of self-repair can be stretched beyond their limits, yet still benefit from medical intervention. The stresses that living in a human environment put on a cat are more than made up for by the benefits, for example, good preventative and curative medicine.

and recognize the damage. They act to cut out the damaged bit of molecule and splice in a fresh, healthy section. Such molecular mistakes occur daily in the renewal of the body's cells. When natural defences are overwhelmed, either by environmental threats or through natural aging, these molecular mistakes are not corrected. The self-repair mechanisms fail, and the new defective cells multiply to create a cancer.

HOW MEDICINE HELPS

Drugs alone do not cure infections or other disorders. The purpose of drugs is to help the body to recover itself; they facilitate self-repair. For example, a cat may develop a bite abscess after a fight, and bacteria from that abscess may escape and enter the cat's general circulation, causing malaise and an associated fever and lack of appetite. Antibiotics can help to destroy the bacteria, but at the same time repairs to the damage caused by the infection must be undertaken. The antibiotics give the body's immune system the time it needs to prepare its counter-attack, to deploy the natural killer cells, scavenging macrophage cells, and other components of the immune system. The real source of repair is the immune system; it is aided and abetted by modern medicines.

A FULL VIEW

Your cat's body is constantly repairing itself, but there are times when these mechanisms are inadequate, and intervention is needed to maintain good health. When this happens, it is important to recognize what the problem really is. Answering these questions help to clarify what can or should be done.

NATURE OR NURTURE?
Genetics play an important role in feline medical conditions. A cat may carry a genetic predisposition to an inherited disease, which is then triggered by environmental factors. This is thought to be the case with many lower urinary tract conditions. Some conditions have also become concentrated in certain pedigree breeds.

RED ABYSSINIAN

DO DESIGN CHANGES CAUSE PROBLEMS?
Through selective breeding we have enhanced the luxuriousness of the coat in some cats, and flattened the face in others. These may be attractive design changes but they may increase the risk of medical conditions, for example, skin irritation or upper respiratory tract infections.

WHAT ROLE DOES THE ENVIRONMENT PLAY?
In their natural environment, unrelated cats seldom meet each other. In the congestion of the human environment, cat-to-cat encounters are far more likely. Infectious microbes have taken full advantage of this environmental change, and the result is a range of "emerging" infectious diseases.

WHAT ARE SIGNS AND WHAT ARE SYMPTOMS?
Signs of illness are what you see as an observer. For example, difficulty breathing is a consequence of blocked nasal passages, lung, or chest damage. While cats also have symptoms, which describe how they feel, we can only guess at what they are.

IS A SIGN THE ILLNESS OR THE DEFENCE?
A cat with an upper respiratory tract infection or with allergies may sneeze. Sneezing, however, is not part of its illness; sneezing is part of the cat's defence. It helps get rid of the irritation in its nose. From an evolutionary viewpoint, microbes have taken advantage of this defensive measure and use it as a method for spreading themselves. Cats, like us, have a variety of these natural defences, which they use to fight infection. For example, a fever is beneficial: it helps destroy certain germs. Fasting is also beneficial: it may starve some germs of vital nourishment. These defensive measures may themselves cause problems if they are excessive.

HOW DOES THE ILLNESS DEFEAT DEFENCES?
Think of infectious diseases as the enemy, always looking for any weakness in a cat's defences. Some microbes, such as feline immune deficiency virus (FIV), have learned how to neutralize the cat's immune system. Others adapt, such as bacteria that are challenged by antibiotics and evolve to become resistant to certain antibiotics.

THE IMMUNE SYSTEM

A CAT'S HEALTH DEPENDS ON AN EFFICIENT IMMUNE SYSTEM THAT defends its body from disease. The immune system can be activated by either physical or psychological triggers and under normal conditions should turn on and off as necessary. If the system does not turn on properly, a cat is "immune-suppressed".

Feline viruses, such as feline immune deficiency virus (FIV) and feline leukemia virus (FeLV), are linked with immune suppression. The immune system may also turn on at the wrong time or fail to turn off, leading to allergy, asthma, or auto-immune diseases, in which the immune system attacks a vital part of the body.

The immune system in all mammals works through almost every system of the body, from the surface of the skin right into the marrow of the bones. A wide range of parts of the body (*see box, below*) contain cells that are integral to a healthy and efficient immune system that defends the body from both internal dangers, such as cancer cells, and external pathogens, such as viruses and bacteria.

Cats may respond to attack from infection in various ways. For example, the brain may set its thermostat at a higher level to induce a fever. Not eating is another defensive strategy. During an infection, a cat releases leucocyte endogenous mediator (LEM). This inhibits the absorption of iron, which is vital for the multiplication of some bacteria, from the gut.

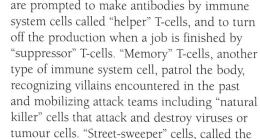

Exercise for immunity
In people, exercise is known to increase the numbers and activity of the immune system's natural killer cells. It is likely there are similar benefits for cats. Mental and physical exercise can only be beneficial for your cat.

HOW THE SYSTEM WORKS

White blood cells called neutrophils are the front line "attack soldiers" of the immune system. Their function is to guard and protect against bacteria and fungi. Other white blood cells work in slower, more complicated ways. One type of white blood cell, a B-lymphocyte, produces antibodies, proteins that neutralize and destroy harmful microbes. B-lymphocytes

are prompted to make antibodies by immune system cells called "helper" T-cells, and to turn off the production when a job is finished by "suppressor" T-cells. "Memory" T-cells, another type of immune system cell, patrol the body, recognizing villains encountered in the past and mobilizing attack teams including "natural killer" cells that attack and destroy viruses or tumour cells. "Street-sweeper" cells, called the

macrophages (literally "big eaters"), are the final part of the system. They arrive and clean up the debris. All of these activities usually occur in a harmonious and balanced way, but sometimes the immune system seems to "forget" how to turn on or when to turn off.

AN OVERACTIVE IMMUNE SYSTEM

The helper T-cells activate and deactivate parts of the immune system, but sometimes they misinterpret the instructions. They may activate the immune system when the cat's body is not being threatened by dangerous microbes or cancer cells, but when it is simply in contact with normally innocuous substances, such as foods, flea saliva, house dust, or plant pollens. These substances can trigger an allergic reaction, which may appear as anything from itchy skin, watery eyes, sneezing, or asthma to vomiting or diarrhoea.

Problems can also arise with immune responses if the suppressor T-cells do not carry out their "turning off" duties effectively. In this case, the immune system remains in overdrive, and it may erroneously start attacking a specific part of the cat's own body – for example, the red blood cells. When this self-destructive reaction occurs, it is called auto-immune disease. Renal amyloidosis, a type of kidney failure seen in cats, is one specific example

THE CAT'S IMMUNE SYSTEM

Elements of the immune system
We often think of the immune system as the white blood cells, tonsils, and lymph nodes, but there are many other elements. In fact, the skin is the largest component of the immune system. The intestines also form another vital part as do the spleen, the thymus, and bone marrow.

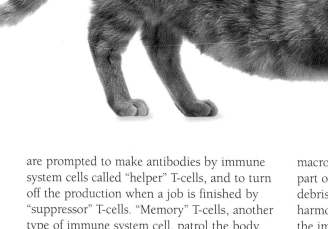

TONSILS

LYMPHATIC SYSTEM

INTESTINES

LYMPH NODES

SKIN

SPLEEN

BONE MARROW

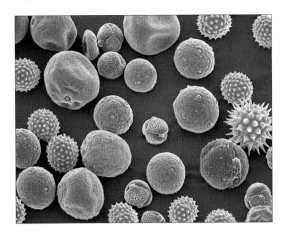

Tiny irritants
Some feline disorders, including dermatitis, eczema, colitis, hay fever, and coughing can all have allergic origins. The substances that cause problems range from flea saliva and dust mite droppings to human hair and skin cells, or tiny pollen grains such as those shown here. The time from exposure to an allergen to the allergic, histamine-releasing reaction is about eight minutes.

In the brain, chemicals called neuropeptides have a powerful effect on pain control, energy, and a sense of well-being. Pain is part of the cat's defence system, prompting it to avoid continuing dangers. During the emotional intensity of a fight, cats produce protective neuropeptides called endorphins that reduce the unpleasantness and intensity of pain. Chronic stress may trigger excesses or deficiencies in neuropeptides: pain and stress can therefore both affect suceptibility to or speed of recovery from illness. Use only feline pain killers: drugs that work in dogs or humans can be lethal to cats.

of an auto-immune disease. Veterinarians are increasingly diagnosing both allergies and auto-immune diseases in cats. While this may be due in part to improved diagnostic methods, many veterinarians feel that both of these problems are increasing in frequency among cats. The corticosteroid group of drugs are used to suppress an overactive immune system.

of its immune system, and so be subject to an increased incidence of infectious diseases, auto-immune disorders, and cancers.

A stress-free life
The sustained release of stress-related chemicals can cause damage to a cat's body, including the immune system, and affect the speed of its recovery from injury as well as illness.

ASTHMA AND ALLERGIES
Allergic reactions in cats to certain foods and chemicals are a relatively recent phenomenon. When a cat inhales, swallows, or is otherwise in contact with any "trigger" substance, the immune system responds by producing an antibody called immunoglobin E (IgE). In allergic cats, IgE binds onto receptor sites on specialized immune cells called mast cells, which are located in the skin and the lining of the stomach, lungs, and upper airways. These mast cells are like primed mines, filled with irritating chemicals, and the IgE causes the mast cells to literally explode, scattering irritating and inflammatory substances, such as histamine. Anti-histamine drugs act to neutralize the released histamine.

CANCER IN CATS
Cancer cells are renegades that have avoided the immune system. They trick "natural killer" cells into regarding the cancer cells as "self" and not attacking and destroying them. Having eluded the immune system, cancer cells, like parasites, embark on producing countless generations of cancer cells. Malignant cancer cells spread to other parts of the body, and some can produce chemicals that actively suppress a cat's immune system. The skin, blood-related organs, mouth, and bones are the most common sites for feline tumours to begin.

AN UNDERACTIVE IMMUNE SYSTEM
There are certain conditions, for example, an infection by feline immune deficiency virus (FIV) or feline leukemia virus (FeLV), in which the cat's immune system is suppressed. Affected cats are susceptible to infections and to cancers. Also, every cat that lives into old age is likely to experience a natural decline in the competence

Cats as allergens
Cats may be allergic to several things, including us, but increasingly we are allergic to cats. It is estimated that as many as 15 per cent of people may be allergic to a specific protein found in cats' skin, and also in their saliva. The suffering cat owner may find some relief if someone else grooms the cat daily, including a wipe with a damp sponge to remove the saliva that is inevitably left on a cat's fur when it grooms itself.

THE CHANGING NATURE OF ILLNESS

JUST AS THE DISEASES THAT WE FACE TODAY DIFFER DRAMATICALLY from those our grandparents had to contend with, so it is for the cat. The cat, however, appears to be more susceptible than many other species, including us, to a wide range of influences. They suffer from "new" infections such as feline immune deficiency virus (FIV) and feline infectious peritonitis (FIP), and also "new" metabolic disorders such as overactive thyroid glands and heart disease. Cats, their owners, and their veterinarians, face new genetic, environmental, and geriatric challenges. The nature of diseases, and their treatments, are constantly changing.

The threats from disease are always changing. The knowledge I gained about feline diseases as a veterinary student in the 1960s is now hopelessly dated. Then, feline infectious enteritis was a common and lethal killer; now, preventative inoculation and other factors have dramatically reduced its incidence. Then, heart disease was rare; today it is relatively common. When I began in practice, hyperthyroidism, an over-production of thyroid hormone, had never been reported in a cat; now I regularly see individuals with this old-age-onset condition. Some infectious agents, such as feline immune deficiency virus (FIV), have only emerged in recent decades. The nature of disease and illness is always changing.

PARASITES AND BACTERIA

The threat from parasites varies enormously, according to where you live and whether or not your cat has access to the outdoors. External parasites such as fleas and ticks are not only irritating in themselves. They act as carriers for other parasites. The common cat tapeworm, *Taenia taeniformis,* is brought into the intestines when a cat grooms itself and swallows a flea. *Toxoplasma* is transmitted when a cat eats meat harbouring the parasite; this meat is most often wildlife, such as birds or rodents, but not always; any undercooked meat is potentially hazardous.

Yet another parasite, the waterborne *Giardia,* has become more widespread recently, and a cat can contract this from drinking apparently clean water outside.

ANTIBIOTICS AND VACCINES

Vaccines have been dramatically effective in reducing threats from viral or bacterial infections in the cat, as in other species. Viral enteritis, a "parvovirus" sometimes called feline distemper, has virtually disappeared in many parts of the feline world.

Today, bacterial infections are treated with antibiotics, and here one problem is that certain bacteria have developed resistance to a range of previously effective antibiotics. When any population of bacteria is confronted with an antibiotic, sometimes not all die. The trait that allowed the survivors to survive and multiply forms the basis of a new bacterial strain resistant to that antibiotic; this new strain's resistance can also potentially be transferred to unrelated bacteria. By using antibiotics we have unwittingly accelerated bacterial evolution.

Taenia taeniformis
This tapeworm attaches itself by hooklets and suckers to the intestines of cats. Its chain-like body, up to 60 cm (24 in) long, consists of egg-carrying units that break off and pass out of the host to be consumed by a new victim.

Irritant parasite
Fleas can be a serious problem for cats. Some cats suffer quite marked allergic reactions to their bites, and fleas can also pass on parasites if they are ingested when a cat grooms itself. Treat your cat regularly with one of the long-lasting modern flea killers for maximum protection.

Invading the home
Fleas were once a seasonal nuisance; they are are now an irritation endured all year round, because our warm homes are an ideal environment for their over-wintering and breeding. Once fleas have established themselves in a warm home, it takes serious effort to remove them.

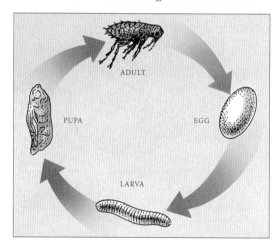

NEW DISEASES

But other infections have emerged. The first case of feline infectious peritonitis (FIP) was seen in the 1950s. Its development coincided with the establishment of large groups of cats in breeding colonies and rescue centres. In this new environment, an innocuous virus called a coronavirus somehow mutated into a lethal form. In early infections, cats suffered a fatal inflammation to the lining of the abdominal cavity, but today in some countries cats have adapted to the presence of this virus and show a variety of different clinical signs of disease. In the 1980s, the first indication that bovine spongiform encephalopathy (BSE) could jump from herbivores to carnivores was seen when cats developed a fatal spongiform condition in the brain after eating beef tainted with BSE prions. Feline spongiform encephalopathy (FSE) became common in the United Kingdom

Swings and roundabouts

The care that we give to cats, protecting them against the diseases and environmental factors that would shorten their lives in the wild, has paradoxically given rise to a host of new problems arising from greater density of feline populations, or simply from living longer.

in the late 1980s, although dogs that ate tainted beef never suffered a similar infection. Also in the 1980s, a neurological condition called autonomic polyganglionopathy, or Key-Gaskell Syndrome, reached epidemic levels in the United Kingdom and to a lesser extent in Sweden, then subsided. In the 1990s, the incidence of cats over 10 years of age suffering from overactive thyroid glands had reached similar epidemic proportions in North America and Europe. The nature of disease in cats is more complicated and changing more rapidly than in any other domesticated species.

ENVIRONMENTAL EFFECTS

So why do we see these changes in the medical problems facing cats? The cat evolved as a lone hunter, seldom meeting other cats. When we moved cats into catteries, rescue centres, even our homes, we changed their environment dramatically: far more than when we began our

EVOLVING TREATMENTS

Just as the nature of disease changes, so does its treatment. New medications are constantly licensed for use in cats: treatments for thyroid management, heart disease, and pain control are all investigated and sanctioned, and even chemotherapy can be highly effective for certain cancers, sometimes offering hopes of permanent remission. These treatments lead to ethical questions. How far should such treatment be taken, and when does treatment cease to be in an individual cat's interest? These are difficult questions.

Ethical issues

Medical treatment is always stressful for a cat. It is important to carefully weigh up the level of stress against the possible benefit to the cat: the equation will not always be the same as it would for a human.

domestication of the dog, which was already gregariously sociable. Mixing cats from different backgrounds increases the risk of infectious disease. Simply living in our centrally heated homes changed the nature of feline disease. Feline asthma, once rare, is now commonly diagnosed both in outdoor and indoor cats. Another effect of this new lifestyle has been a dramatic increase in their risk of trauma. The danger of road-traffic accidents is obvious. Less evident but more worrying is the incidence of air-rifle and gunshot wounds, regrettably common injuries to cats in some localities.

GENETIC PRESSURES

Genetic mutations are always occurring. In the wild, where survival of the fittest determines who will and will not breed, the beneficial mutations are retained while less successful ones die out. The progenitors of the Maine Coon, and the Norwegian and Siberian Forest Cats had to survive cold winters and capture rabbits or hares as well as rodents to survive. These needs predisposed to the genetic adaptation of long, dense coats. When we intervene in cat breeding, we influence genetic pressures as surely as we do with antibiotics and bacteria. Breeding to pedigree standards accelerates evolutionary change. If a small population of cats is used for breeding, deleterious genetic changes in the nature of disease can also occur: recessive genes for medical problems such as kidney failure or cataracts may become concentrated in a single line, and spread through the breed. This has already happened in several breeds.

AN AGING POPULATION

One of the greatest changes in the nature of feline disease is that cats are quite simply living longer than ever before. Parasite and disease prevention, good nutrition, effective veterinary treatments, and increased safety have brought about the need for geriatric cat care. The old calculation that one feline year is equivalent to seven human years ceased to be relevant long ago, and today one feline year is closer to five human years. Cats, especially Siamese, often reach their late teens or even early twenties. As a result, metabolic disorders, such as kidney failure, heart disease, cancers, and general wear and tear, are being diagnosed more than ever before. It is unlikely that the incidence of these conditions has increased; more probably there are now more older cats with owners willing to obtain veterinary advice and treatment for old cat conditions such as cancers. Even senile dementia, similar to Alzheimer's in people, now occurs in cats. Geriatric medicine is a new field in veterinary medicine, and it exists because of cat owners' success in looking after their feline companions.

Downward mobility

The short-legged look of the Munchkin breed found in North America has turned up more than once in feral cat colonies, but the fact that there are no wild cats with this physique strongly implies that it carries disadvantages for the cat.

SIGNS OF ILL HEALTH

IN THE PRESENCE OF SEVERE OR SOMETIMES EVEN DEVASTATING illness, felines will often pretend that nothing is wrong. From an evolutionary viewpoint, this has been a successful adaptation: they hide their illnesses and injuries from predators until they recover. In the newer context of living with us, however, this is unfortunate. Without acute observation on our part, a cat may not be treated for an illness until the condition has reached an advanced stage. This makes a successful outcome less certain. All cats develop natural rhythms to life; be alert to even the slightest changes in your cat's behaviour and contact your veterinarian for advice.

The understanding of feline medical problems has increased dramatically in the last few years. New methods of diagnosis and treatment mean that, increasingly, it is possible to assess the nature of a problem accurately, and then successfully treat and correct it.

VIRAL COMPLICATIONS

A cat may silently carry any of three potentially serious viruses: feline immune deficiency virus (FIV), feline leukemia virus (FeLV), and feline infectious peritonitis (FIP). These slow-acting viruses may be an underlying cause of seemingly small problems, such as gum inflammation and bad breath, and may be associated with poor recovery. Without knowledge about the presence or absence of these viruses, there is always an undercurrent of veterinary uncertainty, even when diagnosing apparently minor conditions.

Maternal danger

A female cat can transmit potentially lethal infections to her young just before, during, or shortly after birth. She may be a symptom-free carrier of feline leukemia virus (FeLV), feline infectious peritonitis (FIP), or feline immune deficiency virus (FIV). All cats, male and female, should be tested for these viruses before being used in breeding programmes.

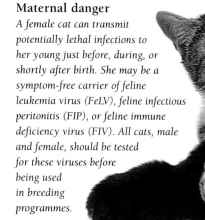

SIGNS OF DISORDERS

CLINICAL SIGNS	CAUSES	ACTION
SCRATCHING/ EXCESS LICKING	External parasites are the most common cause of scratching, but allergies to food or environmental causes may also be responsible. Cats lick most wounds thoroughly. Impacted anal sacs cause a cat to lick its anal region.	Check the skin thoroughly for signs of parasites such as specks of black, glistening flea droppings, and treat accordingly. If there is a small wound, see your vet. It may seem insignificant but there might be hidden damage.
EYE CONDITIONS	Viral, bacterial, and chlamydial infections cause inflammation and discharge. Many of these diseases transmit easily to other cats. Eyes may be injured or become inflamed through environmental irritation or allergy.	Vaccinate your cat preventatively against transmissible infections. Treat affected individuals with appropriate antibiotics or other drugs. Always have your vet examine any physical injuries to the eyes.
EAR CONDITIONS	Ear mites cause dark, gritty wax to build up in the ears. Cats scratch and shake their heads. Infections are common. Many older cats have benign tumours of the ear canal called ceruminomas.	Eliminate parasites with appropriate medications. Break down built-up ear wax with olive oil. Untreated chronic ear infections lead to ceruminomas, which may need surgical removal.
BAD BREATH/ DROOLING	Gum infection is the most common problem suffered by felines. Fetid breath is caused by bacteria multiplying at the tooth margins. Severe gum inflammation, which may be virus-induced, causes drooling.	Prevent gum disease by permitting your cat to exercise its teeth and gums; chewing on bones is the most effective and natural method, but beware of possible splinters. Viral gum disease needs veterinary attention.
SNEEZING/ NASAL DISCHARGE	A variety of viral and bacterial diseases cause sneezing or nasal discharge. Allergies and foreign objects such as blades of grass in the nose may also be responsible. Older cats can develop irritating nasal polyps or tumours.	Vaccinate your cat preventatively against transmissible diseases. If your cat has allergies, avoid allergens when possible or use antihistamines and other treatments to reduce irritation.
COUGHING/GAGGING	Coughing may be caused by a variety of medical conditions, including allergies, parasites, chest and upper respiratory infections, more serious chest complaints, or forms of heart disease.	Although it may simply be an allergic bronchitis, cats can suffer from a wide range of more serious chest complaints. If your cat is coughing, always consult your vet for an accurate diagnosis.
BREATHING DIFFICULTIES	Chest and upper respiratory tract infections may cause breathing difficulties. So too does asthma, a serious, even life-threatening problem. Trauma to the chest or diaphragm seriously affects breathing.	All breathing difficulties are serious. Your cat should be seen immediately by a veterinarian. Intervention with drugs or surgery is often vital to restore ease of breathing.
VOMITING	Occasional vomiting is often caused by diet or eating grass. Cats "regurgitate" hairballs, an action seemingly similar to vomiting. Persistent vomiting often has a metabolic cause elsewhere in the body.	Use cat laxatives for cats suffering from hairballs. Offer grass to eat. All cases of persistent vomiting should be seen by a veterinarian who will make an accurate diagnosis of the cause.

APPETITE CHANGES

Any change in eating and drinking habits is significant. Refusing to eat can indicate pain in the mouth, but it is often of greater significance. Asking to eat but then refusing is always a serious cause for concern. A cat may simply not like what it has been offered, but this behaviour is sometimes associated with tumours in the abdomen's lymph nodes, a fairly common condition called lymphoma. Excess eating with weight loss is frequently associated with an overactive thyroid gland. Affected cats are usually over 10 years old, active, affectionate, and playful. Any increase in thirst should be reported to your vet. It may only be that the environmental temperature has increased, but it is more likely to be a significant sign of illnesses such as kidney or liver problems and different forms of diabetes. All of these conditions are relatively common in older cats.

Age-related weight loss

Senior cats commonly lose weight, but rapid weight loss is cause for concern. Combined with a good appetite, it may indicate an overactive thyroid, kidney problems, or other serious conditions.

ALTERED BEHAVIOUR

With a growing elderly cat population, heart problems are increasingly common. An affected cat often shows only the slightest change in behaviour, possibly sleeping more, being less inclined to patrol its territory, or flopping when lying down rather than reclining gracefully. In severe cases there is coughing, and with accompanying weight loss the heart can be seen beating quickly through the chest wall. As soon as you see any changes in your cat's behaviour, report them to your vet; you may be seeing a problem at an early stage, and it is always easier to treat health conditions sooner rather than later. One of the best methods of preventing disease is to arrange for routine medical examinations for all individuals over nine years old. With skilful diagnosis, age-related illnesses can be caught early and treated more effectively.

Excessive licking

Cats are fastidious groomers and clean their coats by licking in a ritualized fashion. Excessive licking, however, may indicate local injuries or wounds, or the presence of skin parasites such as fleas and ticks.

Listlessness

Cats enjoy "cat-naps" throughout the day, but increased lassitude, listlessness, and sleeping, retiring to unusual places or hiding for no apparent reason might be caused by illness. Consult your vet if you notice any change in your cat's daily routines.

Changed eating habits

Assume that your cat is ill if it eats or drinks less than usual, or refuses food and water. If it eats or drinks more than normal, it may have kidney or liver disease, or an overactive thyroid gland.

SIGNS OF DISORDERS

CLINICAL SIGNS	CAUSES	ACTION
DIARRHOEA/CONSTIPATION	Diarrhoea is often diet-induced, but it may also be caused by parasites, infection, malabsorption conditions, or, in older cats, by masses in the intestinal wall. Do not confuse the straining of diarrhoea with constipation.	Contact your vet for advice. Simple diarrhoea can be treated successfully with a short fast. The cause of persistent diarrhoea needs to be known for effective treatment. Prevent constipation with cat laxatives.
LOSS OF WEIGHT	This is a general sign that a cat is unwell. Starving is the most common cause in young cats. Kidney problems or overactive thyroids are the most common causes in older cats that continue to eat well.	Always see your vet if your cat is eating but still losing weight; it is usually a sign of significant problems. Changes in diet, surgery, and other medical treatments may be useful.
LOSS OF APPETITE	A short loss of appetite is seldom a problem. A cat may be bored or have eaten elsewhere. Loss of appetite for more than a day is an important clinical sign that a cat is unwell; it is associated with many problems.	Consult your veterinarian, who will carry out an examination and possibly tests to find the cause of your cat's loss of appetite. Although many causes are serious, others can be treated easily.
BIRTH PROBLEMS	Most cats experience few difficulties giving birth. Some, however, have poor contractions or are unable to deliver their kittens because of problems in the birth canal.	If your cat is pregnant, make sure that your vet is informed of the expected due date. Prepare a quiet place at home for the birth. If labour lasts for more than two hours, veterinary help is required.
URINATION PROBLEMS	Straining to urinate may be caused by mineral crystals or infection in the urinary tract. Decreased urination is caused by dehydration or a blocked bladder. Increased urination has a variety of significant causes.	Contact your vet if there are any changes in your cat's urination routines. All are significant and some changes are associated with serious conditions that need immediate medical attention.
LOSS OF BALANCE	Injuries are the most common cause of balance problems. Blood loss, strokes, middle-ear infections, viral diseases, and other serious problems can cause a drunken-like walk.	All causes of loss of balance are potentially serious; some are life-threatening emergencies. See your vet immediately. Your cat may look normal outwardly, but have very serious internal problems.
SEIZURES/PARALYSIS	Epilepsy, brain inflammation, and poisons can cause seizures. Paralysis is frequently associated with severe neck injuries, but may also be caused by infections and poisons.	Seek urgent veterinary attention for paralysis, and urgent veterinary advice for seizures. Vaccinate your cat against infectious diseases, including rabies where warranted or necessary.
BEHAVIOUR CHANGES	A variety of physical but also emotional problems may manifest themselves only through moderate changes in your cat's behaviour. Trauma, infection, and metabolic conditions have behavioural consequences.	If your cat's behaviour changes, no matter how slightly, always consult your vet. A minor behaviour change may signify a considerable problem; cats are naturally reticent and seldom complain.

GROOMING

CATS WITH LONG, LEAN BODIES AND SHORT COATS NEED ONLY A little grooming for hygienic reasons: brushing the coat once a week will stimulate the oil glands, remove dead hair, and keep the coat lustrous. Cats with dense or long coats need more frequent grooming, while older cats need more attention to their teeth, nails, and anal regions. From the time your feline arrives in your home, train it to accept combing, dental attention, and nail clipping. Always use food treats and play as rewards for good behaviour. Grooming reinforces your cat's relationship with you, reducing your blood pressure and that of your cat, a potent sign of relaxation.

GROOMING A LONGHAIRED CAT

Routine, simple grooming is an excellent opportunity to bond with your cat; it is an activity that should always be relaxing and pleasurable for both parties. Regular grooming keeps your cat in pristine condition and warns you of possible medical problems. Evolution has equipped the cat with its own grooming tool: the barbed tongue. Consequently, cats are magnificently self-cleaning and self-grooming.

However, this evolutionary grooming tool is inadequate for felines with dense or long coats; these individuals require their owners' aid in the judicious use of a wide-toothed comb and a bristle brush. Older felines with dense coats will require even more grooming assistance from their owners. Longhaired breeds with heavy, predominantly guard-hair coats need frequent brushing, and grooming around the ears and legs is necessary at least weekly.

All cats with soft, Persian-like coats are high-maintenance pets. They have long, fine down hair, which can easily become matted. This requires daily, whole-body grooming in order to prevent serious tangles. These felines are also prone to hairballs, which are potentially dangerous if ingested, therefore preventative grooming is necessary. Although the grooming steps in the sequence below are for a longhaired cat, they can be followed for all types of coats.

1 Use a wide-toothed comb to comb the fur between the hind legs, behind the forelegs, and behind the ears; knots are common in these regions. Next, comb the fur on the abdomen; many cats are sensitive in this area, so be gentle.

2 Tangles develop at the wispy ends of the hair on the top of the neck and back. Working on one section at a time, part the hair and comb gently upwards from the roots. This removes loose hairs at their source and helps to eliminate knots.

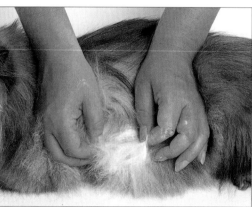

3 As you complete each section, part the hair and ensure that the skin is healthy. Tease out remaining tangles by hand. Check for tiny specks of shiny, black dust. If you find any, your cat and its environment need treatment for flea infestation.

4 Comb the hair upwards behind the ears and over the ruff to enhance this area and remove any tangles. Knots and mats are common in both these regions, because a cat cannot groom them; they are inaccessible to the reach of its tongue.

5 Finally, groom the tail. Do not simply brush along its length; many cats resent this method. Parting the hair and gently brushing it to either side is much less traumatic. Always finish grooming sessions with rewards of food and play.

GROOMING EQUIPMENT

For grooming, use a wide-toothed comb then a bristle brush; a fine-toothed comb can be used to expose fleas. A child's toothbrush fits comfortably into a cat's mouth. "Guillotine" nail clippers cut nails without crushing.

TOOTHBRUSH

WIDE-TOOTHED COMB

BRISTLE BRUSH

"GUILLOTINE" NAIL CLIPPERS

TRIMMING THE NAILS

Outdoor cats benefit from sharp claws: they are part of its natural defences, and assist it in climbing. If you keep your cat indoors, trim its nails routinely. Older cats often grow long nails, which may curl in and penetrate the paw pads. Although nail trimming is painless, cats rarely enjoy it, so start training early. Accustom your feline to having its paws handled and its nails extended. Always use a comfortable "guillotine" nail clipper, but if the nail has grown into the pad, consult your veterinarian. If your feline is a "wriggler", vary the location in your home where you clip its nails so that it does not learn to associate a particular site with the activity. Trim the nails when your cat is relaxed and reward compliance with food treats and play.

1 Cats naturally retract their claws. To extend each nail, press gently behind the claw to expose the brown to white tip and pink central core or "quick" of the nail. The "quick" is living tissue and must be avoided when cutting the nail.

2 Position the clippers over the nail and in front of the "quick". Squeeze firmly, cutting straight across the claw. Speak reassuringly to your cat while you are trimming its nails, and reward it with food treats and play when you are finished.

BRUSHING THE TEETH

Overwhelmingly, the most common reason why cat owners visit the veterinarian is tooth and gum disease. Tooth enamel erodes, exposing the living pulp within; gums become infected, producing disease and foul breath. The most effective way to prevent oral problems is by brushing your cat's gums and teeth. Begin with mock brushing, then graduate to cleaning the teeth several times each week. Always reward good behaviour.

1 A cat usually resents oral attention, so train it to associate teeth and gum cleaning with simple rewards. Using a cotton bud scented with pet toothpaste or even tasty food, gently massage the gums and teeth.

2 Once your cat tolerates the cotton bud in its mouth, graduate to using a soft, child's toothbrush. Gently massage the gums and brush the back teeth, the region where tooth resorption and gum disease most commonly occur.

CLEANING THE EYES, NOSE, AND ANUS

Most felines have superb anatomy but human intervention has led to eye and nose problems in some breeds. In the relatively flat-faced breeds, tears overflow and normal nasal secretions crust around the nostrils. Tear overflow produces a mahogany-coloured discoloration to the hair around the eyes. Cotton wool dampened with antiseptic eye-wash solution can be used to remove excess tears, clean the nasal fold, and remove crust from the nostrils. Consult your vet if your cat has a chronic tear overflow, or suffers from frequent sneezing or nasal discharge. Excessive licking of the anal region indicates that the anal sacs are blocked. Your vet can show you how to empty impacted anal sacs manually.

CLEANING THE EARS

Ear mites are the most common cause of gritty, dark wax in a cat's ears; consult your vet for treatment. An excess of normal but oily and irritating ear wax is common, particularly in indoor cats. Use recommended wax remover, then remove the excess wax with soft cotton wool. Never poke a cotton wool bud into the ears: it pushes the wax into the ear canal.

TRAVELLING AND TRAINING

CATS ARE REMARKABLY GOOD TRAVELLERS AND MOST OF THEM FIND journeys interesting. Even those that prefer the security of home will simply hunker down and withdraw into themselves until the cat carrier is opened and normality returns. Because they are so adaptable, sedation before journeys, even those by aeroplane, is unnecessary, even counter-productive. New vistas will bring new risks, and cats are more likely to stray on new territory. By training your cat to come to you on command or to miaow when it is called, you increase the likelihood of finding it if it disappears in fright. Begin training at an early age to ensure success.

Holidays are always a difficult time for pet owners. While dogs will happily travel long distances with the family, most cats tend to prefer their own territory, and unless your cat is particularly adaptable it is easier to leave it either at home or with an approved cattery. However, it is important that your cat is at ease with travelling, if only for those important visits to the vet. Plan ahead by accustoming your cat to its carrier. Use it as a cat bed at home so that the cat feels secure and content when in it, and considers it part of its own territory. If you do intend to travel frequently with your cat, you should get it used to car travel when it is young. This might entail taking the cat on any short journeys, or even driving around the block occasionally. Of course, never leave your cat in the car in hot or sunny weather. Because it cannot sweat it is at risk of heatstroke, and death can occur very quickly. Once you reach the end of your journey, simply open the cat carrier and leave your cat to get its bearings. Its natural curiosity should soon overcome its initial wariness, and it will emerge from the carrier when it is ready. Because cats can get lost in unfamiliar territory, you should always ensure that they wear a collar and an identity tag bearing up-to-date information.

BOARDING ALTERNATIVES
Alternatives to catteries include leaving your cat with friends, or asking neighbours or house-sitters to keep an eye on it. It is usually best for your cat to stay in its own home with a reliable person visiting each day to feed it, clean its litter tray, and occasionally play with it. Leaving your cat with friends can be problematic if it is an outdoor cat. It will be unfamiliar with the new territory and if close enough to home may try to return there. Others may simply stray.

Wherever your cat stays, make sure that it is protected against contagious diseases and fleas. If travelling abroad, be aware that some of the necessary vaccination documents will need to be dated several months before departure.

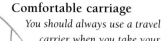

Comfortable carriage
You should always use a travel carrier when you take your cat out of its home. Increase your cat's enjoyment of its carrier by leaving it open at home, filled with soft bedding and a few hidden food treats. If you only use the carrier to take your cat to the vet or the cattery, your cat is likely to disappear each time you fetch the carrier.

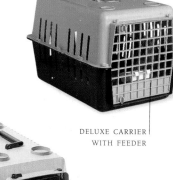

TRADITIONAL WICKER CARRIER

DELUXE CARRIER WITH FEEDER

EASY-TO-CLEAN PLASTIC CARRIER

Hygienic accommodation
Good catteries are designed to permit a cat to move in and out of doors. Some will also have larger units for groups of cats that are used to living together. Cats that are not from the same household should never be kept in the same unit. Each unit should include both perches and hiding places, and central heating should be available when necessary. Always inspect a cattery for cleanliness. Feeding and drinking bowls should be spotless and bedding should be disinfected for each new occupant. Ensure your cat is vaccinated against contagious diseases.

Types of cat carriers
Wicker carriers are attractive to you and your cat but can be extremely difficult to clean when soiled. Moulded plastic cat carriers are easy to disinfect, not attractive to fleas, and easy to disassemble for storage.

TRAINING TO "COME" ON COMMAND

Cats can be trained, and training is fun for both of you, provided that you understand that cats do not respond to touch and kind word rewards in the way that dogs do. Cats really respond well only to food rewards. Needless to say, they never respond to punishment. Successful training involves conditioning your cat, using food treats, to respond to your command. Initially, the cat is rewarded each time it performs the behaviour: coming forward to eat a food treat, for example. As it moves forward you say its name and the word "Come". When your cat understands what it must do to earn its reward, switch to intermittent rewards. Professional trainers often use "clickers". As the cat comes for its food reward, the trainer clicks the clicker. The cat learns to associate the click noise with the reward. After a while the food reward is phased out. The cat simply responds to the click sound. Training your cat to "Come" on command is vital to attract your outdoor cat indoors when you are going out. Train your kitten as soon as it arrives in your home.

CAT RISES UP TO HAND

RAISE HAND ABOVE CAT'S HEAD

1 Through trial and error, discover a food treat that your cat craves. Professional trainers often use pieces of liver, fried to a crisp. It is easy to accustom a young kitten to the flavour of yeast-based vitamin tablets. Begin by feeding treats from your hand, then raise your hand so that your cat rises up to get the treat. You may wish to incorporate a clicker noise as professional trainers do.

STAND BACK AND CALL CAT

CAT COMES FORWARD FOR REWARD

2 Once your cat has learned to love the food treat, show it the treat, then take a step back and say "Come" as it moves towards you. If you are using a clicker, click it as your cat takes the necessary step forward. Over successive training periods, increase the distance until your cat comes from another room. Each training session should never be more than a few minutes long.

CAT IS ATTRACTED TO REWARD

REWARD NOT GIVEN EVERY TIME

CAT MIAOWS IN RESPONSE TO HEARING ITS NAME

"MIAOW" ON COMMAND

Life is endlessly interesting for a cat, but sometimes its curiosity can lead it into situations that it finds frightening. A cat's natural response to fear is to hide, usually in some small, dark corner. Unfortunately, cats sometimes find it easier to get into hiding places than to get out of them, and may need to be rescued from an awkward spot. A cat may also be so frightened that it simply does not want to leave the security of its hiding place. Fortunately, cats recognize human voices and can differentiate known from unknown ones. If your cat has been trained to miaow when it hears its name called, it is likely to do so, even from its hiding place, when it hears a friendly voice that it knows.

1 Just as when training for any other activity, find a treat your cat adores and habituate it to that treat. If you have had success with the clicker, use it here too. Call your cat's name and/or use the clicker as it comes forward. Always train in the quiet of indoors and never with any other animals about, otherwise the cat will be easily distracted.

2 After your cat has learned to associate hearing its name or the click sound with the food reward, tease it a little with the food. Your cat may try to bat the food from your hand, but many cats that already use their voices to demand food will miaow. Give the treat immediately. Reinforce the miaow response by giving the food only intermittently during subsequent training. Gradually increase the distance between you and your cat until you are in separate rooms. Finally, move outdoors to continue training.

BEHAVIOURAL PROBLEMS

MARKING TERRITORY WITH URINE, DIGGING TOILET SITES IN GARDENS, preying upon songbirds, climbing curtains, scratching furniture, and even creating "Grand Prix" racing circuits in the living-room are all natural feline behaviours. Cat owners, however, are likely to consider them problematic. True behaviour "problems", such as compulsive grooming, obsessions with articles, or even psychosomatic illnesses, do exist but are relatively rare, compared with natural behaviours that we want to modify because we find them offensive. Use common sense and originality, never harsh discipline, to change your cat's behaviour.

In most countries, cats historically have had the freedom to roam outdoors; in Australia, fears about the cat's predatory effects on ground-nesting birds and other species have led to laws restricting cats indoors during dusk and dawn, their favourite hunting times. In other countries, such as the United States, half or more of all cats never go outdoors, yet their need to hunt remains.

FORMS OF AGGRESSION

If a cat cannot hunt, it may satisfy its natural predatory needs by stalking, pouncing, and biting its owner's ankles or a resident dog's wagging tail. This unpleasant, anti-social behaviour can be diminished by giving indoor cats an acceptable predatory outlet. Toy manufacturers have created a vast array of stalk-and-capture toys for felines, but even a simple table-tennis ball with a few grains of rice inside will make a satisfying toy.

In Aesop's fable, *Belling the Cat*, mice try to outwit a cat by putting a bell on its collar, but there is no evidence that a "belled" cat is any less successful as a predator. Dr. David Paton at the University of Adelaide in Australia found that belled cats were just as successful as hunters as those without bells, perhaps because rodents and birds have not learned to associate the ringing of a bell with danger.

Predation is not the only form of aggression; others include aggression between two cats in a household (*see page 84*), maternal aggression, dominant aggression, fearful aggression, and, perhaps the most exasperating to owners,

Playing rough
The vast difference in comparative size does not stop this ginger cat from mock-biting its canine companion's leg in a predatorily aggressive but inhibited, playful manner.

Fearful aggression
This feline, ready to roll over and use all its claws, and with its ears folded back flat against its head for protection, shows fearful aggression.

petting aggression: having seemingly enjoyed being stroked, a cat suddenly lashes out with teeth and claws and jumps away, only to return seconds later for more petting. A possible cause of this behavioural quirk is conflict in the feline's mind. Being stroked is pleasurable; it is reminiscent of its mother's grooming. However, cats have evolved to have no body contact with other cats in adulthood other than during mating, an occasion particularly fraught with conflict. Petting aggression can be avoided by restricting your stroking of prone individuals to short episodes.

Jailed frustration (left)
When an indoor cat spots natural prey but cannot stalk and pounce, its teeth may chatter. This odd response does not occur when predation is possible.

Scent marking (right)
Cats spray urine on upright objects such as walls, chairs, and curtains to mark their territory. Neutering can diminish this behaviour.

Scratch happy

All cats scratch, both to unsheathe their claws and visibly mark their territory. Unfortunately, an indoor cat is likely scratch furniture, carpets, or curtains. This natural behaviour can be prevented by training your cat from an early age to use a scratching post, and by trimming its nails regularly. Use blunt, stick-on "artificial nails" on felines that persistently prefer to scratch your furniture rather than their own.

Although most cats willingly eat together in a multi-cat household, feeding can become competitive and potentially intimidating. If your cat eats quickly then vomits its food, feed it smaller quantities more often.

PSYCHOSOMATIC ILLNESSES

Dr. Tony Buffington, a veterinarian at Ohio State University, has described a form of bladder inflammation in cats that is not related to diet or microbes, but responds to "anti-anxiety" drugs. In certain felines, even mild stress can bring on an episode of cystitis (painful urination sometimes accompanied by bleeding into the urinary tract). Some veterinarians believe that cats can exhibit other medical manifestations of stress-induced anxiety, such as itchy skin or a tendency to vomit easily. The behavioural component of these medical conditions has not yet been studied but it is possible that there are various feline behavioural problems that mask themselves as chronic medical conditions.

ELIMINATION PROBLEMS

Overwhelmingly, the most common behaviour problems that frustrate cat owners are those concerned with elimination. Cats might decide to use places other than the litter tray for toileting, possibly because they associate the tray with the pain of cystitis or blocked anal sacs. When medical considerations have been ruled out, examine the circumstances in which the problem occurs. Try different forms of litter to discover which your cat prefers. Keep the litter tray clean and place it in a quiet location.

MARKING TERRITORY

Many cats scent-mark their territory with urine. The urine-marker does not squat in the normal bladder-emptying fashion. Instead, the cat backs up to a vertical object, raises its quivering tail, then pulsates a spray of urine against the object. Urine is intentionally left as a "calling card" for other cats. Scratching furniture serves a similar purpose. A well-scratched piece of furniture is a visible way of marking territory. These problems can be diminished or eliminated using "aversion" methods: a squirt from a water pistol or a sudden loud noise just as your cat takes aim or begins to scratch can be sufficiently offensive to stop the behaviour. Neutering is the most effective way to reduce urine marking.

EATING PROBLEMS

Refusing to eat is the most common feline eating problem. Cats are adapted to go without food longer than we can: in a battle of wills between cat and owner, the owner usually gives in first, offering a *smorgasbord* of tempting delights.

When your vet has concluded that your cat's behaviour is due to stubbornness, not illness, concentrate on the reasons why it will not eat. Never place food too close to the litter tray. Do not position the feeding bowl in busy areas: cats like peace and quiet while they are eating. Ensure that the feeding bowl does not have an unpleasant smell: certain plastics hold odours that cats do not like. Feed two cats in separate areas to avoid rivalry.

Feline food fads (above)

When medical considerations have been ruled out, a refusal to eat tasty and nutritious food should be considered a behavioural problem. It often occurs when owners offer their cats a selection of foods from a large menu. Eventually, the cat may develop a food fad, and refuse to eat anything else except its favourite food.

In the genes (left)

If a kitten is taken away from its mother before completing its 12-week suckling period, it suckles its owner instead, "kneading" with its paws and sucking skin or clothes. Wool sucking in the Siamese breed is slightly different. Restricted to specific genetic lines, this activity begins in adulthood in cats weaned normally, rather than continuing from kittenhood in individuals weaned early.

CARING FOR A SICK CAT

THE CAT'S NATURAL INDEPENDENCE USUALLY SURFACES WHEN IT IS unwell. Although some cats seek the attention of their owners, the majority try to conceal signs of illness, often hiding away. When a cat is sick, other resident house cats may behave spitefully towards it, an anti-social clue to illness. Always contact your veterinarian when you observe changes in your cat's routine, and follow your vet's instructions for giving medicines and for changes in diet. Prevention is always simpler and cheaper than treatment: arrange for routine check-ups to catch old-age problems early, when changes in diet can forestall development of age-related decline.

Good cat care depends upon careful observation of your cat's behaviour to spot any changes in its routine. Just as important is your willingness to take your cat to the vet when necessary, and to carry out treatments that the vet prescribes.

VISITING THE VET
Early training to accept a cat carrier means that when it is necessary to use one for transport to the vet, your cat will not resist being put in it. At the vet's, gently calm your cat and reassure it of your presence. Give the vet as detailed a resumé of the problem as possible. Remember, your vet relies upon you to provide information on how your cat is behaving and responding to treatment. The vet may want to take blood samples or carry out diagnostic tests. If this treatment requires hospitalization, arrange to leave your cat's bedding or any other item that your cat is familiar with, and tell your vet what your cat likes to eat. Follow the vet's routines concerning your possible visits while your cat is hospitalized. Generally speaking, cats should recuperate at home whenever possible.

CONSIDERATE NURSING
The typical cat does not want to be excessively fussed over. Provide it with a warm bed and follow your vet's instructions for feeding and medication. If your cat has had surgery, check the wound carefully, looking for unexpected swelling, redness, or discharge. Bandage changes can be very difficult and are best left to the veterinary staff. A sensible cat hates

Giving liquids
With your cat comfortably relaxed, insert the nozzle of a syringe or a pipette behind the canine teeth, and squeeze in a few drops. Let your cat swallow this liquid before giving more.

being given medicine, which almost always tastes unpleasant, so you may need someone to help you. Unlike dogs, cats are unlikely to eat tablets buried in food. Wrap your cat in a bath towel if necessary and tuck its body under your free arm as you give the pill. An ill cat may also need hand-feeding in the same manner.

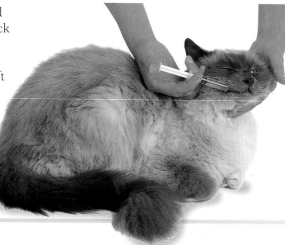

ADMINISTERING TABLETS

Cats do not like being given pills. It is also uncommon for a cat to eat willingly any form of medication placed in its food. The only way to ensure that medicine in pill form is consumed is to give it directly to your cat. Speed and confidence are essential. It is usually best to have the help of extra hands: one person holds, while the other gives the pill. When this is not possible, wrap your cat's body in a bath towel, and hold it under your arm while you give it the pill.

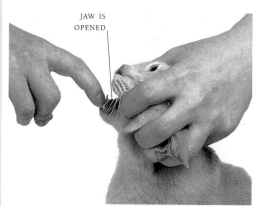

JAW IS OPENED

1 While one person steadies your cat, use one hand to hold its head, tilting it upwards. The jaw muscles may slacken slightly as its head tilts. Use your other hand to pull the lower jaw downwards, gently but firmly. Push your finger on the skin in the space between the jaws.

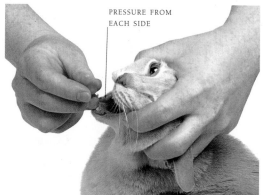

PRESSURE FROM EACH SIDE

2 Your forefinger and opposing thumb will keep your cat's jaws open for a few seconds. With its head tilted upwards, a pill can be dropped over the tongue directly to the back of the throat. Lubricating the pill with butter can be beneficial both because of its taste and its slipperiness.

3 As soon as the pill drops to the back of the throat, close your cat's jaws using one hand and keep the jaws closed. Massage the neck. When your cat licks its lips it has swallowed its medicine.

Comfort and warmth
A sick cat needs to be kept warm. Either place a hot-water bottle under its bedding, or use special cat heating pads that are activated in a microwave oven.

Giving ear drops
Ear mites frequently take up residence in a cat's ear, and sometimes unpleasant microbes thrive, causing infection in the ear canal. Ear drops should be warmed to body temperature by rolling the bottle in your hands for a minute. Invert the bottle, insert into the bottom hole in the ear, and squeeze. Massage below the ear to spread the medicine thoroughly.

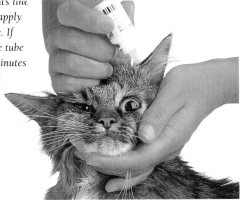

Giving eye drops
Gently hold your cat's head with one hand and approach the eye from above, out of your cat's line of vision. When possible, apply drops at body temperature. If using an ointment, roll the tube in your hands for a few minutes to warm the ointment and improve its flow.

Warm food up to body temperature, and feed small amounts as frequently as possible. Fluids are absolutely essential and should be given by syringe if your cat is not drinking sufficient quantities by itself. Even moderate dehydration is a serious complication in any illness.

A sick cat may appear to be completely better before it really is. Always finish any course of medication. Treatments will sometimes involve altering your cat's feeding routines to prevent a recurrence of the problem. This after-care is just as important as the initial treatment.

AFTER SURGERY
Routine surgery is usually followed by a week of indoor rest, while bone surgery may require four to eight weeks of involuntary restriction. Confinement may be boring for your cat but is essential for its recovery, and allows you to keep an eye on its progress. The surgical site should look clean with little swelling or redness. Your cat's appetite should return rapidly, as should its desire to investigate its territory. If you are concerned about its appetite, mobility, or demeanour, contact your vet for advice.

APPETITE LOSS
A short fast is of no concern but prolonged feline fasting is dangerous, and can even lead to serious liver-related disease. If your cat has lost interest in its food, consult your vet. Stimulate its taste buds with extra-aromatic foods and warm food up slightly, to make it give off stronger aromas. Stay with your cat while it

eats, as encouragement can sometimes help. If your cat refuses to eat, your vet will prescribe a liquid diet that you can feed it by syringe until it is well enough to eat on its own. If the cat becomes dehydrated, your vet will have to feed it either intravenously or by stomach tube.

CARE OF THE ELDERLY CAT
Cats are living longer than ever before, in part because of advances in disease control through vaccination, and in part because so many cats live safely indoors. In feline veterinary medicine, medical conditions associated with old age have increased in frequency. Older cats need special care. Tooth and gum disease remains the most common reason for seeking veterinary attention and routines such as tooth brushing and feeding may need to be altered. With age, many cats fail to groom themselves as well as they once did, leading to a greater likelihood of parasites. Take extra care of your elderly cat's skin and coat. Old skin loses its elasticity, so you will need to be careful not to damage it accidentally when grooming. Clean the ears more routinely for wax, discharge, or debris. Small benign tumours called ceruminomas are fairly common in the ears of elderly cats.

HOW FAR SHOULD WE GO?
Medically speaking, most of the diagnostic aids or therapeutic techniques available for us can be used on cats. Vets routinely use ultrasound, endoscopy, and imaging apparatus to make diagnoses. They have the skill to use advanced

surgery, including transplantation, to treat feline illnesses. The availability of this vast array of equipment and knowledge raises the question of how far cat care should be taken. Cost of care is a financial consideration: what a cat owner will spend is a personal decision. More important is what the patient gains from the procedure. The long-term quality of a cat's life should be the most important consideration, not the medical challenge to the vet or the emotional stress the cat's owner is suffering.

Preventing self-inflicted damage
It is sometimes necessary to prevent a cat from rubbing or scratching at its face. A protective collar permits a cat to eat and drink, and allows air to circulate around the healing region, while protecting it from further irritation.

FIRST AID

THE CAT'S NATURAL CURIOSITY IS WELL DOCUMENTED, AND OFTEN places it in danger. The need for first aid, and then for help from your veterinarian, is more common for outdoor individuals. Your objectives in giving first aid are to save life, prevent any further injuries, control damage, minimize pain and distress, and then to get your cat to the vet for attention. Follow practical steps to assess the situation. Is your cat in danger? Are you in danger if you try to help? Look for life-threatening signs, sustain life by giving emergency first aid, and, if other people are available, they can help by arranging transport to the nearest vet.

When emergencies occur, carry out a physical examination to determine the severity of the injuries. Do not be fooled by any lack of external damage: felines often suffer hidden internal injuries that are potentially life-threatening.

GENTLE RESTRAINT

If a cat is injured, its instinct is to hide and protect itself from further damage. Calm your cat with soothing talk and use as little restraint as necessary to carry out your examination. Eliminate escape routes and, avoiding sudden gestures, place a soft object such as a towel over the cat, wrapping it gently to prevent escape and to minimize the risk of injuries to you. Next, gently unwrap the head so that you can watch the cat's breathing, which will normally be at a rate of 10 to 30 breaths per minute. Flaring nostrils are a sign that your cat is not well.

MONITOR VITAL SIGNS

A cat's resting heart rate varies from 100 to 160 beats per minute. Kittens have heart rates up to 260 beats per minute. To detect the heartbeat, feel the chest firmly behind the elbows. This is simple in lean cats but more difficult in plumper individuals. For those, feel the pulse in the thigh by gently pressing your fingertips into the slight groove inside the hind leg where it joins the body. Examine the gums for colour by gently lifting the upper lip. Normal gums are a healthy pink colour, while bluish, pale, or white gums are all causes for concern.

SHOCK

Clinical shock is a potentially life-threatening event; cats may go into this state hours after an accident. The signs of shock are pale or white gums, a rapid heart rate of over 250 beats per minute, and fast breathing of over 40 breaths per minute. Shock has many causes, including internal or external bleeding, burns, heart failure, vomiting, diarrhoea, urinary blockages, trauma, animal or insect bites, electrocution, diabetes, and poisons. Treating shock takes precedence over most other injuries, including

broken bones. Minimize shock by placing the cat on its side on a warm surface with its head extended, elevating the hindquarters with a pillow or towel to send as much blood as possible to the head. Use manual pressure to stop obvious bleeding. Give artificial respiration and heart massage if necessary, wrap your cat in a warm blanket, and go immediately to the nearest veterinarian for life-sustaining medical treatment.

ASSESSING AN INJURED CAT

After restraining your cat if necessary, monitor its breathing and its heart rate. Always look for signs of clinical shock, the greatest hazard to your cat. Regardless of whether your cat is conscious or unconscious, check the colour of its gums to establish its physical state.

1 Gently lift your cat's upper lip to expose the gums. If they are pink, its heart and lungs are probably functioning well; if they are pale or blue, breathing or circulation has been affected; this is an acute emergency.

2 If your cat is unconscious, use your thumb and fingers around the chest just behind the elbow to feel for its heartbeat, or feel the inside of the hindleg for the pulse. If there is no heartbeat or pulse, start heart massage immediately by grasping the cat's chest between thumb and forefinger and squeezing vigorously.

PHYSICAL WOUNDS

Injuries may be open, where the skin is broken, or closed, where the skin is intact. Do not underestimate the significance of closed wounds: typically, there is pain, swelling, discoloration, or heat under the skin and possibly superficial scratches. Apply a cold compress, or a bag of frozen vegetables wrapped in a towel, as soon as possible.

RECOVERY POSITION

When assessing your cat, if its heart is beating and it is breathing, there is no need for heart massage and artificial resuscitation (CPR). Place the cat in the recovery position by laying it on a warm surface on an incline. Make sure that its head is lying lower than its body so that maximum blood goes to the brain. Unless your cat is suffering from heatstroke, wrap it loosely in a light blanket to prevent the loss of body heat, while you seek immediate veterinary attention.

Open wounds are more apparent, with possible pain, bleeding, and broken skin, even if there is only a small puncture. If a cat is not in shock, it will increasingly lick the affected area. Treat open wounds by applying pressure with clean gauze or absorbent material. Once the gauze or material has been applied to the bleeding wound, leave it there until your vet examines your cat: removing a blood-soaked dressing disturbs the clot and can initiate more bleeding. Do not use disinfectants or antiseptics on gaping open wounds. Never rub open wounds because you may cause further damage.

Cats rarely bleed profusely from minor wounds; blood vessels in their skin clot quickly. Internal bleeding is more common, and very dangerous, because it cannot be seen. Always seek veterinary advice after any trauma. If you see bleeding, apply pressure to the wound using clean gauze. Flush minor wounds with clean water or 3 per cent hydrogen peroxide. Bandage injuries only if absolutely necessary. It can be extremely difficult to apply a bandage to any part of a cat's body other than its torso.

CRITICAL INJURIES

Injured cats are often unable to move. Take extreme care when moving a cat with potentially serious physical injuries. Let the cat find its own most comfortable breathing position, then, keeping its back as still as possible, slip one hand under the cat's rump and the other under its chest, and gently slip it onto an improvised "stretcher", such as a blanket or coat, for transport to the nearest veterinarian. A well-cushioned cardboard box is ideal for transportation. Cover the cat in the box to reduce movement and conserve body heat.

MONITOR THE PATIENT

After giving life-saving first aid to your feline, always have it examined by a veterinarian. If your cat's injuries are not life-threatening, carry out a detailed inspection, looking for potentially serious, painful, or distressing conditions. Watch its behaviour and observe its responses to you. Listen for any unusual breathing sounds and watch your cat's movements. Ask your vet to show you how to take your cat's temperature. A normal temperature is between 38°C (100°F) and 39°C (102°F). If it is above or below normal, seek veterinary attention. Examine the eyes, ears, nose, mouth, head, and neck. Feel the body, limbs, and tail. Check the anal region for injuries or unusual odour. Monitor your cat's eating and toileting habits. If there are any changes to normal behaviour, contact your vet.

FIRST-AID EQUIPMENT

Most cats resent the use of any first-aid equipment, especially bandages. This is unfortunate: bandages keep wounds dry and prevent further injuries, including self-inflicted damage. Bandages also protect open injuries from contamination, absorb seepage, and minimize pain. Unless the cat is subdued or sedated, it is almost impossible to apply an effective bandage. If bandages are required, seek veterinary attention.

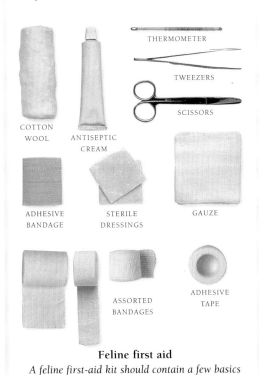

THERMOMETER

TWEEZERS

SCISSORS

COTTON WOOL

ANTISEPTIC CREAM

ADHESIVE BANDAGE

STERILE DRESSINGS

GAUZE

ASSORTED BANDAGES

ADHESIVE TAPE

Feline first aid
A feline first-aid kit should contain a few basics that may be needed in the event of an accident or emergency. Keep all medicines in a safe place.

ARTIFICIAL RESPIRATION

A heartbeat but no breathing means that your cat needs artificial respiration; a cat will die if its brain does not receive sufficient oxygen. Near drowning, electrocution, choking, smoke inhalation, poisoning, diabetic shock, and allergy may all interfere with breathing. If the heart has stopped, combine artificial respiration with external heart massage.

1 Assess the cat's heart rate. If its heart is pumping but it is not breathing, place the cat on its side, clear the airway of debris, and pull its tongue forward. Elevate hindquarters above forequarters.

2 With your hand around the closed muzzle, place your mouth over the nose and blow in to expand the chest. Avert your mouth and let the chest deflate. Repeat 20 to 30 times per minute.

MOVING AN UNCOOPERATIVE CAT

Even the most amiable of felines may lash out with tooth and claw when injured. Talk soothingly to the cat to reassure it, and approach it slowly and cautiously. Drape a lightweight but thick towel or blanket over your cat, to restrain it and to prevent it from running off to hide. If possible, wear gloves to protect your hands from any bites or scratches. Pick up the entire bundle, applying as little pressure as possible. Make sure its face is exposed so that it can breathe easily. Apply gentle pressure to the towel around the scruff of the neck to prevent any attempts to escape.

GLOSSARY

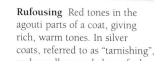

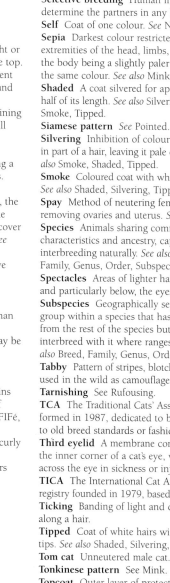

Agouti The lighter areas of fur in the tabby coat; also a term for cats expressing any tabby pattern.

Awn hair Bristly undercoat hair. *See also* Down hair, Guard hair.

Bi-colour Coats consisting of white hair mixed with one other colour.

Boots *See* Gauntlets.

Brachycephalic An abnormally short nose and flattened face.

Breeches Long fur covering the upper part of the hindlegs.

Breed An animal strain visibly different from its parent species, particularly produced by human intervention. *See also* Family, Genus, Order, Species, Subspecies. Other individual interpretations apply within the cat fancy.

Breed clubs Organizations within a registry devoted to one or more particular breeds. Several clubs may exist within a registry for one breed.

Breed standard A description of the ideal characteristics against which each cat in a breed is measured.

Burmese pattern *See* Sepia.

Castrate Method of neutering male cats by removing testicles. *See* Neuter.

Cat fancy Umbrella term for registries and the people who breed and/or show cats worldwide.

CFA The Cat Fanciers' Association, the world's largest feline registry, founded in 1906, primarily based in North America.

Chinchilla Longhair with white coat tipped with black, or such a coat in any cat.

Colourpoint A cat whose face, ears, feet, and tail are a different colour to the rest of its body. *See* Pointed.

Cool Colours as far as possible from red.

Dominant gene A genetic trait that if carried is always expressed. *See also* Recessive gene.

Down hair Soft, insulating hair in undercoat. *See also* Awn hair, Guard hair.

Family Group of animals within an order that share many defining characteristics and evolutionary descent. *See also* Breed, Genus, Order, Species, Subspecies.

Feline Immune Deficiency Virus (FIV) A relative of the HIV virus, which weakens the immune system, causing death. Highly contagious to other cats, but not to humans or other animals.

Feline Infectious Peritonitis (FIP) A viral disease that is usually fatal. Symptoms include fluid accumulation in the abdomen, jaundice, and anaemia.

Feline Leukaemia Virus (FeLV) A virus affecting the lymphatic system, suppressing immunity to disease.

Feral An animal living wild but descended from domestically bred stock.

FIFé Fédération Internationale Féline, an umbrella organization for European cat clubs, founded in 1949.

Flehming Lifting the upper lip to bring a scent into contact with the vomeronasal organ. *See* Vomeronasal organ.

Free-breeding *See* Random-breeding.

Frown lines Dark lines forming a letter "M" on a cat's forehead.

Gauntlets White hindpaws in a coloured or bi-coloured cat, ending below or above the hock. *See also* Mittens.

GCCF The Governing Council of the Cat Fancy, founded in 1910, the governing body for most British cat clubs.

Gene pool The genetic diversity available in any given species, race, or breed.

Genus A group of species within a family that share characteristics and ancestry not shared by other species. *See also* Breed, Family, Order, Species, Subspecies.

Ghost markings Faint tabby markings sometimes seen in the coat of non-agouti cats, usually in kittens.

Guard hair Long, coarse hairs that protect the undercoat and provide a waterproof layer. *See also* Awn hair, Down hair.

Hock A cat's ankle.

Hot Colours that resemble red.

Jacobson's organ *See* Vomeronasal organ.

Kitten In common understanding, a kitten below the age of separation from its mother at about 12 weeks. In shows, cats may still be called kittens for some months after this.

Kitten cap White cats carry the dominant gene W, which masks other colours. White kittens sometimes show a hint of their underlying colour in a "kitten cap" of hair on their heads, which disappears with age.

Locket Small area of white hair on chest of a coloured cat, a fault in many breeds.

Manx Tailless breed of cat. The taillessness is caused by a gene that can cause fatal problems if passed on by both parents.

Mascara lines Dark lines extending from the outer corners of the eyes.

Mink A combination of Pointed and Sepia patterns, with coloured body and moderate pointing. The characteristic pattern of the Tonkinese breed.

Mittens White forepaws in a coloured or bi-coloured cat, typically stopping below the ankle. *See also* Gauntlets.

Mosaicing Random patching or mottling of coat colours, as in tortoiseshell colours.

Mutation Change away from the normal state in a genetic trait.

Neuter To castrate males or spay females to prevent reproduction and unwanted sexual behaviour.

Non-agouti Any solid-coloured cat showing no tabby markings.

Nose break *See* Nose stop.

Nose stop A change in direction, slight or pronounced, seen in profile at the nose top.

Odd-Eyed Cat with eyes of two different colours; in breed standards, one blue and one orange.

Order Group of animals with one defining characteristic, for example carnivora, all carnivorous mammals. *See also* Breed, Family, Genus, Species, Subspecies.

Pedigree A record of ancestry, showing a cat's parentage over several generations.

Pointed Colour restricted to the extremities of the head, limbs, and tail, the body remaining pale, first known in the Siamese breed. Also a generic term to cover other less obvious pointing patterns. *See* Mink, Sepia.

Polydactyly Any number of toes above four on forepaws or five on hindpaws.

Queen Unspayed female cat.

Random-breeding Process of animals choosing their own mates without human intervention. Also called free-breeding.

Recessive gene A genetic trait that may be carried without being expressed.

Registry An authority, national or international, that decides on breed recognition and standards, and maintains records of pedigree breeds. Made up of constituent breed clubs. *See also* CFA, FIFé, GCCF, TCA, TICA.

Rex Term for any mutation causing a curly coat. Also referred to as rexing.

Ruff Longer fur that sometimes appears around the neck and on the chest.

Rufousing Red tones in the agouti parts of a coat, giving rich, warm tones. In silver coats, referred to as "tarnishing", and usually regarded as a fault.

Scent marking A cat marks its territory with urine, or with scent from glands on its face, lips, and ears.

Selective breeding Human intervention to determine the partners in any breeding.

Self Coat of one colour. *See* Non-agouti.

Sepia Darkest colour restricted to the extremities of the head, limbs, and tail, the body being a slightly paler shade of the same colour. *See also* Mink, Pointed.

Shaded A coat silvered for approximately half of its length. *See also* Silvering, Smoke, Tipped.

Siamese pattern *See* Pointed.

Silvering Inhibition of colour production in part of a hair, leaving it pale or white. *See also* Smoke, Shaded, Tipped.

Smoke Coloured coat with white undercoat. *See also* Shaded, Silvering, Tipped.

Spay Method of neutering female cats by removing ovaries and uterus. *See* Neuter.

Species Animals sharing common characteristics and ancestry, capable of interbreeding naturally. *See also* Breed, Family, Genus, Order, Subspecies.

Spectacles Areas of lighter hair around, and particularly below, the eyes.

Subspecies Geographically separated group within a species that has differences from the rest of the species but will interbreed with it where ranges overlap. *See also* Breed, Family, Genus, Order, Species.

Tabby Pattern of stripes, blotches, or spots, used in the wild as camouflage. *See* Agouti.

Tarnishing *See* Rufousing.

TCA The Traditional Cats' Association, formed in 1987, dedicated to breeding cats to old breed standards or fashions.

Third eyelid A membrane concealed in the inner corner of a cat's eye, which draws across the eye in sickness or injury.

TICA The International Cat Association, a registry founded in 1979, based on genetics.

Ticking Banding of light and dark colours along a hair.

Tipped Coat of white hairs with coloured tips. *See also* Shaded, Silvering, Smoke.

Tom cat Unneutered male cat.

Tonkinese pattern *See* Mink.

Topcoat Outer layer of protective hairs, carrying pattern in tabbies. *See* Guard hair.

Undercoat Layer of insulating fur under topcoat. *See also* Awn hair, Down hair.

Vomeronasal organ A sensory organ in the nasal cavity that analyzes smells and tastes. Also known as Jacobson's organ.

INDEX

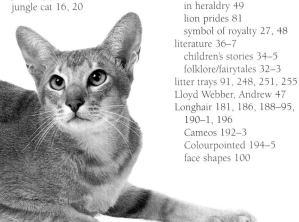

PUBLISHER'S ACKNOWLEDGMENTS

We are indebted to the many owners who gave up their time to allow us to photograph their cats for the breed profiles. Without their assistance, Chapter Five would not have been possible in its present form.

The cats used in the profiles are credited here page by page: page numbers are given in bold, and positions on the page in terms of top (t), bottom (b), left (l), centre (c), and right (r). Each cat's name is followed by that of its breeder and (in brackets) owner.

Any awards the cat holds are given, such as champion (Ch), Grand Champion (GrCh), Premier (Pr), Grand Premier (GrPr), Supreme Grand Champion (SupGrCh), Supreme Grand Premier (SupGrPr), European Champion (EurCh), or International Champion (IntCh). Many of the younger cats that we photographed have gone on to win further awards, but the details remain as they were at the time of photography. We regret that it is beyond the scope of these credits to give the same depth of information for the photographs provided by Chanan and Tetsu Yamazaki, which were not commissioned by Dorling Kindersley. Abbreviations are given for the photographers Tracy Morgan (TM) and Marc Henrie (MH).

108 tr *Pennydown Penny Black* SW McEwen (SW McEwen) MH, cr Yamazaki, bl Yamazaki; **109** tl Yamazaki, tr Chanan, b Chanan; **112** tr *Tameko Tamoshanter* M Simon (R Taylor), c Ch *Sargenta Silver Dan* U Graves (U Graves), bl GrCh *Starfrost Dominic* E Conlin (C Greenal), br GrCh *Maruja Samson* M Moorhead (M Moorhead), all MH; **113** tc Ch & SupGrPr *Welquest Snowman* A Welsh (A Welsh), cr *Susian Just Judy* S Kempster (M Way), br *Miletree Black Rod* R Towse (R Towse), all MH; **114** tr GrCh *Westways Purrfect Amee* A West (GB Ellins), cl *Kavida Kadberry* L Berry (L Berry), cr *Kavida Primetime & Amethyst* L Berry (L Berry), br *Miletree Magpie* R Towse (M le Mounier), all MH; **115** tc GrCh *Miletree Masquerade* R Towse (P Allen), cl Ch *Bartania Pomme Frits* B Beck (B Beck), cr *Cordelia Cassandra* J Codling (C Excell), br *Kavida Misty Daydream* L Berry (L Berry), all MH; **116** tr *Minty* L Williams (H Walker & K Bullin) MH, bl Yamazaki, br Chanan; **117** tc Yamazaki, br *Adrish Alenka* L Price (L Williams) MH; **118** l Yamazaki, r Chanan; **119** all Yamazaki; **120** r Chanan; **121** tl Yamazaki, tr, cr, bc all Chanan; **122** all Chanan; **123** tr Yamazaki, bc Chanan; **124** all Yamazaki; **125** all Chanan; **126** IntCh *Orions Guru Lomaers* (Mulder-Hopma) MH; **127** tl *Eldoria's Crazy Girl*, tr *Eldorias Goldfinger*, bl, *Eldoria's Yossarian* all O van Beck & A Quast (O van Beck & A Quast), br *Aurora de Santanoe* L Kenter (L Kenter), all MH; **128** tr Ch *Comte Davidof de Lasalle*, bl IntCh *Amaranthoe Lasalle*, all K ten Broek (K ten Broek) MH; **129** tr *Donna Eurydice de Lasalle*, bl IntCh *Amaranthoe Lasalle*, all K ten Broek (K ten Broek) MH; **132** tr *Astuhazy Zeffirelli* (M von Kirchberg) MH, bl Yamazaki; **133** tr, br Chanan, cl Yamazaki; **134** tr *Karthwine Elven Moonstock* R Clayton (M Crane), cl GrCh *Emarelle Milos* MR Lyall (R Hopkins), bl Ch *Anera Ula* C Macaulay (C Symonds), bc *Braeside Marimba* H Hewitt (H Hewitt), br Ch *Anera Ula* C Macaulay (C Symonds), all MH; **135** tl *Lionelle Rupert Bear* C Bailey (C Tencor), c *Satusai Fawn Amy* I Reid (I Reid), br *Braeside Marimba* H Hewitt (H Hewitt), all MH; **136** T Straede; **137** tr *Silvaner Pollyanna*, b *Silvaner Kuan*, all C Thompson (C Thompson) MH; **138** tr *Phoebe* (F Kerr) TM, cl & br GrCh *Aerostar Spectre* JED Mackie (S Callen & I Hotten) MH; **139** c Yamazaki, t, b, Chanan; **142** tr *Kartuch Benifer* C & T Clark (C & T Clark), b *Ballego Betty Boo* J Gillies (J Gillies), all MH; **143** t *Lasiesta Blackberry Girl* GW Dyson (GW Dyson), b *Vatan Mimi* D Beech & J Chalmers (J Moore), all MH; **144** tr *Boronga Blaktortie Dollyvee*, P Impson (J Quiddington), bl *Boronga Kreem Kaskuli*, P Impson (J Quiddington), br *Boronga Black Othello*, P Impson (J Thurman), all TM; **145** l *Vervain Ered Luin*, r *Vervain Goldberry*, both N Johnson (N Johnson) TM; **146**

all Chanan; **147** all Chanan; **148** tr Ch *Bambino Seawitch* B Boizard Neal (B Boizard Neal), c *Impromptu Crystal* M Garrod (M Garrod), bl Ch *Hobberdy Hokey Cokey* A Virtue (A Virtue), all MH; **149** tl Ch *Bambino Dreamy* B Boizard Neal (B Boizard Neal), tr *Braeside Red Sensation* H Hewitt (H Hewitt), bl GrCh *Bambino Alice Bugown* B Boizard Neal (B Boizard Neal), all MH; **152** *Romantica Marcus Macoy* (Mrs Davison) TM; **153** tl *Grimspound Majesticlady*, Miss Hodgkinson (Miss Hodgkinson), tr *Tonkitu's Adinnsh Xin Wun*, bl *Tonkitu Mingchen* both D Burke (D Burke), br *Episcopus Leonidas* (Mrs Murray-Langley), all TM; **154** tr Ch *Willowbreeze Goinsolo* Mr & Mrs Robinson (TK Hull-Williams) MH, c Yamazaki; **155** tr Yamazaki, l Yamazaki, br Ch *Pannaduloa Phaedra* J Hansson (J Hansson) MH; **156** tr *Indalo Knights Templar* P Bridham (P Bridham) MH, c Ch *Darling Copper Kingdom* I George (S Mauchline) MH, bl *Mewzishun Bel Canto* A Greatorex (D Aubyn) TM, br GrCh *Dawnus Primadonna* A Douglas (A Douglas) MH; **157** tl Ch *Sisar Brie* L Pummell (L Pummell) MH, cl GrCh *Pannaduloa Yentantethra* J Hansson (J Hansson) MH, cr *Merescuff Allart* (E Mackenzie-Wood) TM, Ch *Rosan Carmen* S Bell (LE Martin) MH; **160** cl *Jasrobinka Annamonique* P & J Choppen (P & J Choppen) MH, bl *Tenaj Blue Max* J Tonkinson (K Iremonger) MH, r *Simonski Sylvester Sneakly* S Cosgrove (S Cosgrove) MH; **161** tl GrPr *Jasrobinka Jeronimo* P & J Choppen (P & J Choppen) MH, c ChPr *Adixish Minos Mercury* A Concanon (A Concanon) MH, br GrCh *Sukinfer Samari* J O'Boyle (J O'Boyle) MH; **162** tr *Scilouette Angzhi* C & T Clark (C & T Clark) MH, bl *Adhuish Tuwhit Tuwhoo* N Williams (N Williams) MH; **163** tl *Sunjade Brandy Snap* E Wildon (E Tomlinson) MH, tr *Scintilla Silver Whirligig* P Turner (D Walker) MH, bl *Saxongate Paler Shades* (D Buxcey) TM, br *Parthia Angelica* MA Skelton (MA Skelton) MH; **164** bl *Ngkomo Ota* A Scruggs (L Marcel) MH, r Yamazaki; **165** all Yamazaki; **166** all Chanan; **167** l Pr *Adkrish Samson* PK Weissman (PK Weissman), tc *Leshocha Azure My Friend* E Himmerston (E Himmerston), cr *Myowal Rudolph* J Cornish (J Compton), all MH; **168** tr GrCh *Ikari Donna* S Davey (J Plumb), bl Pr *Bobire Justin Tyme* IE Longhurst (A Charlton), br GrPr *Bevilleon Dandy Lion* B Lyon (M Chitty), all MH; **169** tl UKGrCh *Nobilero Loric Vilesilenca* AE & RE Hobson (M Reed), tr *Adhuish Grainne* N Jarrett (J Burton), cr *Sailorman Hooray Henry* K Hardwick (K Hardwick), *Myowal Susie Sioux* G Cornish (J & B Archer), all MH; **170** tr *Reaha Anda Bebare* S Scanlin (A Rushbrook & J Plumb), c Yamazaki, bl Yamazaki; **171** all Chanan; **172** all Yamazaki; **173** all Yamazaki; **174** all Chanan; **175** tl Yamazaki, cr Chanan, cl Yamazaki, br Chanan; **176** tr *Gaylee Diablo* M Nicholson (M Nicholson) TM, c Chanan, bl *Gaylee Diablo* M Nicholson (M Nicholson) TM; **177** *Gaylee Diablo* M Nicholson (M Nicholson) TM; **178** all Yamazaki; **179** all Yamazaki; **184** tr name unknown Jane Burton, bl *Friskie* (Bethlehem Cat Sanctuary) MH, **183** tl name unknown, cr *Sinbad Sailor Blue* (V Lew), bl *Crumpet* (Bethlehem Cat Sanctuary), all MH. **188** tl Chanan, bl *Mowbray Tanamera* D Cleford (D Cleford) MH; **189** l, br both Yamazaki; **192** tr *Cashel Golden Yuppie* A Curley (A Curley), cl *Honeymist Roxana* M Howes (M Howes), cr GrPr *Bellra Faberge* B & B Raine (B & B Raine), bl *Casalina Dolly Mixture* E Baldwin (E Baldwin), all MH; **193** l *Watlove*

Hamish H Watson (H Watson), tr *Adirtsa Choc Ice* D Tynan (C & K Smith), cr *Adhuilo Meadowlands Alias* P Hurrell (S Josling), br *Bellrai Creme Chanel* B & B Raine (B & B Raine), all MH; **194** t *Impeza Chokolotti* C Rowark (E Baldwin), c *Anneby Sunset* A Bailey (A Bailey), b *Amocasa Beau Brummel* I Elliott (I Elliott), all MH; **195** Ch *Watlove Mollie Mophead* H Watson (H Watson) MH; **198** tc *Lizzara Rumbypumby Redted* G Black (G Black) TM, tr *Chanterelle Velvet Cushion* L Lavis (G Black) TM, cr *Schwenthe Kiska* FE Brigliadori (FE Brigliadori & K Robson) MH, bl *Panjandrum Swansong* A Madden (S Tallboys) MH; **199** tr *Saybrianna Tomorrow's Cream* A Carritt (A Carritt), bl Ch&GrPr *Panjandrum April Surprice* A Madden (A Madden), br *Aesthetical Toty Temptress* G Sharpe (H Hewitt), all MH; **200** tr *Chehem Agassi* (Christine Powell) TM; **201** tl *Pandapaws Mr Biggs* S Ward-Smith (J Varley & J Dicks) MH, cr *Chehem BryteSkye* (Christine Powell) TM, b *Rags n Riches Vito Maracana* Robin Pickering (Mrs J Moore) TM; **202** all Chanan; **203** Chanan; **204** tr *unnamed kitten*, cl *Ch Keka Ursine Edward*, bl *Keoka Aldebaran*, all D Brinicombe (D Brinicombe) TM; **205** tl Chanan, tr *Ch Keoka Ford Prefect* TM, bl *GrCh Adinnlo Meddybemps*, br *Keoka Max Quordlepleen*, all D Brinicombe (D Brinicombe) TM; **208** bl *Skogens SF Eddan Romeo* AS Watt (S Garrett) MH, r *Lizzara Bardolph* (Ginny Black) TM; **209** tl *Skogens Magni* AS Watt (S Garrett) MH, tr *Tarakatt Tia*, br *Sigurd Oski* both (D Smith) TM; **210** all Yamazaki; **211** tl Yamazaki, tr Yamazaki, br *Olocha* A Danveef (H von Groneberg) MH; **212** tl Yamazaki, br Chanan; **213** all Yamazaki; **214** t Chanan, bl Yamazaki; **215** tc Yamazaki, bl Yamazaki, br Chanan; **216** bl *Bruvankedi Kabugu* B Cooper (B Cooper), r *Champion Cheratons Red Aurora* Mr & Mrs Hassell (Mr & Mrs Hassell), all TM; **217** tl *Cheratons Simply Red* Mr & Mrs Hassell (Mr & Mrs Hassell) TM, cr *Bruvankedi Mavi Bayas* (Mr R Cooper) TM, bl Ch *Lady Lubna Leanne Chatkantarra* T Boumeister (J Van der Werff) MH; **218** tr Chanan; **219** all Yamazaki; **220** Chanan; **221** tl Chanan, tr *Shanna's Tombis Hanta Yo* M Harms-Moeskops (G Rebel van Kemenade) MH, bl *Shanna's Yacinta Sajida* M Harms (M Harms) MH, bc MH; **224** tr *Dolente Angelica* L Brisley (L Brisley) TM, bl *Bealltaine Bezique* T Stracstone (T Stracstone) MH; **225** tl *unnamed kitten*, b *Beaumaris Cherubina*, both A & B Gregory (A & B Gregory) TM; **226** G & T Oraas; **227** tr *Favagella Brown Whispa* J Bryson (J Bryson), b *Kennbury Dulcienea* C Lovell (K Harmon), all MH; **228** tr Pr *Pandai Feargal E Corps* (BV Rickwood), bl *Blancsanglier Rosensoleil* A Bird (A Bird), all MH; **229** tl Pr *Blancsanglier Beau Brummel* A Bird (A Bird), tc Ch *Apricat Silvercascade* R Smyth (E & J Robinson), tr GrPr *Nighteyes Cinderfella* J Pell (J Pell), b *Palvjia Pennyfromheaven* J Burroughs (T Tidey), all MH; **230** tr GrPr *Nighteyes Cinderfella* J Pell (J Pell), cl *Jeuphi Golden Girl* J Phillips (L Cory), br *Mossgems Sheik Simizu* M Mosscrop (H Grenney), all MH; **231** t *Ronsline Whistfull Spirit* R Farthing (R Farthing), cl Ch *Apricat Silvercascade* R Smyth (E & J Robinson), br *Dasilva Tasha* J St John (C Russel & P Scrivener), all MH; **232** tr *Chantonel Snowball Express* R Elliott (R Elliott), bl *Lipema Shimazaki* P Brown (G Dean), all MH; **233** tr *Palantir Waza Tayriphyng* J May (J May) TM, b *Quinkent Honey's Mi-Lei-Fo* IA van der Reckweg (IA van der Reckweg) MH; **234** all Chanan; **235** all Yamazaki; **236** all K Leonov; **237** tr Chanan, b Yamazaki; **240** tr *Maggie* (Bethlehem Cat Sanctuary), bl *White Mischief* (P Brown), all MH; **241** t *Dan-I-Lion* (V Warriner), c *Dumpling* (Bethlehem Cat Sanctuary), br *Stardust* (V Warringer), all MH.

CARTOGRAPHY
James Anderson, David Roberts

ILLUSTRATORS
Main artworks on pp 56–75, Janos Marffy
Additional artworks, Samantha Elmhurst,
Malcom McGregor, Amanda Williams

AUTHOR'S ACKNOWLEDGMENTS

Early in this project, when Candida Frith-Macdonald handed me an armload of argument, gleaned from the Internet, on whether or not cats have one or two diluting colour genes, I knew this would be a fully researched effort, with one "author" seriously underacknowledged. Thank you, Candida, for devoting your self-confessed flypaper brain to this book. My thanks also go to the rest of the team, especially Ursula Dawson for her fluid designs, and Sarah Wilde, Heather Dunleavy, Tracie Lee Davis, Sharon Lucas, Frank Ritter, and Derek Coombes.

As a veterinarian, I have worked with cats for over 30 years but, needless to say, much of the information in this book was new to me. It was not new to those people worldwide who answered my faxes and letters with reams of information. I appreciate the assistance of the Governing Council of the Cat Fancy (GCCF) in Great Britain, and the Cat Fanciers' Association (CFA) in the United States, for providing me with their address lists of cat clubs, and thank all the cat breeders in the United States and Great Britain, unfortunately too numerous to mention here, who responded to my questionnaire. Thanks also to the Breed Club Secretaries in those countries for providing me with such detailed information on their breeds, and to the cat breed Internet website compilers. My thanks also go to individuals around the world who provided regional information about cats. These include Magister Herbert Berthold (Austria), Dr. Marie Rihova (Czech Republic), Kerstin Breyhan (Germany), Ashley McManus (Ireland); Keiko Yamazaki, Tak Muroya, Moto Arima, and Dr. Norio Kogure (Japan), Olga Sizova (Russia), Professor Eric Hurley (South Africa), Dr. Jaume Camps (Spain), Shin Dai, Wu Hung Biksu, and Dr. Du Bai (Taiwan), Dr. Ann Schneider (USA), and Dr. Frances Barr and Joan Moore (UK).

Dorling Kindersley would like to thank the following: Patrick Roberts for the loan of feline stamps; Jan Peder Lamm of the Statens Historiska Museum in Stockholm, Katerina Ågren, and Malcolm Ross-Macdonald for help with Scandinavian research; Irmgard Schmatz for help with German research; Joan Moore of Cat World magazine for help with breed pictures; Mrs L.K. Pring of GCCF; Gloria Stephens; TICA webmaster; the CFA Web Page committee and others involved in the CFA site; Marie Lamb, Orca Starbuck, Laura Gilbreath, Jean Marie Diaz, and Barb French of the Fanciers list and website.

BIBLIOGRAPHY

Alderton, David, *Eyewitness Handbook: Cats*, Dorling Kindersley, London, 1992

August, John, Ed., *Consultations in Feline Internal Medicine*, WB Saunders Co, Philadelphia, 1994

Boylan, Clare, *The Literary Companion to Cats*, Sinclair-Stevenson, London, 1994

Bradshaw, John, *The True Nature of the Cat*, Boxtree, London, 1993

Busch, Heather and Silver, Burton, *Why Cats Paint: A Theory of Feline Aesthetics*, Ten Speed Press, Berkeley, California, 1994

Clutton-Brock, Juliet, *The British Museum Book of Cats*, British Museum Press, London, 1988

Davies, Marion, *The Magical Lore of Cats*, Capall Bann Publishing, Berkshire, 1995

Fogle, Bruce, *First Aid for Cats*, Penguin Books, London, 1995

Fogle, Bruce, *The Cat's Mind*, Pelham Books, London, 1991

Gettings, Fred, *The Secret Lore of the Cat*, Grafton Books, London, 1989

Howard, Tom, *The Illustrated Cat: Animals in Art*, Grange Books, England, 1994

de Laroche, Robert and Labat, Jean-Michel, *Histoire Secrète du Chat*, Casterman, Brussels, 1993

Loxton, Howard, *The Noble Cat*, Merehurst Press, London, 1990

Macdonald, David, Ed., *The Encyclopedia of Mammals*, George Allen & Unwin, London, 1984

Malek, Jaromir, *The Cat in Ancient Egypt*, British Museum Press, London, 1993

Morris, Desmond, *Cat World*, Ebury Press, London, 1996

O'Farrell, Valerie and Neville, Peter, *Manual of Feline Behaviour*, BSAVA Publications, 1994

Kirk's Current Veterinary Therapy XII, Ed. WB Saunders Co, Philadelphia, 1995

The Cat Fanciers' Association Cat Encyclopedia, Simon & Schuster, New York, 1993

Tabor, Roger, *Understanding Cats*, David & Charles, London, 1995

Van Vechten, Carl, *The Tiger in the House*, Dover Publications Ltd, New York, 1996

Wright, Michael and Walters, Sally, Ed., *The Book of the Cat*, Pan Books, London, 1980

PICTURE CREDITS